PREPARATION AND CRYSTAL GROWTH
OF MATERIALS WITH LAYERED STRUCTURES

PHYSICS AND CHEMISTRY OF MATERIALS WITH LAYERED STRUCTURES

Managing Editor

E. MOOSER, *Laboratoire de Physique Appliquée, CH-1003, Lausanne, Switzerland*

Advisory Board

E. J. ARLMAN, *Bussum, The Netherlands*

F. BASSANI, *Physics Institute of the University of Rome, Italy*

J. L. BREBNER, *Department of Physics, University of Montreal, Montreal, Canada*

F. JELLINEK, *Chemische Laboratoria der Rijksuniversiteit, Groningen, The Netherlands*

R. NITSCHE, *Kristallographisches Institut der Universität Freiburg, West Germany*

A. D. YOFFE, *Department of Physics, University of Cambridge, Cambridge, U.K.*

VOLUME 1

PREPARATION AND CRYSTAL GROWTH OF MATERIALS WITH LAYERED STRUCTURES

Edited by

R. M. A. LIETH

Technische Hogeschool, Eindhoven, The Netherlands

D. REIDEL PUBLISHING COMPANY

DORDRECHT-HOLLAND/BOSTON-U.S.A.

Library of Congress Cataloging in Publication Data
Main entry under title:

Preparation and crystal growth of materials with layered structures.

    (Physics and chemistry of materials with layered
structures; v. 1)
    Includes bibliographical references and index.
    1. Layer structure (Solids)—Addresses, essays, lectures.
2. Crystals—Growth—Addresses, essays, lectures. I. Lieth,
R. M. A., 1930–   II. Series.
QD921.P73   541'.042'1   77-888
ISBN 90-277-0638-7

Published by D. Reidel Publishing Company
P.O. Box 17, Dordrecht, Holland

Sold and distributed in the U.S.A., Canada, and Mexico
by D. Reidel Publishing Company, Inc.
Lincoln Building, 160 Old Derby Street, Hingham,
Mass. 02043, U.S.A.

All Rights Reserved
Copyright © 1977 by D. Reidel Publishing Company, Dordrecht, Holland
No part of the material protected by this copyright notice may be reproduced or
utilized in any form or by any means, electronic or mechanical,
including photocopying, recording or by any informational storage and
retrieval system, without written permission from the copyright owner

Printed in The Netherlands

TABLE OF CONTENTS

PREFACE	VII
FOREWORD	IX
J. G. HOOLEY / Elements	1
J. SCHOONMAN and R. M. A. LIETH / Metal Halides	35
H. R. OSWALD and R. ASPER / Bivalent Metal Hydroxides	71
R. M. A. LIETH and J. C. J. M. TERHELL / Transition Metal Dichalcogenides	141
R. M. A. LIETH / III–VI Compounds	225
P. BUCK / IV–VI Compounds	255
INDEX OF SUBJECTS	275

PREFACE

The goal of the series *Physics and Chemistry of Materials with Layered Structures* is to give a critical survey of our present knowledge on a large family of materials which can be described as solids containing molecules which in two dimensions extend to infinity and which are loosely stacked on top of each other to form three-dimensional crystals. Of course, the physics and chemistry of these crystals are specific chapters in ordinary solid state science, and many a scientist hunting for new phenomena has in the past been disappointed to find that materials with layered structures are not entirely exotic. Their electron and phonon states are not two-dimensional, and the high hopes held by some for spectacular dimensionality effects in superconductivity were shattered.

Nevertheless, the structural features and their physical and chemical consequences singularize layered structures sufficiently to make them a fascinating subject of research. This is all the more true since they are met in insulators and semiconductors as well as in normal and superconducting metals. Although for the time being the series is intentionally limited to cover inorganic materials only, the many known organic layered structures may well be the subject of future volumes.

Among the noteworthy peculiarities of layered structures, we mention specific growth mechanisms and crystal habits. Polytypism is very common and it is fascinating indeed to find up to 240 different polytypes in the same chemical substance. In view of this abundance, it is not surprising that the exact determination of the properties of any one polytype and its structural characterization are often difficult and sometimes impossible. Connected with polytypism is the occurrence of stacking faults, which at high densities introduce one-dimensional disorder into layered structures. The low dimensionality and hence the relative simplicity of the phenomenon may well prove useful in coming to a better understanding of amorphous materials.

Easy cleavage and glide along the layer planes have permitted highly interesting and esthetically pleasing electron microscopic studies of the dislocation networks in layered structures. Similar studies in connection with the hitherto not fully understood phase transitions involving charge density waves are still in progress.

The anisotropy of the selection rules governing optical transitions in layered semiconductors often permits absorption measurements well beyond the fundamental absorption edge. Because of this, some of the most spectacular magneto-absorption spectra have been obtained in layered materials.

As far as phonons are concerned, perhaps the most conspicuous feature is the quadratic dispersion observed in some of the acoustic branches. The free charge carriers in layered semiconductors interact strongly with homopolar optical phonons and, unlike in cubic semiconductors, it is this interaction which limits their mobility.

Layered structures may contain many atoms per unit cell, which renders electron band calculations difficult. The preoccupation with layered structures has therefore brought about a considerable development of techniques for such calculations, which was further enhanced by the discovery in some dichalcogenides of charge density waves. In order to come to an understanding of this phenomenon, efficient methods had to be developed to carry out band calculations in compounds containing transition element atoms.

At the time of this writing, the theoretical and experimental investigations on the formation of charge density waves and on the associated phase transitions in layered materials are multiplying. Many interesting results may be expected in the near future, results that will find a rightful place in the present series.

Another rapidly developing field of interest is that of photo-electron-emission spectroscopy. Because the cleavage faces of layered crystals do not contain broken bands, photoemission spectra yield a nearly undistorted picture of their electron density of states. Layered materials have thus become model substances for testing and exploring the photoemission method.

The above survey does not do justice to the many highly interesting aspects of layered materials and is at best only an indication of the numerous subjects which will be treated by specialists in this series. Each volume of the series will cover a particular area of interest, i.e.:

Volume 1: Preparation and Crystal Growth
Volume 2: Crystallography and Crystal Chemistry
Volume 3: Electrons and Phonons in Layered Materials
Volume 4: Electrical and Optical Properties
Volume 5: Structural Chemistry of Layer-Type Phases

Two more volumes are in an advanced state of planning, one on intercalated layer materials and one on photo-electron-emission spectroscopy.

The articles in the different volumes are essentially of two types. On the one hand there are broad surveys of fields of advanced research which aim at informing and stimulating the experienced scientist, and on the other hand there are specialized articles describing new theoretical and experimental techniques. This distinction is not very strict, however, and the character of any one article depends largely on the personality of its author. Once an author has been chosen, he is free to organize his subject at will. This freedom leads to a certain inhomogeneity in the text but assures its spontaneity, which hopefully will make the series the stimulating research tool which I would like it to become. If this goal should be reached, it is above all due to the unfailing efforts of the authors and editors, to whom I extend my sincere thanks.

E. MOOSER

Lausanne, October 1976

FOREWORD

The study of materials with layered structures – their interesting anisotropic behaviour attracts ever increasing attention – has made considerable progress in the past years. Successful growth of single crystals of most of these substances has made it possible to apply new investigation techniques to old problems and to attack new ones. Our knowledge of both the physics and chemistry of these materials has thus rapidly advanced.

Although the large family of layered materials also include silicates (mica, clays), organic compounds, ternary inorganic compounds and the rapidly growing group of intercalation compounds, we have limited ourselves in this volume specifically to binary inorganic compounds and to some elements. Some of the other materials mentioned will be treated in future volumes. Here we include the elements carbon, phosphorus, arsenic, antimony and bismuth, and compounds like the di- and trihalides, the hydroxides, the transition metal dichalcogenides and the compounds of the III–VI- and IV–VI families.

The object of volume 1 is to present a review of the current application of preparative methods and crystal growth techniques to the investigation of the abovementioned compounds and elements. Various aspects of the growth of single crystals are discussed and reference is made to the importance of single crystal X-ray work in connection with the occurrence of polytypism, the influence of phase relations on growth experiments, and attention is given to the problem of purity and purification.

In Chapter 1 the elements and some of the intercalates in graphitic carbon are discussed. Chapter 2 presents the vast group of the di- and trihalides, while in Chapter 3 attention is focused on the less known group of bivalent metal hydroxides. In Chapter 4 another large group is presented, that of the dichalcogenides of the transition metals. Chapter 5 contains the sulfides, selenides and tellurides of gallium and indium and in Chapter 6 the chalcogenides of silicon, germanium, tin and lead are discussed. By compiling a considerable amount of otherwise widely dispersed data in one single volume, it is hoped that the present book will facilitate the access to this information. If, in addition to this, the book helps to establish better contact among different investigators in the field of layered materials, we shall feel that our efforts have been well worthwhile.

As editor I wish to express my gratitude to the authors for their fruitful collaboration. Thanks are due also to all those who have contributed by allowing us to use unpublished results and have critically read and commented on the different chapters.

Eindhoven, The Netherlands   R. M. A. LIETH
1977

ELEMENTS

J. G. HOOLEY

Dept. of Chemistry, University of British Columbia, Vancouver, Canada

Table of Contents
1. Introduction
   1.1. Survey of Topics
   1.2. Survey of the Structure of Elements
2. Varieties of Graphite
   2.1. Natural Graphite
   2.2. Polycrystalline Carbons
3. Intercalation in Graphite
   3.1. General Behavior
   3.2. Intercalation by Halogens and by Metal Chlorides
      3.2.1. Halogens
      3.2.2. Metal Chlorides
   3.3. Theory of Intercalation
      3.3.1. Introduction
      3.3.2. Expansion and Contraction of the Interlayer Spaces
      3.3.3. Structure of Adsorbed Layers
      3.3.4. Function of Cl_2 in the Intercalation of a Metal Chloride
      3.3.5. Long Range Order
   3.4. Intercalation by Other Adducts
      3.4.1. Metals
      3.4.2. Acids
      3.4.3. CrO_2Cl_2
      3.4.4. Oxides and Sulfides
4. Boron Nitride
References

Symbols and Units
1 nm = 10^{-9} m = 10 Å
1 J   = 1 joule = 0.2390 calories
K   = degrees Kelvin
1 atm = 1 atmosphere = 101325 Nm^{-2}
      = 760 mm Hg
1 torr = 1 mm Hg
1 h   = 1 hour

1. Introduction

1.1. SURVEY OF TOPICS

This article will, for the most part, deal with graphite and its intercalation compounds. This is because much more is known about the layered form of carbon than about the layered forms of all the other elements. Furthermore, much of that information is about the intercalation properties of graphite. Thus, it has been known since 1860 that certain elements and compounds would cause graphite to swell and to increase in weight. It is generally agreed that the adduct or reactant diffuses from the periphery of the layer planes into some or all of the interlayer spaces and leaves by the same route on pressure reduction. The other characteristics of layer structures are associated with the anisotropy of their physical properties but again, graphite is the one that has been most studied. Before dealing with graphite, however, this introductory section will discuss the bonding factors that lead to the existence of solids with layer structures in the cases of C, P, As, Sb and Bi. Boron nitride BN is included here because it is isoelectronic with carbon and has a form with a layer structure. Then the many forms of graphite will be described which differ in the size, the degree of perfection and the relative orientation of the layers of hexagonally arranged carbon atoms. Then a detailed description of the intercalation of graphite by halogens and by metal chlorides will serve to introduce and to support a theory of intercalation. This qualitative theory will try to explain the existence of a threshold pressure below which intercalation is not observed and the existence of stages, the stage number being the number of carbon layer planes between layers of reactant in the compound. Following this, the intercalation of graphite by other reactants such as metals, acids, etc. will be described in detail. Finally, contradictory reports of the behavior of the layered form of BN with various reactants will be discussed. These reports, along with some of the work on graphite, illustrate the experimental difficulty of deciding whether or not intercalation does take place with a given reactant. Two of the best criteria are the form of the isotherm of composition against pressure and the direct observation of the expansion of a single flake of graphite.

1.2. SURVEY OF THE STRUCTURE OF ELEMENTS

In every crystal there are planes that consist of ions, atoms or molecules of a given kind. When the bonding energy within a given plane is appreciably greater than that between planes then the structure is said to be layered and will be dealt with in this series. Appreciably greater means that a given set of planes will remain structurally intact while the distance between them is changed by some physical or chemical process. Cleavage into thin flakes is one such obvious process although intercalation or marked anisotropy of some physical property are also the result of this difference in bond energies. Amongst the elements there are five cases – C, P, As, Sb and Bi which will now be discussed in terms of bonding theory. The famous example is graphite with strong sp^2 bonding within each plane of carbon atoms. The internuclear distance of 0.142 nm is intermediate between that for a single and a double bond because the π electron system is delocalized over the two sides of the plane. It is the interaction between the π electron systems that holds this stack of giant planar molecules together. Within a stack the sequence is either the hexagonal *abab*... or the slightly less stable rhombohedral *abcabc*.... These are illustrated in Figure 1 as a plan view of the two ways *B* and *C* in which a layer of carbon atoms can be placed over a similar layer *A* in such a way that in the adjacent layer half the atoms are directly over carbon atoms and half are over the centers of hexagons. In hexagonal graphite either *B* or *C* is used with *A* and in rhombohedral, both *B* and *C* are used, but always in the sequence noted above. These two forms show the same distance between planes of 0.3354 nm but if the stacking order becomes random the distance increases to 0.344 nm. Pauling [1] has suggested that only those layers with the larger separation have the symmetrical resonance structure with all bonds of equal length – 0.142 nm. Those with the smaller spacing have a quinoid structure with one double and two single bonds to each C atom. This, he points out, permits better packing of the superimposed layers and is also more in accord with the compressibility of graphite in a direction parallel to the planes.

A third form of carbon is the diamond with sp^3 bonding around C atoms 0.154 nm apart. At 300 K and 1 atm. it is less stable than graphite: $\Delta H° = 1.9$ and $\Delta G° = 2.9$ kJ mol^{-1}. Other forms of carbon are lonsdaleite, chaoite and carbon VI [2] but only graphite is a layer structure. The other members of Group IV are all diamond-like or metallic even though each has, like carbon, four valence electrons. Their inability to form a layered structure has been related to the fact that p_π orbital overlap and therefore the π bonding characteristic of graphite decreases as the main quantum number increases. This trend is the result of the fact that maximum electron density occurs at a distance from the nucleus which increases with the main quantum number. Hence the p_π overlap decreases much more than p_σ overlap as one goes to higher atomic numbers in the series.

The compound BN is isoelectronic with carbon and has both a graphite and a diamond-like structure. However, the polarity of the bonding leads, in the layered form, to a superimposition of layers such that each atom of one kind is directly above and below an atom of the other kind in adjacent layers as in Figure 2. The interlayer spacing is 0.333 nm and the bond length within a layer is 0.145 nm. The evidence for a change in layer spacing in certain chemical reactions will be dealt with in Section 4 but is not conclusive.

In Group V there are five valence electrons and these, in the elements, are used in three quite different ways. One structure is the diatomic molecule in which a p_σ bond and two p_π bonds hold the nuclei together, leaving a lone pair of electrons on each atom. A second structure is the tetrameric molecule in which four nuclei at the corners of a tetrahedron are held together by three p_σ bonds directed from each atom along the edges of the tetrahedron. Again, a lone pair remains on each atom. The third structure is a double layer of atoms held together by three p_σ bonds from each atom but the bond angles range from 94° to 103°30'. The usual lone pair remains on each atom. Now the p_π bonding in the series of diatomic molecules is weaker for those valence electrons with a higher main quantum number because the nuclei are further apart and hence π overlap is less. This was also observed in Group IV but whereas p_π bonding stabilized a layer structure in that group it stabilizes the diatomic molecule in Group V. That is why the bond energy of the diatomic molecule decreases from a high of 940 kJ mol^{-1} for N_2 to 163 for Bi_2. Hence for nitrogen, N_2 is the only unit in gas, liquid and solid. For the other members of the series the diatomic molecule does exist in the vapor but in the solid the unit is, for P, As and Sb, either a tetramer or a macro layer depending on pressure and temperature. For Bi it is a macro layer. These two forms, as noted above, are held together by three *p* bonds from each atom and, in the absence of distortion, the three bonds should be at right angles. This explains the fact that the layer form, in which the angle ranges from 94° to 103°30′ is more stable than the tetramer with its angle of 60°. The layer form of As, Sb and Bi can be visualized by first considering a plane of atoms in hexagonal array with three bonds to each

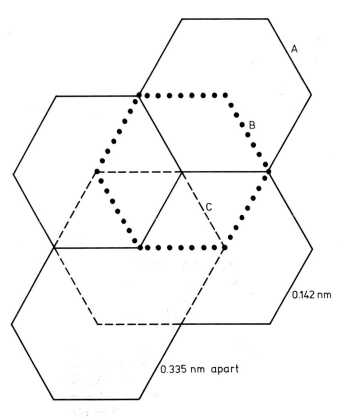

Hexagonal Graphite - any 2 of A,B and C
Rhombohedral Graphite - all 3 of A,B and C

Fig. 1. Graphite, plan view.

atom – as in graphite. Now move along a zig-zag chain of atoms and raise alternate atoms to a new and parallel plane above the original one. Repeat this for adjacent chains to produce the structure shown in Figure 3. This is a double layer and the three bond angles to each atom will be less than the 120° of graphite. Their values for As, Sb and Bi are in Table I and are seen to approach but not reach the 90° of pure *p* bonding as the main quantum number increases. The layer form of phosphorus can also be visualized as the buckling of a hexagonal array of atoms but the structure shown in Figure 4 is produced. This is also a double layer but whereas in, say, grey As all three bonds were between the two halves of the layer, in black P only one of them is

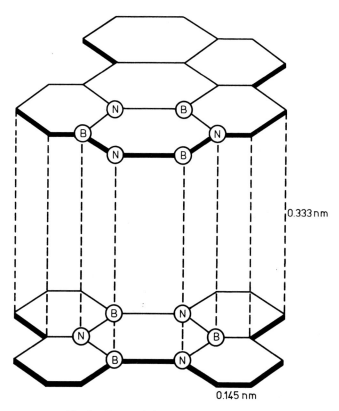

Fig. 2. Boron nitride, layered form.

of this kind. The angle between the two bonds in the plane of the layer is 99° and the angle between either of these two and the third bond is 103°30′ [3]. The bond distances in Table I show that within a double layer the internuclear distance increases with the main quantum number as expected for p_σ bonds. Between the double layers the separation of nearest neighbours is a minimum for grey arsenic. This may be a

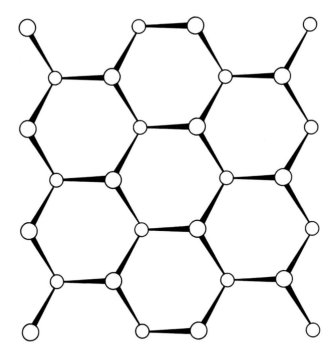

Fig. 3. Arsenic, antimony and bismuth, layered form.

TABLE I

Properties of layered structures

	Bond angle	Nearest neighbor		Conductivity (ohm^{-1} cm^{-1})
		Within layer	Between layers	
Graphite	120°	0.142 nm	0.335 nm	1.2×10^3
P black	102	0.218	0.368	1.4
As grey	97	0.251	0.315	2.6×10^4
Sb grey	96	0.287	0.337	2.7×10^4
Bi metallic	94	0.310	0.347	0.9×10^4
BN	120	0.145	0.333	v. low

reflection of the extra jump in the availability of d orbitals in moving from P to As, or a reflection of the change in the type of layer structure. In any case, the interplanar distance is such that they are all classed as layer structures. Unfortunately the only related property that this reviewer can find is that they are flaky. The conductivities

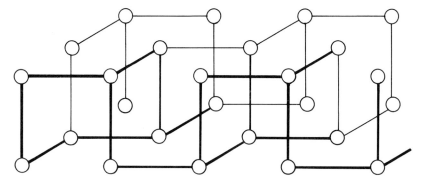

Fig. 4. Phosphorus, layered form.

are given in Table I but they are for polycrystalline material. Although black P is often referred to as a conductor it is a poor one under ordinary conditions. However, it has a high temperature coefficient of conductivity of $+0.0077$ deg^{-1} and a high pressure coefficient. Thus, at 20 katm. the conductivity has risen to 2×10^2 ohm^{-1} cm^{-1} [4]. In addition to these layer forms there are tetrameric species which, as noted above, are less stable because of the 60° angle between the bonds. Their ease of conversion to the layer form increases from P to Bi and indeed for bismuth, only the layer form is known. The first of these, white phosphorus, condenses from the vapor as a colorless solid and remains so if kept in the dark. It is very soluble in benzene and in CS_2 and it catches fire in air at 325 K. Jt was first converted to black P by Bridgman [4a] in 1914 who held the white variety at 500 K and 12 katm for 0.5 h. In 1955 it was prepared at 1 atm by Krebs et al. [4b] who used finely divided Hg and a seed of black P to catalyze the transition at about 650 K. More recently, modern high pressure equipment has stimulated the use of the original Bridgman method. Thus, Wentorf [4c] used the belt type apparatus developed for the conversion of graphite to diamond. At 22 katm he heated white P to 1500 K and then cooled it slowly to obtain a mass of black P crystals. From the partially crushed mass he was able to obtain imperfect crystals of black P. In 1968, Boksha [4d] subjected white phosphorus to an argon pressure of 15–25 katm at 200–300 K above the melting point of black P (1275 K at 20 katm). In 2–3 h. he obtained individual plates 0.15–0.25 mm thick and measuring 4–6 mm across. They were opaque to visible radiation and were metallic in reflected light. The flakes were stable in moist air and will hopefully be used to obtain the sort of data that have been obtained for graphite. Other workers have reported that black P can be ignited with a match and is attacked by cold HNO_3 although not by H_2SO_4. It is insoluble in CS_2. A third form of phosphorus is the red variety produced when white P is held at 520 K or is exposed to the action of light or X-rays. It is not oxidized at 300 K by O_2 and is not appreciably soluble in any known solvent. It has a density of 2.3 g cm^{-3} which is intermediate between those of the white (1.83) and black (2.70) varieties. The structure is not known but is believed to be one or more highly polymerized materials whose structures are intermediate between those of white and black phosphorus.

The next tetramer in the series is yellow arsenic obtained by suddenly cooling hot vapor. It consists of As_4 tetrahedra and changes to the stable grey layer form on heating. Yellow antimony is obtained by the action of O_2 on liquid stibine SbH_3 at 185 K. The Sb_4 units go over to a black amorphous form even in the dark at 185 K and then, on warming, the grey layer structure is produced. Larger crystals up to 8 mm across of the layer forms of As and Sb have been grown hydrothermally in 5–10 M HI [4e]. Other methods of obtaining large layered crystals of As, Sb and Bi from the melt or from vapor are described by Goss [4f]. These three behave as metals which are less electropositive than hydrogen and therefore dissolve only in oxidizing acids. The ease of oxidation decreases from As to Bi.

In Group VI there is, again, a difference between the first member, oxygen, in which the strong π bond overlap of $2p$ orbitals accounts for the stability of the only form O_2, and the other members in which the weaker overlap gives rise to linear polymers such as S_x. The only resemblance to a layered structure is shown by the octomers such as S_8. These are puckered rings which, in the orthorhombic solid, are all parallel to the c axis of the crystal but not to one another. As a result, the crystal has a large positive birefringence with the c axis as the acute bisectrix. The closest distance of approach between the atoms of different molecules is 0.33 nm.

In Group VII, the arrangement of the diatomic molecules of Cl_2, Br_2 and I_2 in the solid can be thought of as layered. Thus, in the Cl_2 molecule the internuclear distance is 0.202 nm. These molecules are arranged in layers between which the Cl—Cl distance is 0.369 nm for some and 0.373 nm for others. These are close to the 0.382 nm separation between the atoms in the solid form of argon, the next element in the Periodic Table. This suggests a van der Waals' attractive force as in argon. However, within a layer, the Cl—Cl distance of 0.334 nm is intermediate between these van der Waals' values and the sigma bond separation of 0.202 nm. Hence, within layers there is interhalogen bonding as in polyhalogens such as Cl_3^-, while between layers the interaction is weaker. Apart from this, however, these solids have not been reported to show layered properties such as flaking or intercalation.

2. Varieties of Graphite

2.1. Natural Graphite

Size varies from the 20 mm × 0.5 mm thick flakes from Ticonderoga, New York to the sub-micron flakes from Mexico and other parts of the world. They consist of the hexagonal and rhombohedral forms and the third stacking sequence which is random or 'turbostratic'. The larger flakes are completely ordered and have only a few percent of the rhombohedral form. Grinding increases the fraction in this form and adds the turbostratic variety. Treatment with hot acid decreases the fraction of rhombohedral, perhaps by reversible intercalation (Section 3.4.2), while heating to 2300–3300 K decreases it to zero. Other defects found in natural graphite are screw dislocations, twinning planes, non parallel layer planes and edge defects. The latter lead to either bonding with foreign atoms or with atoms of neighboring layer planes.

2.2. POLYCRYSTALLINE CARBONS [5]

These are products of the pyrolysis of organic compounds above about 1000 K. They resemble graphite in that they consist of substantially plane condensed aromatic layers stacked in roughly parallel groups but may differ from ideal graphite in the following ways:

(1) Spacing and mutual orientation of the layer planes within each stack.
(2) Defects within these planes.
(3) Dimensions of each stack.
(4) Mutual orientation of the stacks which also determines the pore structure of the solid.
(5) Carbon atoms not in layer planes.
(6) Non carbon atoms.

These differences lead to considerable variations in the physical and chemical properties. To a variable degree the structure and properties approach those of graphite on heat treatment to about 3300 K in vacuum or inert atmosphere. When such change is significant it is referred to as graphitization and the product is a polycrystalline graphitic carbon. It is convenient to distinguish eight structures:

(1) Hard, or non graphitizing carbon in which the stacks are randomly oriented and remain so on heat treatment. These usually come from a solid phase such as cellulose, sucrose or polyvinylidene chloride.

(2) Soft, or graphitizing carbon in which the stacks have some degree of mutual order around pores. These are most likely to be formed from molecules which already exist as large flat polyaromatic molecules and which pass through a fluid phase where there is adequate mobility for ordering to occur. Thus, pitch, petroleum coke and polyvinyl chloride are examples. Above 1300 K the d spacing decreases from the turbostratic value of 0.344 nm towards the graphitic value of 0.3354 nm. The apparent layer diameter and stack height increase.

(3) Carbon blacks in which a gas is pyrolyzed and may pass through a fine liquid drop stage to produce polyhedral particles in which there is some ordering of hexagonal carbon layer planes parallel to the surface. On heat treatment the order between these layer planes improves and the separation decreases. A re-examination of the X-ray diffraction data has led Ergun [6] to suggest that the basic building block is not the crystallite or stack of layers about 2 nm in diameter that had been previously obtained from X-ray line broadening measurements but rather much larger sheets of carbon. This has been confirmed by high resolution electron microscopy [7]. The size distribution is not known but the layers follow a concentric pattern and probably become smaller toward the centre of the particle. It is speculated that the central portion is

composed of the smallest layers having the greatest curvature, accounting for its being the most susceptible to oxidative attack.

(4) Glassy carbon is formed during the pyrolysis of a polymer which retains its original morphology without passing through a plastic phase [8]. One starting material is a phenolic resin and the final brittle product is believed to be a network structure consisting of tangled aromatic ribbon molecules which are cross linked by highly strained carbon-carbon covalent bonds. The development of parallel stacked extensive sheets of graphite is prevented because continuity is preserved along the length of each ribbon by strong C—C bonds which would have to be ruptured to produce extensive areas of graphite. The C—C bonds between ribbons are much weaker and hence the stacks of ribbons can be aligned to some extent by pulling fibers at high temperature. The network structure contains micropores as shown by its low density of about 1500 kg m^{-3} compared with 2200 for single crystals of graphite.

(5) Carbon fibers are formed when the organic material is stretched or drawn into a fiber during pyrolysis [9]. Favorable mechanical properties have led to much work on fibers drawn from rayon and from polyacrylonitrile (PAN). The primary unit is a wavy ribbon of a graphite layer plane cut from a basal plane in any arbitrary orientation. These ribbons are stacked in more or less parallel, but turbostratic sequence and have a preferred orientation parallel to the fiber axis. There are voids between the stacks which are enclosed and long (30 nm × 1–2 nm). The turbostratic stacks increase in size with heat treatment and may reach 10 nm or more by about half that much in thickness.

(6) Pyrolytic Carbon [10] – when relatively low molecular weight hydrocarbons are pyrolyzed, the twin processes of dehydrogenation and condensation produce an entire series of molecular species from aromatics to complex transitional molecules of low hydrogen content and high molecular weight. These lead to two forms of carbon: One is a soft powder or carbon black from the pyrolysis of gas borne droplets. The other is a hard surface deposit from the condensation of large complexes on surfaces in the pyrolysis chamber. The relative amounts formed depend on temperature, time and the geometry of the pyrolyzing chamber. The deposited carbon can have a variety of structures but most of the carbon atoms are in planar hexagonal arrays linked together by strong covalent bonds. These arrays are more or less parallel and are turbostratically stacked. By heat treatment under pressure at 3300 K–3600 K the overall structure approaches that of hexagonal natural graphite. Because the sample can be far larger and thicker (>10 mm) than natural flakes it has been useful in studies of the effect of size on intercalation reported in Section 3.3.2.

(7) Columnar Carbon – when low molecular weight hydrocarbons are pyrolyzed in the presence of smooth surfaces such as glass or ceramic, the deposit is the surface carbon described in the previous section. If certain crystals are present in the surface however, they will initiate the growth of fibrous grains oriented with their major axis more or less perpendicular to the substrate surface. Three forms of this columnar carbon have been reported and are illustrated in Figure 5. In the first two the layers of the carbon network are either in the form of closed cylinders around the column axis

or are rolled up around this axis into tight scrolls. Both forms demonstrate the preference of growth in the direction of the covalent C—C bonds.

The third form has been recently reported to form by the decomposition of CO on twinned SiC crystals above 2100 K [11]. Unlike the first two columnar carbons these fibers have a remarkably constant diameter of 3–6 μm and grow in length to 1 mm or

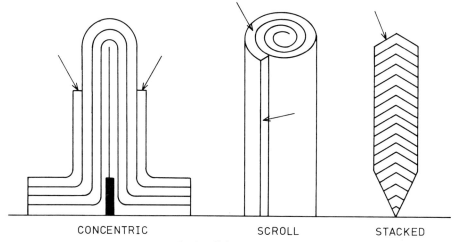

CONCENTRIC   SCROLL   STACKED

Fig. 5. Columnar carbon.

more. The substrate end is conical and the top is a specular cone with an angle of about 141°. The columns cleave very easily parallel to the top cone. These remarkable fibers are believed to be initiated by a cone nucleus from the tip of which circular growth fronts are emitted. The steps on the conical surface are fed by C atoms which reach the step by surface diffusion after having been formed by CO decomposition on the adjacent layer planes. The spacing of 0.340 nm shows a turbostratic ordering of the layer planes that spiral around the axis of the column.

(8) Graphite-Nickel system – at high temperature and pressure an anomalous growth of graphite has been observed [11a] which, according to a microprobe, contains 1.5 ± 0.5 weight %Ni. The electron microscope at $50000 \times$ shows no evidence for a second phase down to 10 nm.

3. Intercalation in Graphite

3.1. General behavior

Of the five elements that are classed here as layered structures all form flakes when mechanically stressed but only graphite has been reported to form intercalation or lamellar compounds on exposure to certain substances. In this process, the C atoms in the layers remain in planes and, to a first approximation, maintain their internuclear distance of 0.142 nm and hence their sp^2 bonding. However, some or all of the interplanar distances are more or less doubled and those spaces now contain the intercalated substance – often called the adduct. The ratio between the number of atoms of carbon and of adduct is sometimes a whole number and there is evidence for charge transfer to or from the adduct – hence the term 'compound'.

The following substances have been reported to intercalate graphite and reviews of their behavior are [12] [13] [14] [15] [16].

(1) Group I metals Li to Cs.
(2) Solutions of the metals of Group I and II and of certain others in liquid NH_3 and in certain other solvents.
(3) Br_2, Cl_2, ICl and other interhalogens, BCl_3.
(4) Various metal chloride vapors but only in the presence of Cl_2.
(5) Various metal chlorides from solutions in nitromethane or in certain other solvents.
(6) Certain Bronsted acids – HF, H_2SO_4, HNO_3, etc.
(7) CrO_2Cl_2 and CrO_2F_2.
(8) Certain oxides and sulfides – CrO_3 etc.

The usual method of preparation has been to hold the reactants at a known temperature and pressure – or concentration – and note the weight increase after removing any excess adduct with a solvent or by sublimation. This final step may remove some intercalated material [17] and hence it is preferable to eliminate it by using a pressure at which no condensation of solid or liquid adduct occurs on the graphite. Reported compositions do vary partly because of this problem and partly because the isotherm is a function of flake size and structure. The general form of an isotherm (of Br_2 around 300 K) is shown in Figure 6 but the exact form varies in the following ways:

(1) The curvature in the region of the first detectable weight increase varies. Part of this may be caused by surface adsorption and indeed, if the specific area is large, the weight increase will begin near zero pressure. Even if this is negligible, the threshold pressure p_t is best taken as the extrapolation of the adsorption curve to zero composition. This p_t has been reported to increase as the degree of perfection of the lattice decreases and as the thickness of the sample in a direction normal to the layer planes increases [18].

(2) Adsorption curves are often reported to have slope changes in the region of certain compositions and these have been shown by X-ray diffraction to be the various 'stages'. The stage number is the number of carbon planes between adjacent layers of adduct. Thus for Br_2 its value near saturation pressure is 2 and below that pressure, values of 3, 4 and 5 have been reported [19]. For other adducts the reported values near saturation range from 1 for ICl to perhaps 5 for WCl_6. Even for a given adduct the composition near saturation will vary with the degree of perfection of the graphite and with the thickness normal to the layer planes [20].

(3) As the pressure approaches saturation capillary condensation in the pores of a powdered sample contributes to the weight increase and causes the rise above 0.90 shown in Figure 6. The extra amount increases with the degree of subdivision of the sample [21].

(4) On desorption, hysteresis is the rule and slope changes have been reported near the compositions for certain stages. At zero pressure, some adduct always remains and

the structure is called a residue compound. The use of the word compound is inappropriate because its composition is a function of the degree of perfection of the graphite, of the composition at which the pressure was reduced to zero, and of the rate of reduction of that pressure [22]. Subsequent heating causes further loss but never to zero composition.

The formation of an intercalation compound involves the transfer of electron charge either to the π bond system by the metals and their solutions, or in the opposite direction for the other adducts. This is shown by the sign of the Hall coefficient and in both cases results in an increase in conductivity. Along the a-axis, parallel to the planes, this increase can be by a factor of 10 which makes the conductivity greater than that of nickel. Along the c-axis there is a remarkable difference between donor and acceptor compounds [24]. For a donor, such as potassium, the increase is 100 which agrees with the hypothesis of metallic bonding along the c-axis in metal compounds.

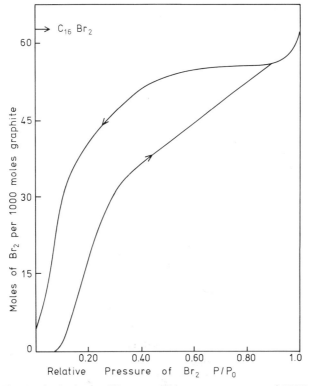

Fig. 6. An isotherm of Br_2 on graphite at a temperature of 300 K.

For an acceptor, such as Br_2 or ICl, the increase is a factor of 2 which agrees with the hypothesis of van der Waals' bonding in that direction. The difference, however, might have been partially caused by a difference in stage number. Thus, the potassium was in stage one whereas Br_2 would be in its minimum value of stage 2 or higher and the authors were not sure whether the ICl was in stage 1 or 2.

Superconductivity has been observed in stage 1 metal compounds such as C_8K [25]. The transition temperatures are low, being 0.56 K for potassium and lower for Rb and Cs. For the stage 2 compound $C_{24}K$, no superconductivfty was found down to 0.011 K. Hannay *et al.* [25] suggest that this is because van der Waals' bonding now plays a role along the *c*-axis whereas in the stage 1 compound it was metallic. Hence all conductivity is in the plane of the layers and theoretically superconductivity cannot exist in a mathematical plane. Salzano and Strongin [26] point out, however, that there may be a transition below 0.011 K and the fact that it is lower than for the stage 1 C_8K could simply be a result of the different number of conduction electrons in the conducting planes. Further studies of these and other compounds in the *c*-direction are obviously needed.

3.2. INTERCALATION BY HALOGENS AND BY METAL CHLORIDES

3.2.1. *Halogens*

Fluorine at 300 K forms the very inert C_4F in the presence of HF gas as a catalyst [13]. It appears that every fourth C atom forms a covalent bond to a fluorine atom and thereby removes one electron from the π bond system. The result is a decrease in conductivity but only to about 1% that of graphite. The carbon layer planes are believed to be still planar and are about 0.534 nm apart. The catalytic effect of HF may be caused by the formation of the intermediate $C_{24}^+HF_2^-$. 4HF from F_2 and a visible liquid film of HF on the surface of the graphite (Section 3.4.2). This is in accord with the observation that above 350 K, no C_4F forms and no liquid film forms. Above about 525 K a new product appears – $(CF_x)_n$, where $x = 0.7$ to 1.1, called poly carbon monofluoride. Every C atom now forms a covalent band to fluorine and the layers are puckered – because of the sp^3 bonding – and about 0.66 nm apart. This nearly white product is remarkably inert, of very low conductivity and has valuable low friction properties at temperatures to 700 K. Finally, if F_2 contacts graphite above about 800 K there is an explosive reaction to form a mixture of gaseous fluorocarbons such as CF_4 and C_2F_6.

Cl_2, Br_2, I_2 and NO_2 have all been shown to lower the magnetic susceptibility of graphite or charcoal and this has been used as evidence for intercalation by Cl_2. However, several workers report that I_2 does not intercalate and there are no claims for NO_2. Hérold [27] measured the weight increase in Cl_2 vapor and concluded that is was caused by capillary condensation between the particles of graphite and not by intercalation. Hooley [28] determined the isotherm for several graphites and reached the same conclusion. He suggested that the decrease in susceptibility was a surface phenomenon. The only X-ray diffraction evidence for intercalation by Cl_2 was done on the residue compound from Br_2 [29] and that may have been detecting the intercalation of BrCl rather than of Cl_2. This is borne out in recent work by Furdin and Hérold [30] who present evidence for a ternary intercalation system of Br_2 and Cl_2 containing a stage 2 $C_8Br_xCl_{1-x}$. A similar product exists for I_2 and Br_2 [31] whereas for I_2 and Cl_2 it is a stage 1 $C_4I_xCl_{1-x}$ [32]. Other interhalogens that intercalate are

ClF$_3$ in anhydrous HF [33] which produces $C_{14}^+H_2F_3^-ClF_2^+HF_2^-$ and also BrF$_3$ and IF$_5$ [33].

Finally, BCl$_3$ was reported by Croft [34] to form a C$_{79}$BCl$_3$ at 285 K. Recently, Freeman et al. [35] determined the isotherms of BCl$_3$ on natural graphite and reported stage numbers 2, 5 and 7 for C$_{24}$, C$_{60}$ and C$_{85}$BCl$_3$.

3.2.2. Metal Chlorides

The first one reported was FeCl$_3$ in 1932 [36] and this was followed in 1956 by the 29 in Table II [34]. Since then a few more have been added but the main effort has been to obtain quantitative data on layer plane separation, stage number and composition.

TABLE II

Metal chlorides reported by Croft [34] to intercalate graphite

YCl$_3$	ZrCl$_4$	ReCl$_4$	CuCl$_2$
SmCl$_3$	HfCl$_4$	FeCl$_3$	AuCl$_3$
GdCl$_3$	TaCl$_5$	RuCl$_3$	AlCl$_3$
YbCl$_3$	CrCl$_3$	CoCl$_3$	GaCl$_3$
DyCl$_3$	MoCl$_5$	RhCl$_3$	InCl$_3$
EuCl$_3$	WCl$_6$	IrCl$_4$	TlCl$_3$
	UCl$_4$	PdCl$_4$	SbCl$_5$
	UO$_2$Cl$_2$	PtCl$_4$	

These have often been elusive for the reasons listed in Section 3.1. The problem is especially acute for those metal chloride vapors with a high threshold pressure relative to saturation. In such cases, the composition is a much more sensitive function of the degree of perfection of the graphite than for a metal chloride with a low threshold. These problems will be discussed in the next Section 3.3 but first, some of the developments since the excellent review article by Rudorff et al. in 1963 [37].

1963 – FeCl$_3$ was shown to intercalate graphite from its solution in nitromethane or nitroethane [38].

1966 – MnCl$_2$, NiCl$_2$ and ZnCl$_2$ vapors in the presence of Cl$_2$ intercalate graphite to give a stage 1 C$_{6.6}$MnCl$_2$, a stage 2 C$_{11.3}$NiCl$_2$ and a stage 3 C$_{16.5}$ZnCl$_2$ [39].

1967 – The isotherm of FeCl$_3$ vapor at 585 K was determined and showed a stage 1 C$_{6.3}$FeCl$_3$ near saturation [40].

1969 – SbCl$_5$ was reported to form a stage 1 C$_{12}$SbCl$_5$ and also stages 2, 3 and 4 [41]. This is to be compared with a C$_{46}$ by Croft [34] and a C$_{25}$SbCl$_5$ by Hooley [20] both near saturation.

1971 – SbF$_3$Cl$_2$ in an inert atmosphere at 363 K gives stage 1 $C_{27}^+SbF_3Cl_2^-$. 2SbF$_3$Cl$_2$ and also stages 2 and 3 [42]. This and UF$_6$ [43] are the only metal halides that have been reported to intercalate graphite in the absence of free halogen.

1972 – PtCl$_4$ was found not to intercalate even in the presence of Cl$_2$ at 625 K [44]. However, at 425 K, H$_2$PtCl$_6$.5H$_2$O formed, in the presence of Cl$_2$, a stage 3

$C_n^+ (HPtCl_6)^- (H_2PtCl_6)_x$. At higher temperatures, HCl was lost and the maximum final composition was $C_{42}PtCl_{4.3}$. This is the first case of the conversion of an acid salt of graphite (Section 3.4.2) to a metal chloride compound.

Nine metal chlorides were reported to intercalate graphite from their solutions in nitromethane [45]. The expansion of a single flake was used as a measure of intercalation and this was found to increase with concentration to a maximum, to vary with the metal chloride and to be less for flakes thicker than about 30 μm. These effects will be discussed in the next Section 3.3.

1973 – $AuCl_3$ liquid at 575 K forms a stage 1 $C_{13}AuCl_3$ [46].

Thionyl chloride was found to be a solvent from which various metal chlorides would intercalate graphite [47]. A stage 1 $C_{19}UCl_5$ did contain some solvent but none was present in a stage 2 $C_{18}AlCl_3$, or in stage 3 $C_{40}NbCl_5$, $C_{32}TaCl_5$ or $C_{41}MoOCl_4$.

The isotherms of 9 metal chloride vapors on graphite were determined [20] and will be discussed in Section 3.3. They are summarized in Table III.

3.3. THEORY OF INTERCALATION

3.3.1. Introduction

The extent to which a given substance will intercalate graphite is determined by the threshold pressure and the stage number near saturation. If either of these is too high, the weight increase associated with intercalation will not be detected. Although there is still no theory that will predict these quantities for a given substance, there is a conjectural theory that will correlate some of the data associated with halogens and metal chlorides [20]. Briefly, it is that adsorption of adduct on the two end basal planes transfers charge and therefore modifies the band structure of the graphite so that bonding in every nth space is weakened sufficiently to allow intercalation in those spaces. The value of n is the stage number and becomes one when sufficient charge has been transferred. This may never happen for some substances, even at saturation pressure. However, n will have some value at all pressures above zero. At the threshold pressure n becomes small enough so that either the increase in weight or in thickness can be detected. If n remains above this critical value at saturation pressure then the substance is classed as a non intercalater. Obviously, a more sensitive method of detecting large values of n is needed before this theory can be tested. However, the data in the following four sections does lend some weight to the theory.

3.3.2. Expansion and Contraction of the Interlayer Spaces

A time sequence study has been reported [18] of the expansion of a pyrolytic graphite cylinder of 3 mm diameter with the carbon planes normal to the length of 6 mm. It was marked at four points along the side and the five distances were measured with a travelling microscope during intercalation by Br_2. The end sections always expanded first thus showing that intercalation starts at opposite basal planes and proceeds towards the center. Furthermore, if the two ends were sealed with 2 mm long glass caps and grease, the weight increase was zero in a pressure of Br_2 that would have intercalated the uncapped cylinder. Apparently, adsorption of Br_2 on or near the end basal planes is necessary for intercalation. If this adsorption does establish the stage number by making every nth space more weakly bonded, then intercalation should start in all such spaces on initial exposure to adduct rather than at the opposite basal planes as observed. One must therefore postulate either (1) the effect of the adsorption is felt in only the nth space from the basal plane and it is the adduct that intercalates in this space which, again by charge transfer, opens the $2n$th space and so on. Or (2) although every nth space has been modified, it is those adjacent to the two basal planes that are initially intercalated because of the greater flexibility of one of the two enclosing layer planes. This would be the outer one. In the central region, both enclosing planes are inflexible. According to this, the process might stop after a certain number of spaces has been intercalated because the enclosing planes are further from the basal plane and therefore less flexible. This indeed has been observed for some adducts. Thus, a nitro-methane (NM) solution of $FeCl_3$ [45] causes an expansion of 60% in an 88 μm thick flake and even after several days there is no further change. If the solution is then replaced by pure NM, there is a shrinkage to a steady thickness followed by expansion when again placed in the $FeCl_3$ solution. However, at the end of the second expansion the total increase is 85% of the original 88 μm and after seven cycles this levels off at 140%. In terms of the theory, the adsorbed adduct gave a certain 'n' value but intercalcation stopped after only a fraction of the spaces had been filled because the enclosing layer planes of carbon became too inflexible. Then on treatment in NM the loss of $FeCl_3$ caused exfoliation or splitting of the two outer portions which had been intercalated. Hence, on reexposure to $FeCl_3$ solution, the intercalation proceeded another fraction of the total and finally, after a number of cyclic treatments which should be and is greater for a thicker flake, the two reaction fronts meet in the region half way between the two basal planes. For a 20 μm thick flake, the initial exposure gave an expansion of 150% and the above cyclic process did not exceed this. Again, the theory would say that 20 μm is sufficiently thin so that flexibility is adequate for complete intercalation during the first exposure.

For other adducts however, this cyclic effect in thick flakes is not observed. Thus, one exposure to Br_2 of a thick natural flake or a well ordered pyrographite is sufficient to complete the intercalation. There is, however, a behavior with a 2475 K pyrographite in Br_2 which may be related to the cyclic process. It is that the sample splits into thin platelets during intercalation [18]. It has been suggested that the expansion of the lattice around the periphery of the carbon planes sets up a tension normal to those planes. This tension is in the central core and is surrounded by a region under compression. In the case of Br_2 in the 2475 K pyrographite the tension splits the sample before the compression stops the intercalation. In the case of the NM solution of $FeCl_3$ in a thick flake, compression stops the intercalation before the sample splits. All this points to a balance amongst threshold pressure, flake thickness and layer plane stiffness. The latter increases with the distance from the basal planes and is less for layer planes that are more nearly parallel as in heat treated pyrolytic graphite or in natural flakes.

The theory would predict that just as intercalation starts at the spaces adjacent to opposite basal planes, so also does the removal of intercalated material. The surrounding pressure of adduct will determine the stage number and indeed, stages are observed along desorption isotherms. Furthermore, just as intercalation of at least certain substances such as $FeCl_3$ in NM has been observed to stop after a certain fraction of spaces has been intercalated, so also the removal of $FeCl_3$ from C_6FeCl_3 to a solvent [17] and the thermal decomposition of C_6FeCl_3 to $C_6FeCl_2 + Cl_2$ [48] have been observed to stop before completion of the process. In all three cases thickness was a factor and it is suggested that the greater flexibility of the outer layer planes was responsible. When either intercalation or its reverse has penetrated to sufficiently rigid layers then the process stops.

3.3.3. Structure of Adsorbed Layers

If charge transfer to or from the basal planes is a factor in determining threshold pressure p_t and the stage number, it should be possible to relate the structure of the adsorbed layer to those quantities. One would expect the large transfer required for a low p_t and low stage number to be associated with a high degree of order and, even more important, of epitaxy with the hexagonal mosaic of C atoms which constitute that basal plane. Well, Lander and Morrison [49] used low energy electron diffraction to examine the degree of order of various adsorbed layers on a basal plane of graphite. They found that some but not all adsorbates gave a two dimensionally ordered layer above a certain pressure. The ordered group contains the three intercalators studied – Cs, Br_2 and $FeCl_3$ – but some others as well. This provides some support for the theory. It is also significant that order appears only above a certain pressure. The value of the stage number might therefore decrease sharply over a narrow pressure range and thereby account for the fairly sharp threshold pressure observed for the first detection of intercalation. Above this p_t the increase in charge transfer per unit of pressure change will determine the composition near the saturation pressure p_0. If all adsorbates were similar in this respect, then a lower p_t would lead to a higher composition near saturation and this is indeed true for most of the 13 adducts whose isotherms have been measured. These are summarized in Table III. For $ZrCl_4$, $SbCl_5$ and $TaCl_5$ however, the coefficient of charge transfer per unit pressure change must be less to account for their relatively lower compositions near saturation.

The case of $MoCl_5$ is of particular interest here. The intercalated material has been shown [50] to be partly in the form of discrete molecules of Mo_2Cl_{10} with a definite orientation relative to the carbon plane and partly in the form of a continuous layer with no particular arrangement of its atoms with respect to the carbon layer. A low rate of intercalation [20] may have been caused by a slow establishment of an equilibrium between these two forms – presumably favoring the epitaxial molecular form. A study of the fraction in each form as a function of time of reaction would be of interest.

TABLE III

Threshold pressure and composition

Adduct	P_t/P_o	Maximum composition at P_o	
		Adduct/Carbon	Stage
K, Rb, Cs	very low	0.125	1
$AlCl_3$	$<10^{-4}$	0.12	1
$GaCl_3$	10^{-4}	0.12	1
$SbCl_5$	<0.01	0.04	
$FeCl_3$	0.01	0.15	1
$ZrCl_4$	<0.02	0.07	
$TaCl_5$	<0.03	0.03	
ICl	0.05	0.12	1
Br_2	0.10	0.06	2
$InCl_3$	0.1	0.07	2
$MoCl_5$	0.3	0.03	3
WCl_6	0.5	0.01	5
$HgCl_2$	0.6	0.04	3

The behavior of the halogens with graphite can be related to their predictable epitaxy with a carbon plane. Thus Br_2, with an internuclear distance of 0.227 nm, has been shown [51] to form chains of Br_2 molecules parallel to the carbon layer planes and the atoms in these molecules are centered over neighboring hexagons. The center to center distance of these hexagons is 0.242 nm which is 0.015 nm greater than that of the atoms in Br_2. For the other two intercalators this difference is even less as can be seen from the internuclear distance of 0.240 nm for ICl and 0.249 nm for IBr. Furthermore, for the two non intercalators the difference is greater than for Br_2. If this is a factor, then Cl_2 does not intercalate [28] because the separation of 0.202 nm between its atoms is too small for epitaxy of both atoms and I_2 does not intercalate because the separation of 0.268 nm is too great. This neat relationship may be fortuitous because there is another factor and that is the amount of charge transferred to or from an atom which is directly over a carbon hexagon.

Finally, the Cl_2 or nitromethane that promotes the intercalation of a metal chloride may act by enhancing the charge transfer. The structure of an adsorbed layer on graphite should be examined in the presence and the absence of these agents to see whether the pressure at which order first appears is changed.

3.3.4. Function of Cl_2 in the Intercalation by a Metal Chloride

A curious fact about all the metal chloride intercalations in Table III is that some Cl_2 must be present. For $AlCl_3$ vapor at 455 K and a pressure relative to saturation of 0.7 the equilibrium composition as a function of Cl_2 pressure (calculated for 293 K) is shown in Figure 7 [20]. The amount of Cl_2 in the compound was calculated from the decrease in pressure during the intercalation. The mole ratio of $AlCl_3$ to C is seen to rise to its top value of 0.106 at a pressure of 50 torr of Cl_2 (293 K). The Cl_2 content also rises with Cl_2 pressure but the data are not sufficiently accurate to establish the form of the curve. However, the amount above 300 torr is 0.25 mole Cl_2 per mole of $AlCl_3$ which is close to the value of 0.17 Cl_2 obtained by Dzurus and Hennig [52] from direct analysis of the chilled product. Note that if one assumes that all the weight increase is $AlCl_3$, one obtains $C_{8.4}AlCl_3$ instead of $C_{9.5}AlCl_3(Cl_2)_{0.25}$. This has been done for all the metal chlorides in Table III because it is only for a few that the

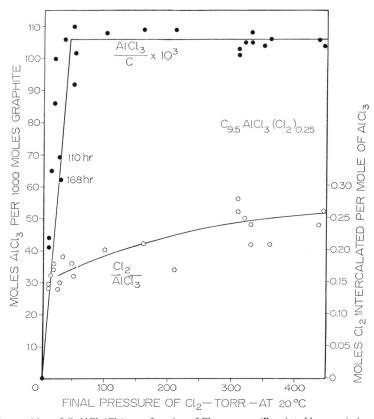

Fig. 7. Composition of $C_m AlCl_3(Cl_2)_n$ as a function of Cl_2 pressure. (Reprinted by permission from [20].)

amount of Cl_2 has been measured. For certain metal chlorides such as $FeCl_3$, $MoCl_5$ and $SbCl_5$ the composition near saturation is independent of the pressure of added Cl_2 from 0–600 torr. These chlorides, however, generate their own Cl_2 by decomposition and the pressure of that Cl_2 is presumably sufficient to promote the maximum composition. The amount of Cl_2 that intercalates along with the metal chloride near its saturation pressure varies from a low of Cl_2/MCl_x of <0.045 for $InCl_3$ to about 0.20 for $AlCl_3$ and $GaCl_3$ to 0.34 for $HgCl_2$ [53]. For the others it is unknown.

There have been two theories about the need for Cl_2. One was that Cl_2 intercalates graphite and therefore 'opens' the layered spaces to the metal chloride. However, a recent report [28] claims that Cl_2 does not intercalate graphite and that the previous work, based on electric and magnetic measurements, was really measuring the effect of surface adsorption of Cl_2. The other theory was based on Hall coefficient data which show that the compounds with $AlCl_3$ and with $FeCl_3$ are p type. An electron acceptor was therefore needed and Cl_2 was given this role. Thus, Dzurus and Hennig [52] wrote $C_n^+ Cl^- \cdot 3AlCl_3$ and $C_n^+ Cl^- \cdot FeCl_2 \cdot 3FeCl_3$. Unfortunately, the amount of Cl_2 intercalated is not well known and the state of the chlorine is in some doubt. Thus, Rudorff and Landel [53] prefer $[AlCl_4]^-$ to Cl^- for the negative product in the electron transfer process. In the case of $FeCl_3$, Mössbauer spectroscopy has been used to look at the intercalated Fe nucleus. Hooley et al. [54] found electron transfer to the Fe(III) which is consistent with the Hall coefficient data. They concluded that at the most, 5% of the iron was in the Fe II state and found a similar trace in 'pure' $FeCl_3$ presumably because of some decomposition during its preparation. Freeman [55] reported that $FeCl_2$ and $FeCl_4^-$ were not present in large proportion, if at all. Grigutsch et al. [56] confirmed these findings and Novikov et al. [57] estimated that 7–18% of the area under the curve was associated with Fe(II). In the original Hennig formula 25% of the iron was Fe(II). All of this throws doubt on the above formulas for the compounds with $AlCl_3$ and $FeCl_3$ and suggests that the charge transfer required by the Hall coefficient is to the metal chloride plus any Cl_2 that does intercalate with it. There is no indication of the distribution amongst the three acceptors except that Fe(III) does receive enough to change its Mössbauer isomer shift.

A related phenomenon is the intercalation of graphite by a solution of a metal chloride in nitromethane (NM) or nitroethane [38] [45]. This occurs for all the metal chlorides in Table III and, at least for $FeCl_3$, some of the NM enters the lattice. Because this NM is rather labile, the final composition is not well known. The extent of intercalation is therefore easier to estimate by measuring the expansion of a single flake placed in the solution and in the field of a travelling microscope. These expansions however, bear no obvious relative relationship to the compositions measured in the chilled products obtained from vapor phase intercalation at high temperatures. Other solvents in which $FeCl_3$ will also cause expansion of graphite are $C_3H_7NO_2$, $C_6H_5NO_2$, CH_3CN, C_2H_5CN and C_3H_7CN [58].

3.3.5. Long Range Order

Sato and Toth in 1961 [59] made X-ray diffraction studies of the alloy CuAu which has been known since 1936 to have two ordered structures. In one of these, there are alternate layers of Cu and Au. In the other, stable over a different temperature range, there is an alternation of the atom in a given layer every 5 unit cells, that is, every 2.0 nm. Now what they found which might be applicable to stages in graphite compounds was that the value of 5 unit cells for the pure alloy could be changed by the addition of a few atom percent of an impurity metal. The change was related to the number of valence electrons in the impurity. They were able to explain this by the effect of the added electrons on the band structure of the alloy. Perhaps in graphite a similar transfer modifies the band structure to weaken the bonding in every nth space.

Another system in which a change in periodicity can be explained by quantum mechanics is that of the polymethines. The free electron molecular orbital (FEMO) model was used by Kuhn [59a] to predict the absorption spectra of these linear, conjugated systems. If the molecule was symmetrical as for the dye cation

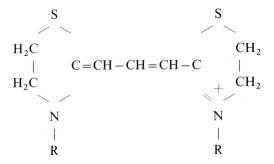

then the two resonating forms, only one of which is shown, are of equal energy and the free electrons along the chain show a periodic potential with a cycle of one bond length. If, however, the molecule is unsymmetrical then the two resonating forms are of unequal energy and a periodicity of two bond lengths is introduced. If graphite is thought of as a giant molecule consisting of carbon sheets held together by the delocalized π electron systems, then it is symmetrical. If the charge on the two basal planes is changed by adsorption, the 'molecule' remains symmetrical. However, real flakes are several thousand layers thick and if, at that distance, one basal plane does not influence the other, then each half of the structure is unsymmetrical and hence a new periodicity may be introduced by adsorption on the basal planes. This is very conjectural because for one thing, the graphite molecule is not a conjugated system normal to the layer planes. However, the interaction of the π bond systems is changed by intercalation as shown by the changes in conductivity described in Section 3.1 and by the change in layer order form the *abab*... of hexagonal graphite to exact registry for the two carbon layers that enclose one layer of intercalated material (Section 3.4.2). Finally, I hope that these conjectural comments will inspire someone to explain the intriguing existence of stages and a threshold pressure.

3.4. INTERCALATION BY OTHER ADDUCTS

3.4.1. *Metals*

From the vapor phase, only the alkali metals Li, Na, K, Rb and Cs have been reported to intercalate graphite. From solution in various solvents, however, intercalation occurs for not only these alkali metals but also for the alkaline earth metals, and certain others. Four groups of metals will be considered:

(a) K, Rb and Cs
It is noteworthy that in the first paper on alkali metal intercalation [60] Fredenhagen and Cadenbach reported that K, Rb and Cs all gave about C_8M with not only natural graphite but with charcoals and lampblacks. In natural graphite it has been shown that every space contains metal and that the atoms are centered over carbon hexagons –

over 25% of them to account for C_8M. Now if the charcoals and lampblacks also give C_8M then it must mean that all the C atoms in these structures are involved in the reaction and that they are close enough to regular hexagons in their arrangements so that metal atoms centered over 25% of them can account for the common C_8M. Otherwise, large variations in composition would occur amongst the various pyrocarbons. This absence of variation has been recently confirmed by Salzano and Aronson [61] for two carbon blacks from the Cabot Corporation – a graphitized Sterling MT-D5 and a non graphitized Spheron 6 – both of which gave a C to Cs ratio of 8.0 ± 0.4 as measured by a tracer technic using radioactive Cs. Their thermodynamic data showed that the cesium had a lower degree of order in the pyrocarbon than in natural graphite. They attributed this to greater 3 dimensional order of the Cs atoms in the well ordered layer plane system of natural graphite as compared with the turbostratic layer planes of a carbon black. This same composition of C_8K has been found for another pyrocarbon, this time in the form of carbon fibers [62]. Again, an absence of 3 dimensional order amongst the K atoms was shown by X-ray diffraction. Still another pyrocarbon – glassy carbon – has been shown by X-ray diffraction [63] to form a stage 2 compound with potassium. The overall composition was not determined so that the fraction of C atoms involved is unknown. However, at least some of the turbostratic stacks of carbon planes are alternately associated with K atoms.

The above behavior is very different from that of Br_2 for which the weight increase near saturation ranges from zero for hard carbons, such as glassy carbon, to C_8Br for natural graphite. The difference may be related to the threshold pressures for intercalation. For graphite, these values, relative to saturation, are about 10^{-4} for K, Rb and Cs and 0.1 for Br_2. For pyrocarbons, they will be higher and hence the isotherms will be shifted to higher pressures. For K, Rb and Cs this does not change the composition near saturation except for glassy carbon. For Br_2 it does change that composition to much lower values.

In addition to the stage 1, C_8M near saturation there are higher stages at lower pressures. For all 3 metals a total of 5 stages has been identified with $n = 8, 10, 24, 36, 48$ and 60 in the formula C_nM. The first 2 are stage 1 and the others range from 2 to 5. The enthalpies and entropies of their interconversion have been measured for well ordered graphite and are summarized by Aronson et al. [64].

The compounds of potassium with well ordered graphite have been used by Nixon and Parry [65] to determine the influence of intercalation on the C—C bond length within the layer planes. They found an expansion which was uniform within a single layer and which was constant for all n carbon layers separating successive intercalate layers in the stage n compound $C_{12n}K (n>1)$. The expansion is consistent with electron transfer from K to the antibonding orbitals in the upper π band of the graphitic region between successive intercalate layers. The discontinuous change of bond length with stage is approximately

$$C—C = [0.14203 + 0.00113/n] \text{ nm}.$$

The fact that the expansion is the same for all n carbon layers implies that the electrons must be delocalized within the unit of graphite structure lying between adjacent layers of potassium.

(b) Sodium

Again, it was Fredenhagen and Cadenbach [60] who first reported that sodium reacted with lampblack but not with hexagonal graphite. Hérold [27] confirmed the lack of reaction with natural graphite at 775 K although he thought there might be some reaction at 1275 K because the product, on exposure to water, produced both H_2 and C_2H_2. This showed that it contained sodium metal, perhaps intercalated, and sodium carbide. He also exposed an artificial Acheson graphite to the alloy CsNa and found that the product gave, with HCl solution, both CsCl and NaCl and hence may have contained intercalated Na. Asher and Wilson [66] were the first to assign a composition to a possible intercalation compound of Na with graphite. At 395 K to 775 K they obtained $C_{64}Na$, a stage 8 intercalation compound, and used X-ray diffraction and magnetic susceptibility to identify it. Hennig et al. [67] however, found reaction at 675 K only in the presence of added contaminants such as O_2, H_2O or H_2 and suggested a ternary system involving the oxide or hydride of sodium. They used X-ray diffraction and electrical conductivity to follow the reaction. Recht et al. [68] used a quartz spring balance to suspend their graphite in sodium vapor at a pressure 70% of saturation. They obtained $C_{64}Na$ only in graphite that had been equilibrated with sodium vapor at a higher temperature and then held at 725 K for seven days. The product was identified by X-ray diffraction as essentially pure $C_{64}Na$. Sleppy [69] separated a spectroscopic graphite electrode from molten $NaNO_3$ by a soda glass tube, and by electrolysis transferred Na to the graphite at 675 K in argon. The weight and magnetic susceptibility were then measured before treating the product with water. Titration with HCl showed the presence of both carbonate and hydroxyl ions. Only the latter was found to be related to the decrease in magnetic susceptibility and was therefore assumed to come from intercalated sodium which would give hydroxyl ion with the water. The diamagnetism was completely eliminated at 2 wt.% of this intercalated Na or $C_{96}Na$. A second form of sodium was believed to be in the form of Na_2CO_3 because the carbonate found in the above titration depended on the amount of oxygen available. The total weight change showed the presence of a third kind of sodium, tightly bound and non reactive with water, and therefore resembling the adduct in the so called residue compounds of Br_2 etc. Aronson and Salzano [70] exposed a well annealed pyrolytic graphite to liquid Na at 925 K in a helium atmosphere in stainless steel for about 20 h. It was opened in a glove box and scraped free of excess Na. X-ray diffraction showed only the $C_{64}Na$ reported by Asher [66] and heating at 1475 K in vacuum removed all sodium as shown by analysis. The e.m.f. of the compound against liquid sodium was measured through a sodium glass at various temperatures to obtain an enthalpy of formation for $C_{64}Na$ of -6.3 kJ mol^{-1} Na and an entropy change close to zero. For the potassium compound nearest to this composition – $C_{60}K$ – the enthalpy change is -50 kJ and this difference, they suggest, accounts for the difficulty in preparing lamellar compounds of sodium. This, indeed, had been proposed by Dzurus et al. [67] on the basis of their plot of the free energy of formation of the K, Rb and Cs compounds against the ionization potentials of those metals. When extrapolated to the potentials of Na and Li, the free energy of formation was positive for stage 1 compounds but did become negative for stage 5 or 6. The subsequent work has confirmed this prediction.

Métrot and Hérold [71] used a Cahn balance to show that hard carbons and hexagonal graphite would not intercalate Na but that soft carbons at 875 K would give stage numbers of 1, 3, 5 and 8, the low ones being for the softer carbons. These, and a $C_{64}Na$ from impure graphite were confirmed by X-ray diffraction. Finally, Pflugmacher and Boehm [72] used a Gerateglas 20 system to distill Na onto various carbon blacks which were held at 575 K for 6 h before distilling off excess Na at 675 K. The product of a given run was either analyzed for Na or examined by X-ray diffraction. They found about $C_{12}Na$ for blacks that had not been heat treated above 1425 K. Above this temperature the Na content was less and became very small for blacks that had been heated above 2275 K. The increase in average interlayer spacing was surprisingly small – only 0.045 nm for the maximum uptake of Na and less for those blacks treated at a higher temperature. They concluded that the sodium is intercalated in an irregular fashion with preference for those parts of the structure where the distance between layers is much wider than the X-ray average.

(c) Lithium

Hérold [27] exposed natural graphite to Li vapor at 775 K in an evacuated system. Reaction was complete in 5 days and the product C_2Li_2 was the carbide because it produced only C_2H_2 on hydrolysis. He found, however, for shorter times of reaction, that products of lower Li content were produced and that these gave on hydrolysis not only C_2H_2, but also H_2. Thus, C_4Li gave about equal amounts of each gas. He concluded that there had been some intercalation. Dzurus et al. [67] used electrical conductivity as a measure of intercalation and found that, as with sodium, there was no intercalation in the absence of O_2 and H_2. In the presence of either of these gases, however, there was reaction and they postulated either catalysis or a ternary system involving Li_2O or LiH. Juza and Wehle [73] used direct analysis and X-ray diffraction to identify intercalation compounds C_6Li, $C_{12}Li$ and $C_{18}Li$ all containing some carbide but gave no details of preparation. Bagouin et al. [74] also used direct analysis and X-ray diffraction to identify intercalation compounds C_6Li, $C_{18}Li$, $C_{28}Li$, $C_{36}Li$ and $C_{72}Li$ from Ceylon graphite. They found that foreign substances such as H_2 or O_2 were necessary to cause intercalation and that the carbide was also produced. Recently, Guérard and Hérold [75] compressed a mixture of Li and Ceylon graphite in stainless steel at room temperature to obtain C_6Li, $C_{12}Li$, $C_{18}Li$ and $C_{27}Li$ free of the carbide. They identified these by chemical analysis, X-ray diffraction and differential heating rates.

(d) Metals in Solvents

Rudorff et al. [76] reported in 1955 that solutions in liquid ammonia of Li, Na, K, Rb,

Cs, Ca, Sr and Ba all gave stage one compounds of general formula $C_{12}M(NH_3)_2$ or, at lower metal concentration, stage two $C_{28}M(NH_3)_{2-4}$. They found that Eu and Yb would behave in a similar fashion and that methylamine or 1,2 diaminoethane could take the place of NH_3. Although Be, Mg and Al did not appreciably dissolve in liquid NH_3, intense blue solutions of these metals were obtained by electrolysis of solutions of the corresponding iodides. These solutions produced with graphite a stage 3 $C_6Be(NH_3)_2$, a stage 4 $C_8Al(NH_3)_2$ and a stage 4 $C_{32}Mg(NH_3)_2$ as well as compounds of La, Ce and Sm [77]. Craven et al. [78] used this same method to add Gd, Tb, Dy, Ho, Er and Tm to the list.

In 1965 Stein et al. [79] introduced a new method for the intercalation of metals which broadened the spectrum of possible solvents. They found that solutions in tetrahydrofuran of the naphthenides of Li, Na and K would intercalate graphite. Ginderow and Setton [80] have shown that benzophenone or benzonitrile can replace naphthalene and that dimethoxyethane can replace THF.

Finally, a solvated proton can be intercalated at a graphite cathode in a solution of an ammonium salt in liquid NH_3. The product is $C_m^- NH_4^+ \cdot nNH_3$ [81].

3.4.2. Acids

That graphite will react with sulfuric acid and an oxidizing agent has been known since 1840 but there are some aspects of the reaction which are still in doubt. It is a complicated collection of three reactions. There is a fairly rapid intercalation to a $C^+ HSO_4^- \cdot nH_2SO_4$ as well as a slow but complete oxidation to CO_2 [82] and a partial oxidation to a graphite 'oxide' which is approximately $C_7O_4H_2$. In the latter, the aromatic layers are further apart than in graphite and are probably flat oxidized sheets kept planar by keto-enol tautomerism [83].

In the intercalation, electron charge can be removed from the carbon layers by HNO_3, CrO_3, $KMnO_4$ etc. or by an anode [13]. Bottomley et al. [84] used the latter method with well oreinted pyrolytic graphite to prepare a series of salts with various acids. The anion concentrations for the various stages were calculated from the position of the slope changes in the plot of voltage vs. coulombs. The content of neutral acid could not be obtained by this method but the sample could be removed for X-ray diffraction studies. In this way they identified stage 3 $C_{72}^+ HSO_4^- \cdot nH_2SO_4$, stage 2 C_{48}^+, and stage 1 C_{24}^+. Chlorosulfonic acid gave stage 1 $C_{24}^+ SO_3Cl^- nSO_3Cl$ and a stage 2 and 3 break in slope. Perchloric acid gave C_{24}^+ and $C_{48}^+ ClO_4^- nHClO_4$. Fluorosulfonic gave a stage 1 and BF_3 – glacial acetic acid gave a stage 3 $C_{32}^+ A^-$. Finally, CF_3COOH, H_3PO_4, H_3AsO_4 and HIO_4 all caused swelling during anodic oxidation but no X-ray data were taken. Other adducts are HF_2^- and $HSeO_4^-$ best prepared with an oxidizing agent.

Hennig [85] studied the electrical properties of the bisulfate compound as a function of composition and concluded that oxidation removes electrons from a nearly full conduction band and that reduction gives a residue 'compound' with about one third of the anions and half the neutral acid originally present. The residue was, he thought, either trapped by collapsing planes or trapped in cracks and imperfections in the process of diffusing out of the structure.

Bottomley et al. [86] have shown for the first 3 stages of the nitrate, bisulfate and perchlorate that the thermal expansion in the c direction is additive to within about 8%. This suggests that the internal vibrations in a filled space are independent of those in neighboring spaces. They also found an anomaly in thermal expansion around 253 K which corresponded to a shrinkage on cooling of 0.005 nm for the first 3 stages of the nitrate. X-ray diffraction showed this to be a transformation from a high temperature liquid form to a low temperature crystalline form in which order now appeared between the nitrate layers.

Nixon et al. [87] determined the stacking sequence in 4 stages of graphite nitrate:

Stage	Stacking sequence
1	A | A| A| A| A| A| A| A| A| A|
2	A | A B| B C| C A| A B| B
3	A | A B A| A C A| A B A
4	A | A B A B| B C B C| C

The letters B and C designate the two possible positions of a carbon hexagonal network relative to a given network A. The vertical lines represent intercalated material. It is apparent that the sequence can be generated from that in hexagonal graphite $ABAB$ if the change $AB \rightarrow AA$ on entry of successive intercalate layers involves a movement of a boundary dislocation through the structure. It was also found that the formation of a stage 2 compound proceeds through the intermediate formation of stage 4 and stage 3 compounds in which the stacking sequences were fully developed. Hence, during progressive nitration, certain layers must be stripped as well as filled. On stripping, the change is from $A|A$ to AB and involves an unexplained memory effect.

Thermodynamic properties of the bisulfate compounds have been measured by Aronson et al. [88]. They prepared the compounds by electrolysis and measured their e.m.f. in a cell of the type Hg, $Hg_2SO_4|H_2SO_4$ conc.| graphite compound in the range 308 K–338 K. Their values in Table IV show that the enthalpy of reaction becomes increasingly positive as the stage number decreases. They suggest that this is caused

TABLE IV

Heats and entropies of formation[a] of C_m^+ HSO_4^- $2 \cdot 5$ H_2SO_4 from graphite and sulfuric acid [88]

m	$\Delta H°$ (kJ mol^{-1} H$_2$)	$\Delta S°$ (kJ^{-1} mol^{-1} H$_2$)
24	30	-15
48	-3	-6.3
72	-11	-5.9

[a] $2mC + 7H_2SO_4 = 2C_m^+ HSO_4^- 2.5 H_2SO_4 + H_2$.

by the greater coulomb repulsion between adduct layers at the higher concentration.

A value of 2.5 moles H_2SO_4 per mole of HSO_4^- anion is the result of some other work by Aronson et al. [89]. They prepared the compound by electrolysis and then used the attached Pt anode wire to suspend it from a balance to obtain the weight increase while still in the solution. A similar method gave a mole ratio of $HClO_4$ to ClO_4^- of 2.0 and they suggest that the lower value is caused by the presence of fewer H atoms and therefore less H bonding to hold neutral molecules.

Recently, Besenhard et al. [90] have used anodic oxidation in non aqueous solvents to prepare new compounds which, in dilute solution at least, contain only the anion. Thus, in 0.1 N $LiClO_4$ in acetonitrile the product was a stage 1 $C_{24}^+ ClO_4^-$ which on analysis contained 0.3%H, 0.25%N and 0.2%Li. Fom 0.01 N solution the Li content was an even lower 0.01%. Other solutes used were $NaBF_4$, KPF_6 and $KAsF_6$ and other solvents were propylene carbonate and nitromethane.

If the oxidation of graphite is continued beyond the lamellar compounds described above, then a so called graphite 'oxide' is produced. Its preparation and properties are dealt with in a series of papers by de Boer and van Doorn [91]. They were able to prepare a $C_7H_2O_4$ from a variety of graphites but did note, as others had found, that the number of successive treatments required to reach this composition increased with the size of the graphite flakes used. This is reminiscent of the behavior of single flakes with $FeCl_3$ in nitromethane described in Section 3.3.2. The product is of uncertain structure but they do believe that a number of double bonds has disappeared to allow the covalent bonding of oxygen and hydroxyl to the C atoms. The electron microscope studies of Carr [83] suggest that the carbon layer planes are still planar. The loss of the delocalized π bond system should and does lead to a lower conductivity – by a factor of 10^5 in the plane of the layers. A great variety of polar molecules such as H_2O, amines, alcohols, etc. will intercalate the structure and in this process there have been no reports of a threshold pressure or stages. In terms of the theory of Section 3.3.1 this could be a result of the loss of the π bond interaction between C layer planes. This would decrease the extent to which charge transfer effects on the end basal planes could influence inter layer bonding within the system. In the absence of such an influence, a threshold pressure and stages would disappear. This change would also explain the observation that the layers are much more flexible than in graphite. Finally, when the layers absorb water and separate, the expansion in the spacing calculated from X-ray diffraction will not account for the amount of water [91] and they have suggested that the spacings may not represent the actual situation but are only average values. Their picture of a water containing graphite oxide particle is shown in Figure 8 and is in agreement with the density of the particles as determined in a pycnometer. This structure is exactly the one proposed in Section 3.3.2 for the structure of graphite during intercalation.

3.4.3. CrO_2Cl_2

In the 5 papers on the intercalation of this tetrahedral molecule by graphite, compositions at or near saturation range over a factor of 10. Croft [34] originally found a

$C_{17}CrO_2Cl_2$ and a $C_{15}CrO_2F_2$ from liquid adduct and natural graphite. He noted the expansion of the c axis from X-ray diffraction data and exfoliation on heating. Rudorff et al. [92] found a $C_{13}CrO_2Cl$ from the liquid but only C_{130} after 3 days in gaseous CrO_2Cl_2. They washed the C_{13} with water and dried it at 395 K to obtain a C_{56} from

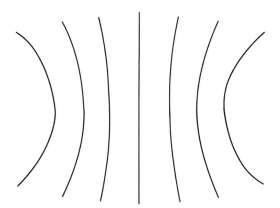

Fig. 8. Graphite oxide plus water – cross section of the layer system.

coarse graphite and a C_{113} from powdered graphite which were stable to 575 K. They suggest that only a fraction of the adduct is decomposed by water and that some remains in the core, stable to 575 K. Hooley [21] used a vacuum system and spring balance to obtain the isotherm of the vapor on natural graphite and found a threshold pressure of 0.3 relative to saturation. His pressure independent composition near saturation did vary in an unexplained way from C_{30} to $C_{45}CrO_2Cl_2$. He also found a C_{40} after exposure to liquid CrO_2Cl_2 at 373 K (B.P. 389 K) for 24 hours. Bach and Ubbelohde [93] also used a spring balance and found a stage 3 $C_{53}CrO_2Cl_2$ from stress annealed pyrolytic graphite. They reported, however, that some specimens of their graphite would not intercalate any CrO_2Cl_2. This same graphite would not intercalate WCl_6 whereas natural flakes show a threshold pressure of 0.5 relative to saturation [20]. This again illustrates the great sensitivity to graphite structure when the threshold pressure is high as for CrO_2Cl_2 or WCl_6.

The intercalated CrO_2Cl_2 shows the curious, and so far unique, behavior of becoming more resistant to removal for longer times spent in the graphite lattice [21]. Thus, if the pressure is reduced to zero immediately after C_{32} is formed, the residue is about C_{80}. If one waits 24 hours at 295 K before reducing the pressure, no CrO_2Cl_2 leaves – there is no weight decrease. This 'fixing' is slower at a lower temperature (272 K) and does not occur for CrO_2Cl_2 held by capillary condensation between the particles of a carbon black for instance. For such material, all is lost whenever the pressure is reduced to zero. The fixing process may be a slow polymerization to a Cl bridged linear polymer. Or it may be a reaction with the π bond system which is slow because of steric requirements. If it is the latter, then the resistance should change with time whereas in the former case it might not.

Takahashi et al. [94] have recently used a vacuum system and spring balance to show that well ordered flakes of Kish graphite (from steel production) produce C_{27} in near saturated vapor and C_{20} in liquid CrO_2Cl_2. Their X-ray diffraction data shows that both are stage 3. They also measured electrical resistance but used a pitch bonded natural graphite heated to 3275 K. In the near saturated vapor the composition was $C_{100}CrO_2Cl_2$ and the conductivity was $3\times$ that of the original sample.

3.4.4. Oxides and Sulfides

Croft [34] heated a number of metal oxides and sulfides with graphite and removed excess reactant with a solvent or by sublimation. He showed that several of these products would swell rapidly on heating and from this fact and from chemical analysis he concluded that intercalation had occurred for Sb_2S_5, TlS_2 (from Tl_2S), CuS, FeS_2, Cr_2S_3, V_2S_3, WS_2, PdS_2, Sb_2O_4, CrO_3 and MoO_3. The compositions ranged from 1.36% of the total weight for Sb_2O_4 to 55% for CrO_3. Of these products, only that with CrO_3 has received further attention. Platzer and Martiniere [95] refluxed graphite with a solution of CrO_3 in anhydrous acetic acid to obtain a product containing 39% CrO_3. For longer times than the 300 seconds to obtain this weight increase, the apparent %CrO_3 decreased because graphite was being oxidized to CO_2. Recently, Lalancette et al. [96] have used Croft's method to prepare a product containing 55% CrO_3 and have shown that it offers a new selective method of preparing aldehydes from the primary alcohol.

Three non metal oxides have been shown to intercalate graphite. Hennig [97] exposed natural flakes to N_2O_5 at 273 K and obtained a maximum weight increase corresponding to $C_9N_2O_5$ which, in a stream of argon at 273 K gave a residue of $C_{12}N_2O_5$. He measured the increase in conductivity and concluded that 2 molecules of N_2O_5 act as 1 electron acceptor. Above 273 K oxidation to CO_2 becomes significant. Fuzellier and Hérold [98] used a spring balance system to expose a natural graphite to N_2O_5 at 273 K and obtained a weight increase equivalent to $C_{16}N_2O_5$. The desorption curve showed breaks at C_{24} and $C_{32}N_2O_5$ and from this and X-ray data they claimed $C_{8n}N_2O_5$ where n is the stage number, minimum value 2.

The second non metal oxide is SO_3 which Bagouin et al. [99] found would form a stage 1 $C_{4.7}$ to $C_{6.5}SO_3$ at 318 K. From the desorption curve and X-ray data they claim evidence for stages 1 to 4.

Finally, Boehm et al. [100] exposed a natural graphite to 0.04 M solution of Cl_2O_7 in CCl_4 at 225 K and at 290 K. Analysis showed 5.8% and 3.5% chlorine in the product. Their main criterion for intercalation, however, was the disappearance of the diffraction lines for the 30% rhombohedral graphite in the original material. Fuzellier and Hérold [101] found that pure Cl_2O_7 explodes in the presence of graphite but when diluted with air gave a stage 1 $C_{12}Cl_2O_7$ which decomposes with evidence for stages 2 to 7.

4. Boron Nitride

Because this material is isoelectronic with carbon, various attempts have been made to intercalate the layer form (Section 1.2). Croft [34] reported that $SbCl_3$, $AsCl_3$, CuCl, $FeCl_3$, $AlCl_3$ and N_2H_4 would all intercalate BN. The % reactant present after washing ranged from 2% for $SbCl_3$ and N_2H_4 to 13% for $AlCl_3$. When the compounds with $FeCl_3$, $AlCl_3$ and N_2H_4 were heated, a very large expansion is obvious in the published photographs. Rudorff et al. [102] found no change in BN after exposure to the vapors of $FeCl_3$, $AlCl_3$, $CoCl_2$, $NiCl_2$ or $NiBr_2$. Hooley [21] found no intercalation of Br_2, ICl or CrO_2Cl_2 on the basis of isotherms at 295 K. Freeman et al. [103] treated BN with $FeCl_3$ vapor and removed excess with acid. He found a 7% weight increase, new X-ray lines and a Mössbauer spectrum that showed electron transfer away from the iron which is opposite to that observed with graphite. They concluded that intercalation also occurred for CuCl, $CuCl_2$, Hg_2I_2 and $AsCl_3$ but not for $AlCl_3$, Hg_2Cl_2, $SbCl_3$, Cl_2, Br_2, I_2 or ICl, etc.

Rudorff and Freeman both suggest that the above discrepancies may be caused by the use of different starting materials. Thus, Croft prepared a black lustrous product easily cleaved into flexible flakes. Analysis showed <0.75% impurities. Rudorff used a white, finely divided material prepared in different ways at 1075 K. Analysis for B and N showed it to be 98% BN. Freeman used B. D. H. Reagent grade BN, a white powder of quoted composition B 43.9%, N 55.6%, B_2O_3 <0.05 and C<0.01%.

I have exposed 5 sources of BN to $FeCl_3$ vapor at 585 K for 5 days in the two temperature reactor tube method [20]. This eliminates the need to use a solvent to remove excess reactant because the pressure was 0.9 relative to saturation. The weight increases as a % of the final weight were:

(1) Norton Co. of Niagara Falls, Canada – white powder, bulk density 100 kg m^{-3} with particles 1 μm in average diameter plus thin plates up to 26 μm across and rods 0.5 μm diameter × 20 μm long. Fired at 2275 K and 98% BN from N analysis. Increased 1%.
(2) Alpha Inorganics, Beverly, Mass. U.S.A. – white powder 99+% BN. Increased 0.5%.
(3) Eimer and Amend, white powder. Increased 70%.
(4) Fisher Scientific Co., white powder. Increased 53%.
(5) Norton Co., a hard, white, smooth plate 1 mm thick. Increased nil.

In a study of BN, Finlay et al. [104] state that some sources are very impure and may contain 9% C or even 40% MgO.

References

1. Linus Pauling: *Proc. Nat. Acad. Sci. (U.S.A.)* **56** (1966), 1646.
2. A. G. Whittaker and G. M. Wolten, *Science* **178** (1972), 54.
3. R. Hultgren, N. S. Gingrich, and B. E. Warren: *J. Chem. Phys.* **3** (1935), 351.
4. P. W. Bridgman: *Proc. Nat. Acad. Sci. (U.S.A.)* **21** (1935), 109.
4a. P. W. Bridgman: *Phys. Rev.* **3** (1914), 187.

4b. H. Krebs, H. Weitz, and K. H. Worms: *Z. anorg. allg. Chem.* **280** (1955), 119.
4c. R. H. Wentorf Jr.: in J. J. Gilman (ed.), *The Art and Science of Growing Crystals*, J. Wiley, 1963, p. 187.
4d. S. S. Boksha: *J. Cryst. Growth* **3/4** (1968), 426.
4e. R. Rau and A. Rabenau: *J. Cryst. Growth* **3/4** (1968), 417.
4f. A. J. Goss: in J. J. Gilman (ed.), *The Art and Science of Growing Crystals*, J. Wiley, 1963, p. 323.
5. J. M. Hutcheon: in L. C. F. Blackman (ed.), *Modern Aspects of Graphite Technology*, Academic Press, London, 1970, Ch. 1.
6. S. Ergun: *Carbon* **6** (1968), 141.
7. D. F. Harling and F. A. Heckman: *Materie Plastiche ed Elastomeri* **35** (1969), 80.
8. G. M. Jenkins and K. Kawamura: *Nature* **231** (1971), 175.
9. D. J. Johnson, p. 52 and A. Fourdeux, R. Perret, and W. Ruland, p. 57, in: *Proc. Int. Conf. on Carbon Fibers*, The Plastics Institute, London, 1971.
10. J. C. Bokros: in P. L. Walker, Jr. (ed.), *Chemistry and Physics of Carbon*, Vol. 5, Dekker, N.Y., 1969.
11. H. B. Haanstra, W. F. Knippenberg, and G. Verspui: *J. Cryst. Growth* **16** (1972), 71.
11a. H. M. Strong and R. E. Hanneman: in H. Steffen Peiser (ed.), *Proc. of an Int. Conf. on Crystal Growth*, Boston, 1966, p. 579.
12. G. R. Hennig: in F. A. Cotton (ed.), *Progress in Inorganic Chemistry*, Vol. 1, Interscience, N.Y., 1959.
13. W. Rudorff: in H. J. Emeleus and A. G. Sharpe (eds.), *Advances in Inorganic Chemistry and Radiochemistry*, Vol. 1, Academic Press, N.Y., 1959.
14. A. R. Ubbelohde and F. A. Lewis: *Graphite and its Crystal Compounds*, Oxford Press, 1960.
15. A. Hérold, R. Setton, and N. Platzer: *Les Carbones*, Vol. 2, p. 465, Masson, Paris, 1965.
16. J. G. Hooley: in P. L. Walker, Jr. (ed.), *Chemistry and Physics of Carbon*, Vol. 5, Dekker, N.Y., 1969.
17. J. G. Hooley and R. N. Soniassy: *Carbon* **8** (1970), 191.
18. J. G. Hooley, W. P. Garby, and J. Valentin: *Carbon* **3** (1965), 7.
19. T. Sasa, Y. Takahashi, and T. Mukaibo: *Carbon* **9** (1971), 407.
20. J. G. Hooley: *Carbon* **11** (1973), 225.
21. J. G. Hooley: *Can. J. Chem.* **40** (1962), 745.
22. J. G. Hooley: *Carbon* **2** (1964), 131.
23. G. A. Saunders: in L. C. F. Blackman (ed.), *Modern Aspects of Graphite Technology*, Blackman, Academic Press, London, 1970, Ch. 3.
24. A. R. Ubbelohde, L. C. F. Blackman, and J. F. Mathews: *Nature, London* **183** (1959), 454.
25. N. B. Hannay *et al.*: *Phys. Rev. Letters* **14** (1965), 225.
26. F. J. Salzano and M. Strongin: *Phys. Rev.* **153** (1967), 533.
27. A. Hérold: *Bull. Soc. Chim. Fr.* (1955), 999.
28. J. G. Hooley: *Carbon* **8** (1970), 333.
29. R. Juza and H. Seidel, *Z. anorg. allg. Chem.* **317** (1962), 73.
30. G. Furdin and A. Hérold: *Bull. Soc. Chim. Fr.* (1972), 1768.
31. G. Colin et A. Hérold: *ibid.* (1972), 3345.
32. B. Bach et A. Hérold: *ibid.* (1968), 1978.
33. A. A. Opalovskii, A. S. Nazarov, and A. A. Uminskii: *Russian J. Inorg. Chem.* **17** (1972), 1366 and 2350.
34. R. C. Croft: *Australian J. Chem.* **9** (1956), 184.
35. A. G. Freeman and J. H. Johnston: *Carbon* **9** (1971), 667.
36. H. Thiele: *Z. anorg. allg. Chem.* **207** (1932), 340.
37. W. Rudorff, E. Stumpp, W. Spriessler, and F. W. Siecke: *Angew. Chem. Int. Ed.* **2** (1963), 67.
38. D. Ginderow and R. Setton: *Compt. Rend.* **257C** (1963), 687.
39. E. Stumpp and F. Werner: *Carbon* **4** (1966), 538.
40. J. G. Hooley and M. Bartlett: *Carbon* **5** (1967), 417.
41. J. Melin and A. Hérold: *Compt. Rend.* **269C** (1969), 877.
42. A. Boeck and W. Rudorff: *Z. anorg. allg. Chem.* **384** (1971), 169.
43. J. Mairé: *Proc. U.N. Int. Conf. Peaceful Uses At. Energy*, Geneva (1958), 392.
44. A. Boeck and W. Rudorff: *Z. anorg. allg. Chem.* **392** (1972), 236.
45. J. G. Hooley: *Carbon* **10** (1972), 155.
46. R. Vangelisti and A. Hérold: *Compt. Rend.* **276C** (1973), 1109.
47. A. Boeck and W. Rudorff: *Z. anorg. allg. Chem.* **397** (1973), 179.
48. J. G. Hooley, J. R. Sams, and B. V. Liengme: *Carbon* **8** (1970), 467.
49. J. J. Lander and J. Morrison: *Surface Sci.* **6** (1967), 1.
50. A. W. Syme Johnson: *Acta Cryst.* **23** (1967), 770.

51. W. T. Eeles and J. A. Turnbull: *Proc. Roy. Soc.* **A283** (1965), 179.
52. M. L. Dzurus and G. R. Hennig: *J. Am. Chem. Soc.* **79** (1957), 1051.
53. W. Rudorff and A. Landel: *Z. anorg. allg. Chem.* **293** (1958), 327.
54. J. G. Hooley, M. W. Bartlett, B. V. Liengme, and J. Sams: *Carbon* **6** (1968), 681.
55. A. G. Freeman: *Chem. Commun.* **193** (1968).
56. F. D. Grigutsch, D. Hohlwein, and A. Knappwost: *Z. Physik. Chem.* **65** (1969), 322.
57. Y. N. Novikov et al.: *Zh. Strukt. Khim.* **11** (1970), 1039.
58. J. G. Hooley: unpublished.
59. H. Sato and R. S. Toth: *Phys. Rev.* **124** (1961), 1833.
59a. H. Kuhn: *J. Chem. Phys.* **17** (1949), 1198.
60. K. Fredenhagen and G. Cadenbach: *Z. anorg. allg. Chem.* **158** (1926), 249.
61. F. J. Salzano and S. Aronson: *J. Inorg. Nucl. Chem.* **30** (1968), 2317.
62. C. Herinckx, R. Perret, and W. Ruland: *Nature* **220** (1968), 63.
63. M. K. Halpin and G. M. Jenkins: *Nature* **218** (1968), 950.
64. S. Aronson, F. J. Salzano, and D. Bellafiore: *J. Chem. Phys.* **49** (1968), 434.
65. D. E. Nixon and G. S. Parry: *J. Phys. C (Solid St.)* **2** (1969), 1732.
66. R. C. Asher and S. A. Wilson: *Nature* **181** (1958), 409.
67. M. L. Dzurus, G. R. Hennig, and G. L. Montet: *Proc. Fourth Carbon Conf.*, Pergamon, N.Y., 1960, p. 165.
68. H. L. Recht, G. M. Wolten, and D. E. Gilmartin: *J. Inorg. Nucl. Chem.* **23** (1961), 275.
69. W. C. Sleppy: *J. Inorg. Chem.* **5** (1966), 2021.
70. S. Aronson and F. J. Salzano: *Nucl. Sci. Eng.* **38** (1969), 187.
71. A. Métrot and A. Hérold: *J. Chim. Phys., Physico Chim. Biol.* **71** (1969), 73.
72. I. Pflugmacher and H. P. Boehm: *Proc. 3rd Conf. Ind. Carbon and Graphite*, London (1970), 62.
73. R. Juza and V. Wehle: *Natw.* **52** (1965), 560.
74. M. Bagouin, D. Guerard, and A. Hérold: *Compt. Rend.* **262C** (1966), 557.
75. D. Guerard and A. Hérold: *Compt. Rend.* **275C** (1972), 571.
76. W. Rudorff, E. Schulze, and O. Rubisch: *Z. anorg. allg. Chem.* **282** (1955), 232.
77. W. Rudorff: *Chimia* **19** (1965), 489.
78. W. E. Craven and W. Ostertag: *Carbon* **4** (1966), 223.
79. C. Stein, L. Bonnetain, and J. Gole: *Bull. Soc. Chim. Fr.* (1966), 3166.
80. D. Ginderow and R. Setton: *Compt. Rend.* **270C** (1970), 135.
81. M. L. Dzurus and G. R. Hennig: *J. Chem. Phys.* **27** (1957), 275.
82. K. Carr: *Carbon* **8** (1970), 155.
83. K. Carr: *Carbon* **8** (1970), 245.
84. M. J. Bottomley, G. S. Parry, A. R. Ubbelohde, and D. A. Young, J. Chem. Soc. (1963), 5674.
85. G. Hennig: *J. Chem. Phys.* **19** (1951), 922.
86. M. J. Bottomley, G. S. Parry, and A. R. Ubbelohde: *Proc. Roy. Soc.* **279A** (1964), 291.
87. D. E. Nixon, G. S. Parry, and A. R. Ubbelohde: *Proc. Roy. Soc.* **219A** (1966), 324.
88. S. Aronson, C. Frishberg, and G. Frankl: *Carbon* **9** (1971), 715.
89. S. Aronson, S. Lemont, and J. Weiner: *Inorganic Chem.* **10** (1971), 1296.
90. J. Besenhard and H. P. Fritz: *Z. Naturforsch.* **27b** (1972), 1294.
91. J. H. de Boer and A. B. C. van Doorn: *Proc. K. Nederl. Akad. Wetensch.* **B61** (1958), 12, 17, 160, 242; **B64** (1960), 34.
92. W. Rudorff, V. Sils, and R. Zeller: *Z. anorg. allg. Chem.* **283** (1956), 299.
93. B. Bach and A. R. Ubbelohde: *J. Chem. Soc.* (1971), A 3669.
94. Y. Takahashi, H. Yamagata, and T. Mukaibo: *Carbon* **11** (1973), 19.
95. N. Platzer and B. Martiniere: *Bull. Soc. Chim. Fr.* (1961), 177.
96. J. M. Lalancette, G. Rollin, and P. Dumas: *Can. J. Chem.* **50** (1972), 3058.
97. G. Hennig: *Nucl. Sci. Eng.* **21** (1965), 34.
98. H. Fuzellier and A. Hérold: *Compt. Rend.* **267C** (1968), 607.
99. M. Bagouin, H. Fuzellier, and A. Hérold: *Compt. Rend.* **262C** (1966), 1074.
100. H. P. Boehm and J. N. Meussdoerffer: *Carbon* **9** (1971), 521.
101. H. Fuzellier and A. Hérold: *Compt. Rend.* **276C** (1973), 1287.
102. W. Rudorff and E. Stumpp: *Z. Naturforsch.* **13b** (1958), 459.
103. A. G. Freeman and J. P. Larkindale: *J. Chem. Soc.* (1969), A 1307.
104. G. R. Finlay and G. H. Fetterley: *Ceramic Bull.* **31** (1952), 141.

METAL HALIDES

J. SCHOONMAN*

Center for Materials Research, Stanford University, Stanford, California, U.S.A.

and

R. M. A. LIETH

Chemical Physics Division, Dept. of Physics, University of Technology, Eindhoven, The Netherlands

Single crystal of PbFBr, grown according to the method described in 5.2.16.

Table of Contents
1. Introduction
2. Crystal Structures and Polytypism
3. Compound Preparations
4. Thermochemical Properties
5. Crystal Growth
    5.1. General Techniques
        5.1.1. Sublimation of the Substance
        5.1.2. Reaction in the Vapour Phase
        5.1.3. Chemical Transport
        5.1.4. Growth from a Melt
        5.1.5. Growth from Solvents

* On leave from Solid State Chemistry Department, Physical Laboratory, University of Utrecht – De Uithof, The Netherlands.

5.2. Growth of Specific Compounds
    5.2.1. Ag_2F
    5.2.2. CaI_2
    5.2.3. CdI_2
    5.2.4. $MnBr_2$
    5.2.5. MnI_2
    5.2.6. $FeBr_2$
    5.2.7. $CoBr_2$ ($CoCl_2$ and CoI_2)
    5.2.8. PbI_2
    5.2.9. $CdCl_2$ and $CdBr_2$
    5.2.10. $FeCl_2$ and $FeCl_3$
    5.2.11. $NiCl_2$ and $NiBr_2$
    5.2.12. AsI_3
    5.2.13. SbI_3 and BiI_3
    5.2.14. SrFCl
    5.2.15. BaFCl
    5.2.16. PbFCl and PbFBr
    5.2.17. PbFI
6. Conclusions
References

1. Introduction

A considerable number of more or less ionic halides with chemical formulae MX_2, or MX_3 (X = Cl, Br, I) crystallize with layer-type structures. In general, those halides with small, or highly polarizing cations in combination with polarizable anions adopt layer structures. The divalent metal halides can adopt either the structure of $CdCl_2$, CdI_2, or PbFCl, the type being determined by the polarizability of the constituent ions. The trivalent metal halides reflect the influence of polarization and covalency by adopting the BiI_3 or the $CrCl_3$ structure. These types are closely related to the CdI_2 and the $CdCl_2$ arrangements, respectively [1–5]. In general the more ionic compounds crystallize in the $CdCl_2$ or $CrCl_3$ structure, whereas the more covalent compounds prefer the CdI_2 or BiI_3 arrangement.

The majority of these halides are extremely reactive towards atmospheric moisture. Many transition metal halides, for example, deliquesce in a matter of seconds. Nevertheless, improvement in preparation, handling and materials characterization techniques over the past two decades has led to a substantial increase in knowledge and applications of the layered metal halides.

A vast number of papers have been published, which describe the formation of organometallic compounds using anhydrous di- and trihalides as starting materials. Several transition metal halides are used as solid catalysts for chlorination of hydrocarbons in the gaseous phase [6].

In view of the difficulties in handling the extremely hygroscopic halides it is not surprising that detailed studies of optical and electrical properties are limited to those members of this family of compounds which are relatively stable: CdX_2, PbI_2, SbI_3, and BiI_3.

The optical properties of the layered halides have been reviewed extensively by Tubbs [7]. As can be expected for these layered crystal structures the electrical conductivity is highly anisotropic, the electrical conductivity parallel to the layers in, for example, CdI_2, Ag_2F, and PbFCl being some 10^2 to 10^3 times larger than that normal to the layers. Several layered halides are sensitive to actinic radiation. Particularly sensitive compounds like CdI_2, PbI_2, and BiI_3 have been investigated thoroughly, and their photochemical behaviour is presently well understood [8–11]. The unique photochemical properties of these materials have led to applications in microrecording, holography, printed circuits, and digital information storage systems [12, 13].

In the present chapter the preparation of anhydrous layered metal halides and their thermochemical properties will be reviewed. In addition to compound preparation, attention will be focussed on techniques utilized for growing single crystals. Only an introductory discussion will be given of the crystal structures and the occurrence of polytypism, since in Volume II of this series the crystallography, crystal chemistry, and the phenomena of polymorphy and polytypism will be treated in full detail.

2. Crystal Structures and Polytypism

In the crystal structures to be discussed, the coordination number of the metal ions is 6 or 8. In most of the layered di- and trihalides the metal ions are octahedrally surrounded by the halide ions. When octahedra share three pairs of opposite edges so that no two octahedra are in contact by vertices only, MX_2 layers can be formed. A layer may be regarded as a two-dimensional molecule. In the crystal the layers are stacked so that all X ions are close-packed and held together by weak van der Waals forces. The metal ions occupy one-half of the octahedral interstices. The *hcp* structure is typified by CdI_2, and the *ccp* structure by $CdCl_2$. In both structures the unit cell contains one molecule.

Perspective drawings of these structures are shown in Figures 1 and 2. They can be represented diagrammatically by showing the type of layer, *A, B* or *C*, and their method of stacking (Figures 3 and 4).

The cadmium iodide structure occurs predominantly in those halides where marked polarization effects are to be expected. The iodides of moderately polarizing cations and the bromides and iodides of d^{10}, some transition element cations, and magnesium adopt this structure; i.e. (Ca, Cd, Ge, Pb, Th, Tm, Yb) I_2, Mg (Br_2, I_2), Mn (Br_2, I_2), Fe (Br_2, I_2) and Co (Br_2, I_2). Several chlorides of more highly polarizing cations crystallize in this arrangement also, i.e. Ti (Cl_2, Br_2, I_2) and V (Cl_2, Br_2, I_2). A well known exception is Ag_2F with an anti-CdI_2 structure.

The chlorides, and occasionally the bromides and iodides of moderately strongly polarizing divalent cations crystallize with the cadmium chloride structure, i.e. $MgCl_2$, $MnCl_2$, $FeCl_2$, $CoCl_2$, $Ni(Cl_2, Br_2, I_2)$, $Zn(Br_2, I_2)$, $Cd(Cl_2, Br_2)$ and PbI_2. In these compounds the extent of polarization is less pronounced than in compounds adopting the cadmium iodide structure. The alkaline earth halides with the CdI_2 or $CdCl_2$ structure show ionic radius ratios, $r_{M^{2+}}/r_{X^-}$ in the range 0.30 (MgI_2) – 0.46 (CaI_2). $MgCl_2$ with $r_{M^{2+}}/r_{Cl^-} = 0.36$ exhibits the $CdCl_2$ structure, whereas MgI_2 and

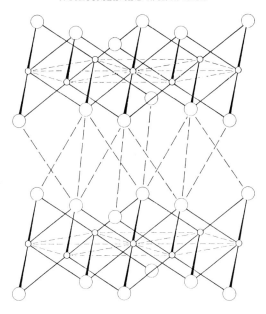

Fig. 1. The crystal structure of CdI$_2$: small circles Cd, large circles I.

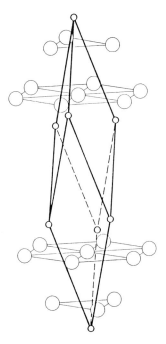

Fig. 2. The rhombohedral unit cell of CdCl$_2$: small circles Cd, large circles Cl.

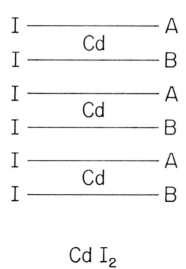

Fig. 3. Method of stacking of the layers in CdI$_2$.

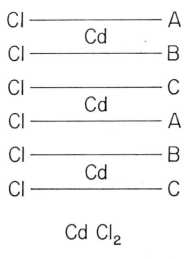

Fig. 4. Method of stacking of the layers in CdCl$_2$.

CaI_2 have the CdI_2 structure. As mentioned in the introduction the $CdCl_2$ structure can be associated with more ionic character than can the CdI_2 structure. It seems that the adoption of either of these structures is more governed by the polarizability of the constituent ions than by the ionic radius ratios [5].

The c/a ratio of the MX_2 compounds appears to decrease as the intercore $X-X$ distance increases [7, 18].

Mercuric chloride has an orthorhombic structure in which sheets of $HgCl_2$ molecules are closely stacked above each other along the c-axis. Mercuric bromide and the yellow form of mercuric iodide have the same general arrangement, but the ions are in such positions that the structure can be considered as a layer structure like CdI_2 and $CdCl_2$ with the halide ions in positions that are a moderate distortion of a mixed cubic and hexagonal close-packing [1]. The red polymorph of HgI_2, being an ionic compound, has iodide ions in a *ccp* arrangement. The metal ions occupy one-quarter of the tetrahedral interstices between close-packed layers of halide ions. The γ-forms of $ZnCl_2$ and $ZnBr_2$ as precipitated from aqueous solution at about 348 K also possess this structure [1, 15]. Several modifications of $ZnCl_2$ are known, and there is even some doubt whether $ZnCl_2$ ever crystallizes in the $CdCl_2$ structure [7].

Layers with stoichiometry MX_3 are obtained when one third of the metal ions are removed from the layers present in the aforementioned structure types. The structure then obtained with the X^--ions in a hexagonal close-packing is typified by BiI_3, whereas $CrCl_3$ represents the structure type with X^--ions in cubic close-packed arrangement [14].

The trihalides which crystallize in the BiI_3 structure are, (As, Sb, Bi, V) I_3, (Ti, Cr, Fe) Br_3 and (Sc, α-Ti, V, Fe) Cl_3.

The trihalides adopting the $CrCl_3$ structure are (Al, Y, In, Tl, Lu, Cr, Rh, Ir) Cl_3.

The triiodides of arsenic, antimony and bismuth form an interesting series [16, 17]. Arsenic and antimony are significantly displaced from the centers of the octahedra to give three near and three far M—I distances, but for Bi no irregularity can be detected. The bond distances are given in Table I together with the melting points.

TABLE I

Bond distances and melting points of the triiodides of arsenic, antimony and bimuth

Compound	Short (Å)	Long (Å)	$T_m(K)$
AsI_3	2.556	3.56	415
SbI_3	2.868	3.32	444
BiI_3	3.1	3.1	681

The degree of ionicity for a compound is usually inferred indirectly from data such as conductivity, bandgap, melting point, bond length or susceptibility. Recently, Tubbs [19] proposed an empirical definition of ionic character, or ionicity, i, in terms of parameters available from optical spectra. The ionicity can be calculated using the equation

$$i = \frac{E_{av}}{\hbar \omega_p}, \quad (1)$$

where $\hbar \omega_p$ is the valence electron plasmon energy, and E_{av} the average energy gap for the valence electrons in the uppermost valence band. Partially ionic compounds exhibit ionicities in the range 0.35–0.5, whereas compounds with ionicities smaller than 0.3 are classified as covalent. Ionic compounds exhibit i values larger than 0.55 For CdI_2, BiI_3 and PbI_2 values of 0.52, 0.41 and 0.35, respectively, are given for the ionicity [7, 19]. So the type of bonding in the di- and trihalides ranges from relatively ionic, like in CdI_2, through the more covalent ones like PbI_2 to those with molecular units like AsI_3 and $HgCl_2$.

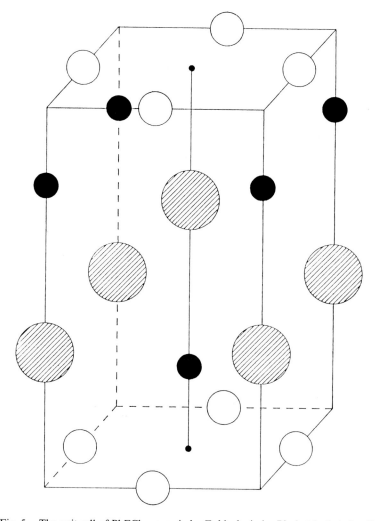

Fig. 5. The unit cell of PbFCl: open circles F, black circles Pb, hatched circles Cl.

A number of the compounds mentioned so far are known to exhibit structures with stacking sequences other than the hexagonal close-packing $ABABAB...$, and the cubic close-packing $ABCABCABC....$ These modifications can be regarded as derived from the common CdI_2 or $CdCl_2$ structure by the insertion of periodic stacking faults. Differently faulted structures have almost equal free energies, and one should therefore expect a random distribution of stacking faults. However, while such structures are not infrequent, there is a marked tendency towards an orderly arrangement of the stacking faults leading to long period structures, which are called polytypes. Polytypes, unlike polymorphs, do not have a definite thermodynamic stability range and are often observed to grow together under identical growth conditions. As compared to the common unit cell, two dimensions of their unit cell are the same while the third one, usually the c-axis, can vary from about 10 Å in small period polytypes to over 1000 Å in some of the long period polytypes [20–22].

The compounds known to show polytypic behaviour are CdI_2, PbI_2, $CdBr_2$ and $NiBr_2$. Detailed studies have been carried out on CdI_2, PbI_2 and $CdBr_2$ only. The largest number of polytypes has been discovered in CdI_2, where more than 160 different modifications have already been found [22] and more appear nearly every day [23–25]. PbI_2 and $CdBr_2$ form a much smaller number of distinct polytypes.

Of the mixed metal halides with layer structures, CdBrI adopts a structure which is related to the $CdCl_2$ structure, whereas mixed strontium, barium and lead halides can crystallize in a layered tetragonal structure known as the PbFCl structure. Only those mixed halides which adopt the PbFCl structure will be included in this chapter.

The unit cell of PbFCl, which contains two molecules, is shown in Figure 5. Lead ions are surrounded by four fluoride ions in a symmetry which resembles that found in the fluorite modification of PbF_2, and by four chloride ions at distances which are the values calculated from the ionic radii [3]. The PbFCl structure is expected for compounds of weakly or moderately polarizing cations with two anions of different sizes, exhibiting low or moderate polarizabilities, i.e. SrFCl, BaFCl, PbFCl and PbFBr. Recently, it was confirmed that PbFI crystallizes also in the PbFCl structure [26].

3. Compound Preparations

In this section the preparative methods commonly employed in the synthesis of anhydrous halides MX_2, MX_3 and MFX will be considered. The examples of preparation cited are intended merely to illustrate the current methods. No attempt is made to record all examples for a particular method.

In general the following techniques can be employed to synthesize the halides under consideration from the elements, or from their salts.

(a) Reaction of the metal with halogen or with the HX vapour.

Preparations are carried out in either flow systems or in sealed tubes. The resulting oxidation state often depends on reaction temperatures. An extremely useful preparative method for dihalides is the direct interaction of the elements in alcohol or in ether at room temperature. Oxides of the first row transition elements can be chlorinated directly. Often the oxide is mixed with carbon, which probably reduces the oxide to the metal, or a carbide, which is then chlorinated [27].

(b) Reduction of a higher halide with hydrogen, or with the particular metal, or by thermal decomposition of the higher halide.

(c) Precipitation of the compounds from aqueous solution, and subsequent dehydration of the formed hydrate in a stream of the appropriate hydrogen halide, or by reaction of hydrated chlorides with thionyl chloride under reflux conditions.

Oxide impurities are easily formed by partial hydrolysis in the dehydration step. They can be removed by subliming for example the chloride in a stream of hydrogen chloride, or in some cases in chlorine.

(d) Chlorination of the oxides.

In addition to the method mentioned under (a), oxides are frequently chlorinated with carbon tetrachloride, ammonium chloride, or thionyl chloride. With carbon tetrachloride, flow system arrangements are used because of the high temperatures. Oxides and sulfides can readily react with aluminium triiodide to give binary iodides. The preparations are carried out in sealed tubes [28–30].

The preparation techniques of the layered halides MX_2 and MX_3 are gathered in Tables II–V. The hygroscopic character of the compounds is also indicated in these tables; +denotes hydroscopic, −degree of hygroscopic character unknown, and ○ means not hygroscopic.

A considerable number of papers dealing with X-ray diffraction studies on these halides describe preparative and handling procedures. They are included separately in the tables under structure references, and the majority were taken from Wyckoff [1, 2].

Most of the di- and trihalides can be obtained commercially, although in many cases probably only on request[†]. The commercially available compounds – marked [a] in the tables – often have purities better than 99.9%, however, additional purification to satisfy specific demands is often needed.

In addition to the literature cited in this section the reader is referred to a review by Corbett [162], describing the preparative methods employed for metal halides in low oxidation states.

The mixed halides, MFX, which crystallize in the PbFCl structure, are in general prepared by reaction of equimolar mixtures of the constituent lead halides, either in aqueous solution, or through melting the mixture. Reaction of an aqueous solution of $PbX_2 (X=Cl, Br, I)$ with an aqueous solution of an alkali fluoride leads only to the formation of PbFCl and PbFBr. PbFI cannot be prepared in this way [26]. The preparative methods are gathered in Table VI.

[†] CERAC/PURE Inc. BOX 703, Butler, Wisc. 53007, U.S.A. specializes in rare fine inorganic chemicals. Its pricelist contains almost all the compounds discussed, with their purity.

TABLE II

Metal halides, MX_2, with the CdI_2 structure and their preparative methods

Compound	Hygroscopic character	Preparative reaction	Reference	Structure reference
Ag_2F	+	$AgF + Ag$ (323–363 K)	31	32
$MgBr_2$ [a]	+	$Mg + Br_2$ (ether)	33	34
		$MgCl_2 \cdot 6H_2O + HBr$	35	
MgI_2 [a]	+	$Mg + I_2$ (873 K)	33	36
		$MgI_2 \cdot 8H_2O + HI$ (N_2)	35	
CaI_2 [a]	−	$CaCO_3 + HI$	35	36
		$Ca + I_2$	35	
CdI_2 [a]	+	$CdSO_4 \cdot xH_2O + KI$	37	38, 39
		$Cd + I_2$ (373 K)	35	
$TiCl_2$	+	$TiCl_4 + Ti$ (excess, 1073–1173 K)	40	40, 41
		$TiCl_4 + Ti$ (1323 K)	42	
		$TiCl_3$ (decomp. 773 K)	43	
$TiBr_2$	+	Ti (excess) $+ Br_2$ (1223 K)	44	44
		$TiBr_4 + Ti$ (1073 K)	44	
		$TiBr_3$ (disprop. >623 K)	45	
TiI_2	+	$TiI_4 + Ti$ (stoich.)	35, 46	40
		TiI_3 (decomp. 623 K)	47[b]	
VCl_2	+	VCl_3 (disprop. 1123 K)	48, 49	48
		$VCl_3 + H_2$ (948 K)	35, 50	
VBr_2	+	$V + Br_2$ (stoich.)	35	
		$VBr_3 + H_2$ (853 K), (723 K)	51, 54	
		VBr_3 (decomp. 853 K)	49, 51	
VI_2 [a]	−	$V + I_2$ (433–443 K)	40[b]	40
		VI_3 (decomp. 573–773 K)	52, 53	
$MnBr_2$ [a]	+	$MnBr_2 \cdot 4H_2O$ (dehydr. 463 K)	55	34
		$MnCO_3 + HBr$ (373, 998 K)	56	
MnI_2 [a]	+	MnO_2, or $MnS_2 + AlI_3$ (503 K)	28–30[b]	57
		$Mn + I_2$ (298 K, ether)	58	
$FeBr_2$ [a]	+	$Fe_2O_3 + HBr$ (473–598 K)	59	34
		$Fe + HBr$	60	
FeI_2 [a]	+	$Fe + I_2$ (gradient, 803/453 K)	61[b]	57
		$Fe + HI$	60	
$CoBr_2$ [a]	+	$Co + HBr$ (red heat)	62	34
		$CoBr_2 \cdot 6H_2O + HBr$ (873 K)	63	
CoI_2 [a]	+	$CoCO_3 + HI$ (H_2/N_2, 473 K)	63[b]	57
		Co_3O_4, or $CoS_2 + AlI_3$ (503 K)	28–30	
ThI_2	−	$ThI_4 + Th$ (excess)	64, 65	66, 67
		ThI_3 (disprop. 823 K)	68	
TmI_2	+	$TmI_3 + Tm$ (excess, 823 K)	69	69, 70
		$HgI_2 + Tm$ (573–673 K)	71	
YbI_2	+	$YbCl_3 + HI$ (863–873 K)	72	69, 73
GeI_2	+	$GeS + HI$	74	75
		$GeO_2 + HI$	76	
PbI_2 [a]	+	Pb-acetate $+ KI$	77	78
		$Pb + I_2$	79	

[a] Commercially available.
[b] See also Reference 233.
+ : Hygroscopic.
− : Degree of hygroscopic character unknown.
○ : Not hygroscopic.

TABLE III

Metal halides, MX_2, with the $CdCl_2$ structure and their preparative methods

Compound	Hygroscopic character	Preparative reaction	Reference	Structure reference
$MgCl_2$ [a]	+	$MgO + NH_4Cl$ (excess)	80	81, 82
		$MgCl_2 \cdot NH_4Cl \cdot 6H_2O$ (decomp. 673–773 K)	83	
		$MgCl_2 \cdot 6H_2O$ (dehydr.)	84	
$ZnBr_2$ [a]	+	$Zn + HBr/Br_2$	35	85, 86
ZnI_2	+	$Zn + I_2$ (298 K, ether)	35	87, 88
$CdCl_2$ [a]	+	$CdCl_2 \cdot 2\frac{1}{2}H_2O + SOCl_2$	89	90, 91
		$Cd + HCl$ (723 K)	92	
$CdBr_2$ [a]	+	$Cd + Br_2$ (723 K)	92	93, 94
$MnCl_2$ [a]	+	$Mn + Cl_2$ (ethanol)	95	81
		$MnO + NH_4Cl$ (excess)	80	
		$MnCl_2 \cdot 4H_2O$ (dehydr. 673 K)	62	
		$Mn + HCl$	96	
$FeCl_2$ [a]	+	$Fe + HCl$ (red heat)	60, 97	91, 98
		$FeCl_3 + Fe$	99	
		$FeCl_3 + H_2$ (573–623 K)	35	
		$FeCl_3$ (decomp. 573 K)	100	
$CoCl_2$ [a]	+	$Co + Cl_2$ (ether, or alcohol)	95	98, 102
		$CoCl_2 \cdot 6H_2O + SOCl_2$	89	103
		$Co_3O_4 + CCl_4$ (573–1173 K)	104	
		$CoCl_2 \cdot 6H_2O$ (dehydr. 673 K)	63	
$NiCl_2$ [a]	+	$Ni + Cl_2$ (flow system)	105, 106	82, 103
		$Ni + Cl_2$ (293 K, ethanol)	95	
		$Ni, C + Cl_2$ (>573 K)	107	
$NiBr_2$ [a]	–	$Ni + Br_2$ (ether)	35	108
		$NiBr_2 \cdot 6H_2O$ (dehydr. 413 K)	35	
		$NiO + HBr$ (353 K)	109	
NiI_2 [a]	+	$NiI_2 \cdot 6H_2O$ (dehydr.)	35[b]	108
		Ni-salt + NaI (ethanol)	110, 111	
		$NiO + AlI_3$ (503 K)	29, 30	

[a] Commercially available.
[b] See also Reference 233.

4. Thermochemical Properties

Following Barin and Knacke [157] we have tabulated for the layered halides the following molar temperature functions: the enthalpy, $H°(T)$, in kcal mol^{-1}, the entropy, $S°(T)$, in cal mol^{-1} deg^{-1}, the free energy, $G°(T)$, in kcal mol^{-1} (Tables VII–X) and the molar heat at constant pressure, c_p, in cal mol^{-1} deg^{-1} (Tables XI–XIV). The molar enthalpy, $H°(T)$, of a pure substance at 1 atmosphere pressure and temperature T is given by

$$H°(T) = \Delta H_f + \int_{298.15}^{T} dH°, \qquad (2)$$

TABLE IV

Metal halides, MX_3, with the BiI_3 structure and their preparative methods

Compound	Hygroscopic character	Preparative reaction	Reference	Structure reference
AsI_3 [a]	+	$AsCl_3 + KI$	112	16, 17
		$As + I_2$ (CS_2)		113, 114
SbI_3 [a]	+	$Sb + I_2$ (benzene)	115, 116	16, 17, 114
BiI_3 [a]	+	$Bi + I_2$ (423–453 K)	117	16, 17, 114
				118
$ScCl_3$ [a]	+	$Sc + HCl$	119, 120	121
		$(NH_4)_3 ScCl_6$ (decomp.)	120, 122	
$TiCl_3$	–	$TiCl_4 + H_2$ (1073–1273 K)	124, 125	121, 123
		$TiCl_4 + $Al-alkyl	123	
		$TiCl_4 + H_2$ (glow discharge)	126	
$TiBr_3$	–	$TiBr_4 + H_2$ (W-filament)	125	123, 127
		$Ti + Br_2$	121	
		$TiBr_4 + H_2$ (gradient, 1023/453 K)	128	
VCl_3 [a]	+	VCl_4 (decomp. 423–443 K)	129, 130	121
		$VO_2 + SOCl_2$	131	
VI_3	+	$V + I_2$ (gradient, 822/516 K)	52	52
		$VI_2 + I_2$ (573 K)	53	
		$V_2O_3 + AlI_3$ (603 K)	29	
$CrBr_3$ [a]	–	$CrBr_3 \cdot 6H_2O$ (dehydr. 673 K)	55	132
		$Cr + Br_2$ (1023 K)	133	
$FeCl_3$ [a]	+	$Fe_2O_3 + CCl_4$ (673 K)	134	135
		$Fe + Cl_2$ (573–623 K)	136–138	
		$FeCl_3 \cdot 6H_2O + SOCl_2$	89	
$FeBr_3$ [a]	+	$Fe + Br_2$ (723 K)	139	140
		$Fe + Br_2$ (gradient, 473/393 K)	141	

[a] Commercially available.

TABLE V

Metal halides, MX_3, with the $CrCl_3$ structure and their preparative methods

Compound	Hygroscopic character	Preparative reaction	Reference	Structure reference
$AlCl_3$ [a]	+	$Al_2O_3 + CCl_4$ (673 K)	134	142
YCl_3 [a]	+	$Y_2O_3 + HCl$ (823 K)	143	143
$InCl_3$ [a]	+	$In + Cl_2$ (473 K)	143	143
		$In_2O_3 + SOCl_2$ (573 K)	144	
$TlCl_3$ [a]	+	$Tl_2O_3 + HCl$	143	143
		$TlCl_3 \cdot 4H_2O + SOCl_2$	35	
$LuCl_3$ [a]	+	$Lu_2O_3 + HCl$	143	143
$CrCl_3$ [a]	+	$Cr_2O_3 + CCl_4$ (923 K)	145, 146	147, 148
		$Cr + Cl_2$ (1233–1273 K)	14	
		$CrCl_3 \cdot 6H_2O + SOCl_2$	89	
$RhCl_3$	O	$Rh + Cl_2$ (673 K)	149	150
		$RhCl_3 \cdot xH_2O$ (dehydr. 473 K)	35	
$IrCl_3$	–	$(NH_4)_2 IrCl_6$ (Cl_2, 713–723 K)	35	
		$Ir + Cl_2$ (873 K, $h\nu$)	35	

[a] Commercially available.

TABLE VI
Mixed halides, MFX, with the PbFCl structure and their preparative methods

Compound	Hygroscopic character	Preparative reaction	Reference	Structure reference
SrFCl	+	$SrF_2 + SrCl_2$ (melt)	222, 223	
BaFCl	+	$BaF_2 + BaCl_2$ (melt)	151	151
PbFCl	+	$PbCl_2$ (excess) + NaF (KF)	152, (26)	153
		$Pb(NO_3)_2 + HCl/HF$	79	
		$PbF_2 + PbCl_2$ (melt)	154	
PbFBr	+	$PbBr_2$ (excess) + KF	155	155
		$PbF_2 + PbBr_2$ (melt)	79	
PbFI	−	$PbF_2 + PbI_2$ (aq. sol.)	26, 156	26

TABLE VII
Thermochemical properties of metal halides, MX_2, with the CdI_2 structure

Compound	M.P.	B.P.	S.P.	$-H°_{298.15}$	$S°_{298.15}$	$-G°_{298.15}$
$MgBr_2$	984	1503		124.0	28.5	132.497
MgI_2	923			86.0		
CaI_2	1013	1373		128.0	34.72	138.352
CdI_2	663	1069		48.0	38.5	
$TiCl_2$	1308		1299	123.2	20.88	129.425
$TiBr_2$	500, d		1208	96.9	25.9	104.622
TiI_2	873	1273	1301	64.5	35.3	75.025
VCl_2	±1623		1300	110.0	23.2	116.917
VBr_2				83.0	30.0	
VI_2			1023/1073	63.0	35.0	73.435
$MnBr_2$	971	1300		92.0	33.0	101.839
MnI_2	911	1290		58.0	36.0	68.733
$FeBr_2(I)$				59.5	33.62	69.542
$FeBr_2(II)$	964	1207		52.324	49.355	84.404
$FeI_2(I)$				25	40	36.926
$FeI_2(II)$	860	1366		17.727	55.965	54.104
$CoBr_2$	951			51.0		
CoI_2	793			21.0		
ThI_2	1137, d					
TmI_2	1029					
YbI_2	1045	1573				
GeI_2	d		513			
PbI_2	675	1135		41.86	41.869	54.343

d: decomposes
$FeBr_2(I)$: 298–650 K
$FeBr_2(II)$: 650–964 K
$FeI_2(I)$: 298–650 K
$FeI_2(II)$: 650–860 K

$H°_{298.15}$ in kcal mol^{-1}
$S°_{298.15}$ in cal. mol^{-1} deg^{-1}
$G°_{298.15}$ in kcal mol^{-1}
M.P., B.P., and S.P. in degrees K

TABLE VIII

Thermochemical properties of metal halides, MX_2, with the $CdCl_2$ structure

Compound	M.P.	B.P.	S.P.	$-H°_{298.15}$	$S°_{298.15}$	$-G°_{298.15}$
$MgCl_2$	987	1691		153.3	21.4	159.680
$ZnBr_2$	675	923		78.3	32.7	88.05
ZnI_2	719	1003		50.0	38.5	61.479
$CdCl_2$	841	1234		93.6	27.6	101.829
$CdBr_2$	840	1136		75.2	33.2	
$MnCl_2$	923	1504		115.2	28.25	123.623
$FeCl_2$	950	1285		81.8	28.7	90.357
$CoCl_2$	1013	1298		74.7	26.1	82.482
$NiCl_2$	1303		1243	73.0	23.35	79.962
$NiBr_2$	1236	1192		51.8		
NiI_2	1070			23.0		

TABLE IX

Thermochemical properties of metal halides, MX_3, with the BiI_3 structure

Compound	M.P.	B.P.	S.P.	$-H°_{298.15}$	$S°_{298.15}$	$-G°_{298.15}$
AsI_3	415	697		13.7		
SbI_3	444	673		23.0	51.5	38.355
BiI_3	681	(814), d		24.0		
$ScCl_3$	1233	1240		215.0	29.0	223.65
$TiCl_3$	d	933	1104	172.5	33.4	182.46
$TiBr_3$			1067	131.5	42.2	144.01
VCl_3	d			134.0	31.3	143.3
VI_3			533, s	67.0	48.5	81.460
$CrBr_3$			s			
$FeCl_3$	577	605		95.46	34.019	105.6
$FeBr_3$	d			63.5		

d: decomposes
s: sublimes

TABLE X

Thermochemical properties of metal halides, MX_3, with the $CrCl_3$ structure

Compound	M.P.	B.P.	S.P.	$-H°_{298.15}$	$S°_{298.15}$	$-G°_{298.15}$
$AlCl_3$	465.6		460	168.58	26.45	176.466
YCl_3	994	1783		232.7	32.7	242.450
$InCl_3$	859		573			
$TlCl_3$	298	d				
$LuCl_3$	1196	1753				
$CrCl_3$	1150		1218	132.0	29.4	140.766
$RhCl_3$	723/773, d	1073				
$IrCl_3$	936, d					

d: decomposes

TABLE XI
Molar heats, c_p, of metal halides, MX_2, with the CdI_2 structure

Compound	Temperature range (K)	A	B	C	D
$MgBr_2$	298–984	16.092	5.190		
MgI_2					
CaI_2	298–1013	16.33	5.59		
CdI_2					
$TiCl_2$	298–1300	16.339	4.308	−0.826	
$TiBr_2$	298–1209	18.185	2.571	−0.826	
TiI_2	298–1301	20.09	1.740	−0.001	
VCl_2	298–1300	17.25	2.72	−0.71	
VBr_2					
VI_2		17.29	2.0		
$MnBr_2$	298–971	16.23	5.93		
MnI_2	298–911	16.06	6.53		
$FeBr_2$	298–964	17.591	5.318		
$FeI_2(I)$	298–650	19.824	0.589		
$FeI_2(II)$	650–860	−3.100	37.00		
$CoBr_2$					
CoI_2					
ThI_2					
TmI_2					
YbI_2					
GeI_2					
PbI_2	298–680	18.00	4.70		

TABLE XII
Molar heats, c_p, of metal halides, MX_2, with the $CdCl_2$ structure

Compound	Temperature range (K)	A	B	C	D
$MgCl_2$	298–987	18.9	1.42	−2.06	
$ZnBr_2$	298–675	12.6	10.4		
ZnI_2	298–719	12.6	10.4		
$CdCl_2$	298–841	16.0	7.7		
$CdBr_2$					
$MnCl_2$	298–923	18.04	3.16	−1.37	
$FeCl_2$	298–950	18.94	2.08	−1.17	
$CoCl_2$	298–1013	14.41	14.60		
$NiCl_2$	298–1303	17.50	3.16	−1.19	
$NiBr_2$					
NiI_2					

where ΔH_f denotes the formation enthalpy of the substance from the stable modification of the elements at 298.15 K. The formation enthalpy of the elements in their stable modifications is zero by definition. The temperature dependence of the enthalpy is given by

$$c_p(T) = dH°/dT. \tag{3}$$

TABLE XIII

Molar heats, c_p, of metal halides, MX_3, with the BiI_3 structure

Compound	Temperature range (K)	A	B	C	D
AsI_3					
SbI_3	298–444	17.0	21.2		
BiI_3					
$ScCl_3$	298–1233	20.25	7.15		
$TiCl_3$	298–1104	22.900	2.644	−0.428	
$TiBr_3$	298–1067	−2.582	67.959	8.169	−28.510
VCl_3	298–900	22.99	3.92	−1.68	
VI_3		23.24	2.0		
$CrBr_3$					
$FeCl_3$	298–577	14.9	27.5		
$FeBr_3$					

TABLE XIV

Molar heats, c_p, of metal halides, MX_3, with the $CrCl_3$ structure

Compound	Temperature range (K)	A	B	C	D
$AlCl_3$	273–466	18.431	11.431		
YCl_3	298–994	21.00	8.08		
$InCl_3$					
$TlCl_3$					
$LuCl_3$					
$CrCl_3$	298–1218	19.44	7.03		
$RhCl_3$					
$IrCl_3$					

The relationship between entropy and enthalpy is given by

$$dS° = dH°/T \tag{4}$$

and

$$S°(T) = S°_{298.15} + \int_{298.15}^{T} dH°/T. \tag{5}$$

By using molar heats these equations become

$$dS° = [c_p(T)/T]\,dT \tag{6}$$

and

$$S°(T) = S°_{298.15} + \int_{298.15}^{T} [c_p(T)/T]\,dT. \tag{7}$$

The molar free energy is related to the enthalpy and the entropy through the relation

$$G°(T) = H°(T) - T \cdot S°(T). \tag{8}$$

For the enthalpy, the entropy and the free energy, the values at 298.15 K are tabulated, i.e. the formation enthalpy and the standard entropy.

In addition the melting points, boiling points, and normal sublimation points (in degrees K) are included in Tables VII–X. The constants A, B, C and D in the captions of Tables XI–XIV represent the temperature coefficients in the expression for $c_p(T)$,

$$c_p(T) = A + B\ 10^{-3}T + C\ 10^5 T^{-2} + D\ 10^{-6} T^2. \tag{9}$$

The enthalpy and the free energy are approximately linear functions of the temperature and, therefore, permit linear interpolations. The temperature ranges as presented in Tables XI–XIV represent the ranges for the interpolation formulae.

The majority of the presented data are taken from the work of Barin and Knacke [157], a few data are from Kubaschewski and coworkers [158] and from the work of Brauer [35]. For a complete discussion of the molar temperature functions and numerical data at higher temperatures the reader is referred to the literature.

For the mixed halides with the PbFCl structure thermochemical data are very scarce. The known melting points are; 1235 K (SrFCl[223]), 1281 K (BaFCl), 874 K (PbFCl), and 834 K (PbFBr [221]).

The decomposition potentials of solid electrolytes can be calculated from their thermodynamic data. The theoretical decomposition voltage would correspond to the reversible electromotive force (emf) of galvanic cells composed of the electrolyte and two electrodes, one of which is reversible to the anion and the other to the cation of the electrolyte. The emf of the cell

$$M(s)/MX_2(s)/X_2(g), \tag{10}$$

is related to the free energy change for the chemical reaction

$$M(s) + X_2(g) \to MX_2(s), \tag{11}$$

and according to convention this free energy change is the free energy of formation, $\Delta G°(= \Sigma v_i G_i°$, where v_i denote the stoichiometric numbers in the chemical reaction), of the metal halide, when the halide, the metal and the gas are in their standard states. The reversible electromotive force is related to the free energy of formation of the elektrolyte through the relation:

$$E° = -\Delta G°/nF, \tag{12}$$

where $E°$ is the standard emf, F the Faraday constant, and n the number of faradays involved in the electrochemical reaction.

Although experimental measurements of decomposition voltages usually do not agree with theoretical values for several reasons (including chemical interaction of the electrodes with the electrolyte, the production of electrodes in a non-standard state, deviation from isothermal conditions and electrode polarization) it is important to know the theoretical values. They aid in elucidating mechanisms and interpretations of experimental observations on solid electrolytes [159–161].

The theoretical standard emf's for the layered metal halides at 298 K are gathered in Tables XV and XVI [159, 160]. It is assumed that the solids are anhydrous.

TABLE XV

Standard electromotive forces for metal halides with the CdI_2, and the $CdCl_2$ structure

Compound (CdI_2)	$E°(298 K)$ (volts)	Uncertainty at 298 K (volts)	Compound ($CdCl_2$)	$E°(298 K)$ (volts)	Uncertainty at 298 K (volts)
$MgBr_2$	2.606	±0.01	$MgCl_2$	3.070	±0.004
MgI_2		±0.01	$ZnBr_2$	1.624	±0.02
CaI_2	2.845	±0.01	ZnI_2	1.185	±0.01
CdI_2	1.141	±0.02	$CdCl_2$	1.775	±0.01
$TiCl_2$	2.255	±0.22	$CdBr_2$	1.537	±0.02
$TiBr_2$	1.995	±0.22	$MnCl_2$	2.287	±0.02
TiI_2	1.409	±0.22	$FeCl_2$	1.565	±0.02
VCl_2	2.103	±0.43	$CoCl_2$	1.464	±0.02
VBr_2	2.016	±0.43	$NiCl_2$	1.412	±0.04
VI_2	1.453	±0.43	$NiBr_2$	1.106	±0.04
$MnBr_2$	1.908	±0.04	NiI_2		
MnI_2	1.388	±0.04			
$FeBr_2$	1.247	±0.01			
FeI_2	0.759	±0.04			
$CoBr_2$	1.149	±0.02			
CoI_2	0.646	±0.02			
ThI_2					
TmI_2					
YbI_2					
GeI_2	1.084	±0.33			
PbI_2	1.008	±0.01			

TABLE XVI

Standard electromotive forces for metal halides with the BiI_3, and the $CrCl_3$ structure

Compound (BiI_3)	$E°(298 K)$ (volts)	Uncertainty at 298 K (volts)	Compound ($CrCl_3$)	$E°(298 K)$ (volts)	Uncertainty at 298 K (volts)
AsI_3	0.318		$(AlCl_3)_2$	2.200	±0.01
SbI_3	0.452		YCl_3	3.163	±0.04
BiI_3	0.549	±0.07	$InCl_3$	1.641	±0.04
$ScCl_3$	2.946	±0.01	$TlCl_3$		
$TiCl_3$	2.154	±0.14	$LuCl_3$	3.049	±0.01
$TiBr_3$	1.843	±0.14	$CrCl_3$	1.706	±0.07
VCl_3	1.735	±0.43	$RhCl_3$	0.593	±0.04
VI_3			$IrCl_3$	0.665	±0.04
$CrBr_3$					
$FeCl_3$	1.197	±0.06			
$FeBr_3$	0.888	±0.06			

5. Crystal Growth

5.1. GENERAL TECHNIQUES

Although there are several ways of growing single crystals, the techniques being adequately described in textbooks and reviews [163–173], data on growth techniques applied to the layered metal halides are scarce. For a number of thoroughly investigated substances, like the cadmium halides, bismuth iodide and lead iodide various growth methods have been reported. For other members of this family, especially the very hygroscopic ones, almost no information can be found in the literature.

In general there are four different ways of growing single crystals, i.e. from the vapour, from a liquid phase, from a supercritical or 'fluid' phase, and from one or more solid phases [173]. The first two techniques are commonly employed for growing crystals of layered halides.

In growing crystals from the vapour phase one may distinguish between three techniques.

5.1.1. Sublimation of the Substance

If the substance is sufficiently volatile and does not decompose during evaporation, crystals can be grown by the sublimation technique. The procedure is such that the substance, placed in the hot zone (source) evaporates and condenses in a region of lower temperature in the reaction tube, owing to the fact that the vapour becomes supersaturated in the colder part of the set-up. There are various varieties of experimental conditions which can be employed. For detailed descriptions the reader is referred to the existing literature [173, 174, 178–182].

Often as a suitable alternative for thin crystals, thin films are used in studies of optical properties of the halides. Commonly, the optical absorption coefficient rises to values of 10^5–10^6 cm^{-1} in the intrinsic absorption region of these layered crystals and it is for this reason that very thin samples with a thickness in the range of 10^{-8}–10^{-7} m are necessary.

Such thin films can be obtained by thermal evaporation – in vacuum – of the substance under consideration. Deposition has to take place onto a suitable transparent substrate, usually silicate. Combined optical- and X-ray diffraction studies of di- and trihalide films show that layers deposited at such low temperatures as 4 to 20 K normally have an amorphous structure which gives broad Debye-Scherrer patterns. Sometimes spontaneous crystallization takes place during annealing at elevated temperatures, sometimes the crystallization process is a gradual one. From the work of Tubbs [7] it appears that crystalline layers can be obtained directly by evaporating the substance onto substrates that are heated above the so called crystallization temperature for the substance involved. It must be noted that for perfect films the substrate-temperature should not exceed the re-evaporation temperature. As a rule two-thirds of the melting point in degrees K seems usually to be an adequate substrate temperature.

Specific crystallization temperatures are reported by Best [176] for a number of halides. Tubbs [177] discusses electron-microscopic and diffraction studies carried out on CdI_2 and PbI_2 films ($\sim 5 \times 10^{-8}$ m). It appears that such a film deposited onto a substrate kept at room temperature is continuous and gives a single crystal diffraction pattern from regions as large as 5 μm in diameter. Thinner films ($\sim 2.5 \times 10^{-8}$ m) give single crystal diffraction patterns from regions of about 2.5 μm in diameter. Surface crystallites in PbI_2 films seem to be oriented with the c-axis parallel to the substrate.

5.1.2. Reaction in the Vapour Phase

When the constituent atoms of a crystal and/or volatile compounds of the constituent atoms are introduced into a hot zone, reaction may take place. In the case that the temperature of the hot zone is so high that the vapour is unsaturated with regard to the solid to be formed, crystallization occurs when the vapour reaches the cold region of the system.

The conditions of growth are now very similar to those described in Section 5.1.1. Well known examples of the successful employment of this procedure can be found in the literature on the growth of II–VI compounds like ZnS, ZnSe and CdS. In most cases the metal vapour is carried by a carrier gas (argon or hydrogen) into the hot zone where it reacts with the nonmetal hydride gases and crystals or crystal-layers are deposited [231, 232].

5.1.3. Chemical Transport

Transport by interaction with solvent or carrier molecules in the vapour phase has been used for the growth of a variety of crystals [175, 183–185]. The source material to be crystallized reacts with the carrier gas to form volatile compounds. From the vapour containing the volatile compounds crystals can grow either by thermal decomposition, or by disproportionation, depending on the temperature dependence of the equilibrium between the crystal to be grown, and the vapour phase containing the volatile species [174]. Commonly, transfer between the source material and the crystal is brought about by a temperature difference between the end zones of the containing tube. The liberated carrier gas recirculates and continues the process.

The vapour transport techniques offer the advantage of growth at temperatures below the melting point. However, the process is usually slow, and the products are often small.

Apparatus for vapour growth ranges from the very simple sealed ampoule, placed in a furnace which has the desired temperature gradient, to elaborate systems including flows of different highly purified gases over sources and deposition areas under precisely specified conditions.

The growth of crystals from a liquid phase may be divided into the following categories [173]: growth of the solid from its own melt, from solution in a molten salt, or from a liquid solvent.

5.1.4. Growth from a Melt

Except for the growth from aqueous solution, crystal growth by cooling a melt in a suitable container until it solidifies from one end upward (directional freezing) was the earliest studied technique. It can be carried out either horizontally – moving a molten zone through the substance in a boat (Pfann) – or vertically – lowering a sealed ampoule with the substance through a temperature gradient (Bridgman and Stockbarger), pulling a single crystal from the melt in an open crucible by means of a seed attached to a holder (Nacken, Czochralsky and Kyropoulos), or moving a temperature gradient relative to a stationary crucible (Stöber).

The growth by controlled freezing is in comparison to other procedures an uncomplicated, readily controllable process. Two important economic advantages are that the basic apparatus is simple and inexpensive, and little operator attention is required. Disadvantages arise from contact between container wall and the molten substance, and from the differential thermal contraction upon cooling. Furthermore, many materials can not be grown from their own melt, since they either decompose before melting, melt incongruently, sublime before melting, or the vapour pressure is too high at the melting point. In addition, the growth of thick single crystals from a thin plate-like seed in, for instance, the Stockbarger method can be difficult in that a high anisotropy of the growth rates will either prevent adequate thickness or will cause sticking to the walls of the crucible. An additional ring-furnace close to the growing interface can overcome these problems [151].

Detailed information on various apparati and sophisticated methods can be found in References 169, 173 and 186.

5.1.5. Growth from Solvents

Growth from a solvent is widely practised, and the main differences between the various methods are of technique rather than of principle; growing from solution in a melt (flux growth), electrolysis of molten salts, or growing from volatile solvents.

There are two reasons why precipitation from a liquid phase is favourable for growth of a wide variety of crystals. Firstly the concentration of many compounds can be high, and secondly the ions often posses high mobilities in solution.

In growing crystals from volatile solvents the object is to supersaturate the solution without causing spontaneous nucleation, and to make the supersaturation and hence the rate as high as commensurable with controlled growth on a seed of material of the requisite perfection. The methods hinge essentially on the manner in which supersaturation is brought about. There are several techniques to cause supersaturation: cooling, if the solubility decreases with decreasing temperature, evaporation of the solvent, decomposition of complexes or chemical reaction [187].

Supersaturation provided by slow cooling is an easy method, although there are disadvantages. The fact that it is a nonisothermal procedure means that any growth-temperature-dependent property of the crystal will not be uniform throughout. If the temperature is lowered stepwise, the growth rate will not be uniform, and rate-dependent properties of the crystal will not be uniform.

In growth by evaporation, an isothermal process, supersaturation is provided by evaporation of the solvent. In all evaporation growth excessive nucleation at the interface is a problem. The ring of spurious nucleation often observed in the container, just above the interface, is caused by the solution moving up by capillarity through the crystallites, and then evaporating at the top of the ring of nucleation. It is worth noting that in evaporation growth, the growing crystal is at a fixed temperature. When a dopant addition is required, which has a temperature-dependent distribution constant, more homogeneous crystals will result from such an isothermal process than from processes involving temperature variations.

In the case where supersaturation is achieved by chemical reaction a stationary seed is used, whereas in the other solvent growth techniques seed crystals are suspended in the solution, and are often rotated to ensure symmetrical growth. Stirring may be used to promote single crystal growth.

The art of growing single crystals in gels was known at the end of last century and continued into the early twentieth century. The gel technique, which is based on chemical reactions in aqueous media, has been revived through the work of Henisch and coworkers [188–194], and through the publication of his text [170]. This technique helps to overcome the difficulty of growing single crystals with low solubilities, and/or low dissociation temperatures. The procedures used in gel growth are variants of two basic methods. A solution of sodium silicate of certain specific gravity is mixed with an acid and allowed to gel at a constant temperature. One of the variants uses a gel containing a concentrated acid. On top of this gel the solution of a salt is placed. The salt solution supplies a reaction component and prevents the gel from drying out. The gel containing the acid supplies the other reaction component. In due course, crystals of the desired material, which are only slightly soluble in water, form in the gel and also to some extent on the surface of the gel, within the salt solution.

The other method utilizes a gel with the weakest possible acid addition in a U-shaped tube. This gel separates the concentrated acid solution in the one leg of the U-shaped tube from the concentrated salt solution in the other leg of the tube. Cations and anions diffuse towards each other through the gel from opposite sides, and crystal nucleation will take place at a site somewhere between both concentrations.

A particular advantage of this technique is that the growing crystal is held in a strain-free manner in the gel. The rate of crystal formation depends on the gel density in that the density affects the diffusion constants, on the concentration of reagents in the gel, temperature, diffusion path length etc. For detailed information on this topic the reader is referred to Henisch's text [170].

5.2. Growth of specific compounds

This chapter covers 57 layered metal halides. For less than half of these substances specific growth experiments have been reported in the literature. For those com-

TABLE XVII
Methods utilized in the crystallization of layered metal halides

Compound	Crystallization method	Reference
Ag_2F	electrolysis of a saturated AgF solution	195, 224
CaI_2	sublimation: crystallization of thin films	176
CdI_2	sublimation: crystallization of thin films	7, 10, 176, 177
	sublimation	196, 203
	liquid phase: melt	196, 197, 198
	liquid phase: aqueous solution	199, 200, 202
$MnBr_2$	liquid phase: melt	201
MnI_2	sublimation: crystallization of thin films	177
$FeBr_2$	sublimation	35
$CoBr_2$	sublimation	35
PbI_2	sublimation: crystallization of thin films	176, 177, 209
	sublimation	20, 210, 215
	liquid phase: melt	77, 198, 211–213
	liquid phase: aqueous solution	209, 214
	liquid phase: gel	170, 213, 216
$CdCl_2$	sublimation	203, 204
	liquid phase: melt	204
$CdBr_2$	sublimation	203, 205
	liquid phase: melt	204
	liquid phase: aqueous solution	205
$FeCl_2$	sublimation, melt	206, 234
$NiCl_2$	sublimation	207, 208
$NiBr_2$	sublimation	207, 208
AsI_3	liquid phase: melt	198
SbI_3	sublimation: crystallization of thin films	177
	sublimation	217
BiI_3	sublimation: crystallization of thin films	177
	sublimation	217, 218
$FeCl_3$	sublimation	206
SrFCl	liquid phase: melt	222
BaFCl	liquid phase: melt	151, 219, 220
PbFCl	liquid phase: melt	154
PbFBr	liquid phase: melt	79
PbFI	liquid phase: aqueous solution	26

pounds the crystallization techniques have been gathered in Table XVII. In addition to this tabulation a brief description of the growth experiments will be presented for each substance.

5.2.1. Ag_2F

Silver subfloride single crystals can be obtained from the electrolysis of a saturated silver fluoride solution in a Leclanché-element. Fairly large crystals, $0.5 \times 0.5 \times 0.1$ cm^3, were grown on a platinum or silver cathode for about two hours as shiny hexagonal crystals [195, 224].

For high current densities silver is deposited on the cathode, whereas low current densities yield silver subfluoride crystals on the cathode. The compound is stable in aromatic hydrocarbons such as benzene, but is very reactive towards water. The compound decomposes in air giving AgF and Ag.

5.2.2. CaI_2

Crystallization of thin films is the only reported technique for the growth of CaI_2 crystals, and both thin films (2×10^{-6}–6×10^{-6} cm) as well as thicker films (10^{-5}–2×10^{-5} cm) of a number of halides have been investigated by optical absorption and X-ray methods [176, 177]. In the procedure followed, the substance is dehydrated in high vacuum (10^{-5}–10^{-6} torr), melted and subsequently evaporated onto a suitable substrate. Both low temperature as well as high temperature substrates have been used. The low temperature substrates [176] were slowly heated after condensation of CaI_2 had taken place. Heated substrates were kept at 350–400 K during the evaporation process [177].

From the X-ray work of Best [176], it appears that the CaI_2 layers condensed on cold substrates (20–90 K) are amorphous, and that the heating procedure following the deposition process does not show a sharp crystallization temperature. Above 260 K crystallization seems to occur gradually. The temperature of 260 K is, therefore, used to indicate the crystallization temperature of CaI_2.

5.2.3. CdI_2

Being one of the most thoroughly investigated compounds of this halide family, it is not surprising that a wealth of information is available on CdI_2. Virtually all the commonly known techniques have been applied to obtain this halide in single crystal form.

Thin film crystallization has been applied [7, 10, 176, 177]; the reported X-ray experiments show that CdI_2 layers deposited at condensation temperatures of 290 K or less are amorphous, and that spontaneous crystallization takes place at temperatures above 335 K. Melt growth has been reported by Kleber and Fricke [196] in their study on the occurrence of polytypism in cadmium iodide and the correlation between preparation technique and polytype structure. With a melting point of 663 K, the maximum temperature in the furnace was kept at about 680 K in their experiments. The sealed and evacuated ampoule was not lowered through a temperature gradient; the whole system was cooled with a rate of $50° h^{-1}$. The crystals produced in this process were platelets of about 1 cm in diameter, and about 0.01 to 0.05 cm thick.

Attempts to grow cadmium iodide with incorporated impurities have also been reported. Lead-doped cadmium iodide was prepared by Goto and Ucta [197] in a Bridgman set-up. From the ingots, slices could be cut normal to the c-axis with dimensions 0.5 cm $\times$ 0.6 cm $\times$ 0.8 cm.

Vapour phase growth experiments have also been successful. The vertical set-up utilized in their melt growth experiments was used by Kleber and Fricke [196] for sublimation experiments. The source zone was kept at 680 K, and the colder tip at about 573 K. Deposition of crystals occurred in a small temperature region: 600–610 K.

Growth from an aqueous solution is carried out by slow evaporation of a saturated ($\sim 5\,N$ CdI_2) solution in complete darkness [199, 200].

5.2.4. $MnBr_2$

Wollan and coworkers reported the use of $MnBr_2$ single crystals in neutron diffraction studies on the antiferromagnetic structure of this compound [201]. Their crystals were grown by slowly moving a sealed quartz crucible with the molten substance through a temperature gradient. Apart from the fact that the best results were obtained when the inner walls of the crucible were coated with aquadag, no further information such as pulling rate, temperatures, or crucible shape is given.

5.2.5. MnI_2

For this compound optical studies using films have been reported by Tubbs [177]. However, detailed information on the crystallization technique is not presented in his report.

5.2.6. $FeBr_2$

$FeBr_2$ is formed by reaction of iron with hydrogen bromide at 1073 K in a stream of nitrogen, and this sublimation process seems to produce the substance in crystalline form in the cooler part of the reaction tube which is placed in a horizontal furnace [35].

According to Best [176] it seems that, owing to decomposition of the vapour, $FeBr_2$ cannot be evaporated onto substrates.

5.2.7. $CoBr_2$, ($CoCl_2$ and CoI_2)

Information on crystal growth experiments concerning these cobalt halides is scarce. Preparative procedures to synthesise the compound [35] indicate the possibility of obtaining $CoBr_2$ crystals by sublimation at elevated temperatures in a stream of hydrogen bromide.

The work reported by Ferguson and coworkers [225] does not give enough information about the growth techniques for $CoCl_2$ and $CoBr_2$. The existing literature on thin film preparation indicates attempts to evaporate the substances onto substrates and mentions the occurrence of decomposition during evaporation. A report by Feseveldt [226] discusses experiments utilizing evaporation onto substrates placed very close to the heated source container.

5.2.8. PbI_2

This compound has been studied thoroughly, and a number of growth techniques have been reported, i.e. thin film preparation, vapour phase growth, melt growth, solution and gel growth.

Thin film crystallization studies have been discussed by Best and Tubbs [176, 177, 209]. At condensation temperatures as low as 20 and 90 K the layers are amorphous.

Spontaneous crystallization occurs at 220 K. Verma and Krishna [20] mention the existence of the polytype 6H, which has been reported to be the result of sublimation growth experiments.

Small lead iodide platelets were obtained in sublimation experiments [210] carried out in sealed, evacuated capsules of various diameters. In all cases the source region in the horizontal furnace was kept at about 660 K, a temperature which is a little lower than the melting temperature. Crystals had diameters from 0.2 to 0.5 cm, and covered a range of thicknesses.

Kato and coworkers [215] grew crystals a few microns in thickness from the vapour phase at 643 K in a CO_2 ambient at a pressure of half an atmosphere.

Gähwiller and Harbecke [213] studied excitonic effects in electroreflectance of lead iodide. Their crystals were grown from a melt, and in a gel.

Dirksen [79] and Petkov [77] grew single crystals of PbI_2 in a Bridgman set-up. The substance was subjected to a zone-refining procedure before crystal growth [77]. The purified material was molten in a sealed, evacuated ampoule and subsequently lowered through a temperature gradient. Petkov [77] discusses several shapes for the crystallization ampoules and stresses that in order to avoid cracking of the crystal it should be allowed to cool at a rate of $5°\,h^{-1}$. Solution growth is reported by Tubbs [209], who prepared crystals by slowly cooling an aqueous solution.

Gel growth has been studied by Henisch and coworkers [170] and others [213]. The anion source was KI, the cation source lead acetate, and their experiments were carried out at 273 K. Reagent grade sodium metasilicate forms a good gel, and the crystals found in this metasilicate are purer than those prepared in other gels. The gel concentrations are of importance for the growth of high quality crystals; a specific gravity of 1.035 to 1.040 $g\,cm^{-3}$ gives optimal results. Henisch and Srinivasagopalan [216] measured p-type semiconductivity in their gel-grown PbI_2 crystals and ascribed this electronic conductivity to a deviation from stoichiometry (excess I) in their crystals.

5.2.9. $CdCl_2$ and $CdBr_2$

Single crystals of both compounds have been grown by the melt growth technique and by sublimation in vacuum. Thin film crystallization has also been applied successfully.

Growth in a Bridgman set-up has been reported by Nakashima and coworkers [204]. The substances are dehydrated and sealed under vacuum in a Pyrex tube which is then lowered through a temperature gradient. These authors also used the sublimation technique to obtain crystals of both compounds. Again, the dehydrated compounds were used as starting materials. Mitchell [205] reported the growth of $CdBr_2$ single crystals both from aqueous solution and by sublimation. His starting material was $CdBr_2 \cdot 4H_2O$. HBr was added to the solutions to prevent hydration. The crystals obtained were up to 0.2 cm across and 0.05 cm thick. Anhydrous cadmium bromide crystals were formed by subliming reagent grade $CdBr_2 \cdot 4H_2O$ in covered crucibles.

5.2.10. $FeCl_2$ and $FeCl_3$

Crystals of both substances have been grown by the sublimation technique. Hydrogen is used as a carrier gas [206]. The starting material was commercially obtained $FeCl_3$. If necessary this material can first be dehydrated. Thereafter, it is reduced at about 598 K in a ceramic boat in a stream of dry hydrogen in a horizontal furnace. At higher temperatures the trichloride is completely reduced to metallic iron. The flow rate of the hydrogen gas, which not only reduces $FeCl_3$ but also acts as a carrier, was determined experimentally to give an optimal yield of $FeCl_2$ at a velocity of 60–70 cm^3 min^{-1}. With an increasing flow rate of hydrogen an increasing amount of $FeCl_3$ was transported to both cold ends of the reaction tube.

A typical run showed the following distribution of $FeCl_2$ and $FeCl_3$ in the reaction tube: on both sides of the boat which contained the starting material, $FeCl_2$ powder was deposited, while crystal platelets grew on the upper tube wall directly above the powder. $FeCl_3$ powder had been transported towards both cold ends of the process tube, while $FeCl_3$ plate crystals had grown on the upper tube wall above this powder. Crystals with an area of 0.5 cm^2 could be obtained in this manner.

5.2.11. $NiCl_2$ and $NiBr_2$

In their study of growth spirals and more recently of surface morphologies called surface spikes, Sickafus and coworkers [207, 208] prepared single crystals of $NiCl_2$ and $NiBr_2$ by sublimation in a slow flow of an inert gas like helium or argon. Commercially obtained material, and material synthesized from the metal and the halogen were used.

Sublimation temperatures usually ranged from 898 to 973 K and the crystals (dendrites, whiskers and platelets) grew on the inside surface of the silica tube that was placed in a region of the furnace where the temperature varied from 753 to 890 K. Crystals grew radially inward in a cluster. The growth rate [207] was greatly decreased by removal of the water of hydration from the growth chamber, and was increased by about an order of magnitude upon insertion of a reservoir of water into the growth chamber in a region where the temperature was near 373 K.

5.2.12. AsI_3

Zerfoss and coworkers [198] have grown a variety of crystals in a Bridgman set-up using temperature gradients in the range 22–26° cm^{-1}. They obtained AsI_3 single crystal boules of about 2 cm. Specific details on the growth of this compound were, however, not presented.

5.2.13. SbI_3 and BiI_3

Both compounds have been obtained in single crystalline form by sublimation [217, 218] and thin film crystallization [177].

In sublimation growth experiments the growth ampoule containing SbI_3 is filled with argon, and in the case of BiI_3 the ampoule is evacuated since the introduction of an inert gas like argon results in partial decomposition of the triiodide vapour. The temperature gradient along the ampoule is such that the starting material in the source region is molten. After heating for some hours crystals had grown in the cooler region of the ampoule. The platelet crystals of both triiodides covered a range of thicknesses, all had their optic axis (c-axis) perpendicular to the plane of the crystal plate. BiI_3 crystals were stored in vacuum, since this compound is slowly converted to bismuth oxyiodide when exposed to the atmosphere.

The growth of platelet crystals of BiI_3, 1 to 10 μm thick, by an open flow sublimation technique is described by Curtis and Brunner [218]. The best crystals were prepared in a Pyrex glass apparatus by passing argon, at a flow rate of 40 cm^3 min^{-1} over iodine at 298 K, and reacting this gas mixture with elemental bismuth heated to 583 K. BiI_3 crystals, suitable for optical absorption measurements grew in a temperature gradient ranging from 483 to 363 K. The optical properties of these crystals were dependent on the growth temperature.

Tubbs [177] reports on the preparation of thin films of both triiodides. It seems that SbI_3 films sublime and possibly decompose in high vacuum at room temperature. For further handling, such films have to be removed from the vacuum directly after preparation.

5.2.14. *SrFCl*

In their report on paramagnetic resonance results on a tetragonal Eu^{2+} center in SrFCl crystals De Siebenthal and coworkers [222] mention the use of a modified Bridgman set-up for the growth of their crystals. Details of this set-up were not published, nor were growth data presented.

5.2.15. *BaFCl*

Patel and Singh [219] have grown single crystals of this substance by the indirect flux method. They achieved platelets of $0.6 \times 0.2 \times 0.01$ cm^3, where 0.01 cm is probably the thickness in the c direction.

NaF and $BaCl_2$ were mixed in stoichiometric proportions in a platinum crucible and heated in a muffle furnace at about 1173 K for 7 h. The furnace was then allowed to cool at a rate of 50° h^{-1}. Crystals were washed in boiling alcohol. The possible reaction that might have occurred in the molten state may be given by

$$NaF + BaCl_2 \rightarrow BaFCl + NaCl. \tag{13}$$

Fong and Yocom [220] have grown BaFCl in a Bridgman set-up, but do not report on the size of their crystals, nor on growth temperatures. Larger and especially thicker crystals were grown by Nicklaus and Fischer [151] in a modified Kyropoulos set-up. A plate-like seed with the c-axis along the growth direction was contacted with the surface of the melt. In order to avoid the prevailing growth in the *a* direction, they have placed an additional ring-furnace around the growing crystal and about 0.5 cm above the surface of the melt. The ring-furnace stopped growth in the *a* direction before the crucible walls were reached, whereas the crystal still grew in the c direction.

In this way they achieved a thickness of 2 cm in the c direction. A large concentration of OH$^-$ ions was detected in their crystals, and ascribed to reaction with water vapour during growth in air.

5.2.16. PbFCl and PbFBr

Single crystals of these compounds have been grown from the melt [79, 154]. Commercially available PbCl$_2$ and PbF$_2$ were used as starting materials. PbBr$_2$ was precipitated from an aqueous solution of lead nitrate with concentrated HBr. Lead fluoride was not further purified, whereas lead chloride [227] and lead bromide [228] were subjected to a zone-refining procedure. An equimolar mixture of PbCl$_2$, or PbBr$_2$ and PbF$_2$, placed in a platinum boat in a zone-refining equipment, was dried for about one day at elevated temperature under dry, oxygen free nitrogen. The temperature of the furnace around the boat was then raised above the melting temperature in order to allow the mixture to react. After reaction and solidification, the substance was subjected to a zone-levelling and a subsequent zone-refining procedure. Both procedures were carried out at a rate of 1 cm h^{-1} under purified nitrogen. In order to grow single crystals the last zone pass was carried out at a rate of 0.1 cm h^{-1}. The furnace was then cooled steadily to room temperature over a period of 24 h. Single crystals with surface areas between 0.3 and 1.0 cm^2 and a thickness between 0.1 and 0.3 cm could easily be cleaved from the crystalline parts of the ingots [154].

5.2.17. PbFI

Aqueous solutions of PbFI are stable up to 373 K [26]. On slowly cooling a saturated PbFI solution Rulmont [26] obtained small platelets oriented perpendicular to the c-axis.

6. Conclusions

Although the literature presents a wealth of information on preparative methods and crystal structures of the layered metal halides, detailed information on crystal growth of many members of this family is meager.

This is reflected by the lack of data on the mechanisms of mass and charge transport in these substances. Adequate data can be obtained from a study of the electrical and photoelectrical properties of these halides. Since their crystal structures suggest the electrical properties to be anisotropic, electrical conductivity studies require large single crystals. For compounds of which large crystals are available such studies confirm indeed an anisotropy in the electrical conductivity [154, 195, 229, 230].

As regards optical absorption studies, much more information is available in the literature [7]. The use of crystallized films in optical studies is influential. Moreover, crystals grown by the sublimation technique, commonly used in optical studies, have in general inadequate dimensions to perform electrical conductivity experiments. Yet vapour phase growth seems the only possible method to grow crystals of a considerable number of these halides. In that respect some of the preparative methods gathered in Tables II-V could in principle be used for the simultaneous growth of crystals as well.

Optical and resonance studies on impurity centers in mixed halides have become the subject of current research. Substitutional replacement of the metal ions by impurity ions in these structures provides an interesting crystallographic surrounding for the impurity center. Such studies have initiated the crystal growth of these compounds. It may well be possible to extend the series with the substances SrFBr, BaFBr, and some mixed halides of the rare earths.

It is hoped that this compilation will stimulate further research into the properties of these layered metal halides, of which a considerable number have received as yet little attention, and that it will aid in selecting the preparative and crystal growth methods for members of this family of compounds.

Note added in proof. Meglino and Kostiner [235] and Belotskii et al. [236] report on the optimum conditions for the Bridgman-growth of centimeter long crystals of MCl_2 (M = Mg, Mn, Fe, Co, Ni), and BiI_3, respectively. Recent reports of Tanguy et al. [237], Lambrecht et al. [238] and Brixner and Bierlein [239] describe the melt-growth of EuFCl (m.p. 1230 K), the compound being prepared from the binary constituents EuF_2 and $EuCl_2$, Tanguy et al. [237] obtained grey-green crystals, while Brixner and Bierlein [239], using the same procedure, obtained colorless crystals. EuFCl is hygroscopic.

EuFBr (m.p. 1088 K) was for the first time prepared by Brixner [240] from EuF_2 and $EuBr_2$. A melt of EuFBr in a Pt-ampoule was cooled at a rate of about $30°\ h^{-1}$, and platy large-grained crystals were obtained. EuFBr is also hygroscopic.

EuFX (X = Cl, Br, I) can be prepared by solid-state reaction under argon between 873 and 1023 K [241]. In the same way MFX (M = Ca, Sr, Ba. X = Cl, Br, I) could be prepared. Beck [241] also reports on fluorination with manganese fluoride according to the reaction

$$MnF_2 + 2MX_2 = MnX_2 + 2MFX.$$

MFX compounds with large cations M and small anions X are fairly stable against water, e.g. BaFCl will only be decomposed by hot water, while the fluorohalides of Ca are quite hygroscopic. Moreover, the fluoroiodides quickly darken in air due to iodine formation [241]. The preparation of MFCl (M = Sr, Ba) has further been reported by Yuste et al. [242, 243].

Interesting phosphors result when small amounts of divalent europium are incorporated in alkaline earth fluorohalides. Preparative methods are presented by Tanguy et al. [244] for $M_{0.97}Eu_{0.03}FCl$ (M = Ca, Sr, Ba), and by Sommerdijk et al. [245] for $M_{1-x}Eu_xFX$ (M = Sr, Ba. X = Cl, Br). Samples were prepared by firing intimate mixtures of the starting materials SrF_2, $SrCl_2$, BaF_2, $BaCl_2 \cdot H_2O$, $SrBr_2 \cdot 6H_2O$ and $BaBr_2 \cdot 2H_2O$ between 973 and 1173 K in a N_2/H_2 atmosphere. The concentration of Eu, added as EuF_3, was 5% of the total cation concentration [245].

Acknowledgements

One of the authors J. S. would like to thank Prof. R. A. Huggins and Stanford

University for providing the time and facilities necessary for the completion of this article, and Dr L. E. Nagel for critically reading the manuscript.

References

1. R. W. G. Wyckoff: *Crystal Structures*, Interscience, New York, Vol. I, 2nd edition, 1965.
2. R. W. G. Wyckoff: *Crystal Structures*, Interscience, New York, Vol. II, 2nd edition, 1965.
3. N. N. Greenwood: *Ionic Crystals, Lattice Defects, and Nonstoichiometry*, Chem. Publ. Co., New York, 1970.
4. A. F. Wells: *Structural Inorganic Chemistry*, Oxford University Press, 3rd edition, 1962.
5. G. M. Clark: *The Structure of Nonmolecular Solids*, John Wiley and Sons, New York, 1972.
6. T. Shiba (ed.): *Catalytic Engineering*, Chijinshokan & Co., Tokyo, Vol. 9, 1965.
7. M. R. Tubbs: *Phys. Stat. Sol. (B)* **49** (1972), 11.
8. J. Malinowsky: *Phot. Sci. Eng.* **15** (1971), 175.
9. W. C. de Gruijter and J. Schoonman: *Phot. Sci. Eng.* **17** (1973) 382.
10. M. R. Tubbs: *J. Photogr. Sci.* **17** (1969), 162.
11. R. A. Fotland: *J. Chem. Phys.* **33** (1960), 956.
12. M. R. Tubbs: *Phot. Sci. Eng.* **18** (1974), 156.
13. A. Buroff, P. Simidchieva, R. Stoycheva, and J. Malinowsky: *Model Investigations of the Photographic Process*, Slunchev Bryag, Bulgaria, 1973, p. 207.
14. B. Morosin and A. Narath: *J. Chem. Phys.* **40** (1964), 1958.
15. G. A. Jeffrey and M. Vlasse: *Inorg. Chem.* **6** (1967), 396.
16. J. Trotter: *Z. Krist.* **121** (1965), 81.
17. J. Trotter and T. Zobel: *Z. Krist.* **123** (1966), 67.
18. J. A. Wilson and A. D. Yoffe: *Adv. Phys.* **18** (1969), 193.
19. M. R. Tubbs: *Phys. Stat. Sol.* **41** (1970), K61.
20. A. R. Verma and P. Krishna: *Polymorphism and Polytypism in Crystals*, John Wiley and Sons, New York – London, 1966.
21. A. R. Verma and P. Krishna: 'Periodic Faulting in Close-packed Structures', in S. Balakrishna, M. Krishnamurthi, and B. Ramachandra Rao (eds.), *Physics of the Solid State*, Academic Press, London-New York, 1969, p. 93.
22. G. C. Trigunayat and G. K. Chadka: *Phys. Stat. Sol. (A)* **4** (1971), 9.
23. V. K. Agrawal and G. K. Chadka: *Z. Krist.* **137** (1973), 179.
24. R. S. Tiwari and O. N. Srivastava: *Z. Krist.* **137** (1973), 184.
25. G. Lal and G. C. Trigunayat: *J. Solid State Chem.* **9** (1974), 132.
26. A. Rulmont: *Compt. Rend. (Ser. C)* **276** (1973), 775.
27. R. Colton and J. H. Canterford: *Halides of the First Row Transition Metals*, Wiley-Interscience, London-New York-Sydney-Toronto, 1969.
28. M. Chaigneau and M. Chastagnier: *Bull. Soc. Chim. France* (1958), 1192.
29. M. Chaigneau: *Bull. Soc. Chim. France* (1957), 886.
30. M. Chaigneau: *Compt. Rend.* **242** (1956), 263.
31. L. Poyer, M. Fidder, H. Harrison, and B. E. Bryant: *Inorg. Synth.* **5** (1957), 18.
32. H. Ott and H. Seyfarth: *Z. Krist.* **68** (1928), 239.
33. W. Biltz and G. F. Huttig: *Z. anorg. allg. Chem.* **119** (1921), 115.
34. A. Ferrari and F. Giorgi: *Rend. Accad. Lincei* **9** (1929), 1134.
35. G. Brauer: *Handbuch der präparativen anorganischen Chemie I and II*, Ferdinand Enke Verlag, Stuttgart, 1962.
36. H. Blum: *Z. phys. Chem.* **22B** (1933), 298.
37. V. K. Agrawal and G. C. Trigunayat: *Acta Cryst.* **A25** (1969), 401.
38. R. S. Mitchell: *Z. Krist.* **108** (1956), 296.
39. G. Hägg and E. Hermansson: *Arkiv. Kemi Mineral. Geol.* **17B**, No. 10 (1943).
40. W. Klemm and L. Grimm: *Z. anorg. allg. Chem.* **249** (1941), 198.
41. N. C. Baenziger and R. E. Rundle: *Acta Cryst.* **1** (1948), 274.
42. P. Ehrlich and H. J. Hein: *Z. anorg. allg. Chem.* **292** (1957), 139.
43. D. G. Clifton and G. E. McWood: *J. Phys. Chem.* **60** (1956), 311.
44. P. Ehrlich, W. Gutsche, and H.-J. Seifert: *Z. anorg. allg. Chem.* **312** (1961), 80.

45. E. H. Hall and J. M. Blocher: *J. Phys. Chem.* **63** (1959), 1525.
46. R. N. Lowrey and R. C. Fay: *Inorg. Synth.* **10** (1967), 6.
47. R. F. Rolsten and H. H. Sisler: *J. Am. Chem. Soc.* **79** (1957), 5891.
48. P. Ehrlich and H.-J. Seifert: *Z. anorg. allg. Chem.* **301** (1959), 282.
49. R. E. McCarley and J. W. Roddy: *Inorg. Chem.* **3** (1964), 54.
50. R. C. Young and M. E. Smith: *Inorg. Synth.* **4** (1953), 126.
51. S. A. Amirova, V. A. Rupcheva, and T. V. Romanova: *Russ. J. Inorg. Chem.* **15** (1970), 173.
52. K. O. Berry, R. R. Smardzewski, and R. E. McCarley: *Inorg. Chem.* **8** (1969), 1994.
53. T. Atolmacheva, V. M. Tsintsius, and E. E. Yudovich: *Russ. J. Inorg. Chem.* **11** (1966), 249.
54. R. E. McCarley and J. W. Roddy: *Inorg. Chem.* **3** (1964), 60.
55. V. A. Rupcheva, T. V. Romanova, and S. A. Amirova: *Russ. J. Inorg. Chem.* **15** (1970), 170.
56. W. B. Hadley and J. W. Stout: *J. Chem. Phys.* **39** (1963), 2205.
57. A. Ferrari and F. Giorgi: *Rend. Accad. Lincei* **10** (1929), 522.
58. J. W. Mellor: *A Comprehensive Treatise on Inorganic and Theoretical Chemistry*, Vol. 12, Longmans, London, p. 384.
59. J. D. Christian and N. W. Gregory: *J. Phys. Chem.* **71** (1967), 1583.
60. G. Winter: *Inorg. Synth.* **14** (1973), 101.
61. H. Schäfer and W. J. Hones: *Z. anorg. allg. Chem.* **288** (1956), 62.
62. R. C. Schoonmaker, A. H. Friedman, and R. F. Porter: *J. Chem. Phys.* **31** (1959), 1586.
63. W. E. Hatfield and J. T. Yoke: *Inorg. Chem.* **1** (1962), 463.
64. D. E. Scaife and A. W. Wylie: *J. Chem. Soc.* (1964), 5450.
65. R. J. Clark and J. D. Corbett: *Inorg. Chem.* **2** (1963), 460.
66. J. S. Anderson and R. W. M. D'Eye: *J. Chem. Soc.* (1949), S 244.
67. L. J. Guggenberger and R. A. Jacobsen: *Inorg. Chem.* **7** (1968), 2257.
68. E. Hayek, Th. Rehner, and A. Frank: *Monatsh.* **82** (1951), 575.
69. L. B. Asprey and F. H. Kruse: *J. Inorg. Nucl. Chem.* **13** (1960), 32.
70. F. H. Spedding and A. H. Daane: *Metallurgical Rev.* **5** (1960), 297.
71. D. Brown: *Halides of the Lanthanides and Actinides*, John Wiley and Sons, London-New York-Sydney, 1968, p. 227.
72. G. Jantsch, N. Skalla, and H. Jowurek: *Z. anorg. allg. Chem.* **201** (1931), 207.
73. W. Doll and W. Klemm: *Z. anorg. allg. Chem.* **241** (1939), 239.
74. E. A. Flood, L. S. Foster, and E. W. Pietruschka: *Inorg. Synth.* **2** (1946), 106.
75. H. M. Powell and F. M. Brewer: *J. Chem. Soc.* (1938), 197.
76. L. S. Foster: *Inorg. Synth.* **3** (1950), 63.
77. K. Petkov: *Izv. Otd. Khim. Nauki, Bulg. Akad. Nauk.* **4** (1971), 51.
78. R. S. Mitchell: *Z. Krist.* **111** (1959), 372.
79. G. J. Dirksen: unpublished results.
80. J. B. Reed, B. S. Hopkins, and L. F. Audrieth: *Inorg. Synth.* **1** (1939), 29.
81. G. Bruni and A. Ferrari: *Rend. Accad. Lincei* **4** (1926), 10.
82. L. Pauling: *Proc. Natl. Acad. Sci.* **15** (1929), 709.
83. D. Bryce-Smith: *Inorg. Synth.* **6** (1960), 9.
84. Y. Sensui: *Bull. Chem. Soc. Japan* **46** (1973), 3324.
85. S. Yamaguchi: *Sci. Papers Inst. Phys. Chem. Res. (Tokyo)* **39** (1942), 291.
86. B. Brehler: *Fortschr. Mineral.* **39** (1961), 338.
87. H. R. Oswald: *Helv. Chim. Acta* **43** (1960), 77.
88. M. Balconi: *Rend. Soc. Mineral. Ital.* **5** (1948), 49.
89. A. R. Pray: *Inorg. Synth.* **5** (1957), 153.
90. Z. G. Pinsker and I. Tatarinova: *J. Phys. Chem. U.S.S.R.* **15** (1941), 1005.
91. L. Pauling and J. L. Hoard: *Z. Krist.* **74** (1930), 546.
92. G. Brauer, J. O. Honigschmid, and R. Schlee: *Z. anorg. allg. Chem.* **227** (1936), 184.
93. J. M. Bijvoet and W. Nieuwenkamp: *Z. Krist.* **86** (1933), 466.
94. Z. G. Pinsker: *J. Phys. Chem. USSR* **16** (1942), 1.
95. R. C. Osthoff and R. C. West: *J. Am. Chem. Soc.* **76** (1954), 4732.
96. R. A. Butera and W. F. Gianque: *J. Chem. Phys.* **40** (1964), 2379.
97. R. C. Schoonmaker and R. F. Porter: *J. Chem. Phys.* **29** (1958), 116.
98. A. Ferrari, A. Celeri, and F. Giorgi: *Rend. Accad. Lincei* **9** (1929), 782.
99. M. Hawthorne, T. D. Andrews, P. M. Garrett, F. P. Olsen, M. Reintjes, F. N. Tebbe, L. F. Warren, P. A. Wegner, and D. C. Young: *Inorg. Synth.* **10** (1967), 112.

100. H. Schäfer: *Z. anorg. allg. Chem.* **266** (1951), 269.
101. W. Kangro and E. Petersen: *Z. anorg. allg. Chem.* **261** (1950), 157.
102. H. Grime and J. A. Santos: *Z. Krist.* **88** (1934), 136.
103. A. Ferrari: *Rend. Accad. Lincei* **6** (1927), 56.
104. Y. I. Ivashentsev: *Tr. Tomskogo Gos. Univ., Ser. Khim.* **157** (1963), 77.
105. J. D. McKinley and K. E. Schuler: *J. Chem. Phys.* **28** (1958), 1207.
106. J. D. McKinley: *J. Chem. Phys.* **40** (1964), 120.
107. Y. I. Ivashentsev and G. G. Bodunova: *Tr. Tomskogo Gos. Univ., Ser. Khim.* **154** (1962), 63.
108. J. A. A. Ketelaar: *Z. Krist.* **88** (1934), 26.
109. L. G. L. Ward: *Inorg. Synth.* **13** (1972), 154.
110. A. B. P. Lever: *J. Inorg. Nucl. Chem.* **27** (1965), 149.
111. D. M. L. Goodgame and L. M. Venanzi: *J. Chem. Soc.* (1963), 616.
112. J. C. Bailar: *Inorg. Synth.* **1** (1939), 103.
113. Y. Morino, T. Ukagi, and T. Ito: *Bull. Chem. Soc. Japan* **39** (1966), 71.
114. L. Vegard: *Skrifter Norske Videnskaps. Akad. Oslo I, Mat. Naturv.*, Klasse No. 2, 83 pp. (1947).
115. J. C. Bailar and P. F. Cundy: *Inorg. Synth.* **1** (1939), 104.
116. M. S. Sneed and R. C. Brasted: *Compreh. Inorg. Chem.* **5** (1956), 121, Van Nostrand, New York.
117. G. W. Watt, W. W. Hakki, and G. R. Chopin: *Inorg. Synth.* **4** (1953), 114.
118. H. Bräkken: *Z. Krist.* **74** (1930), 67.
119. R. W. Stolz and G. A. Melson: *Inorg. Chem.* **11** (1972), 1720.
120. B. C. McCollum, M. J. Camp, and J. D. Corbett: *Inorg. Chem.* **12** (1973), 778.
121. W. Klemm and E. Krose: *Z. anorg. allg. Chem.* **253** (1947), 209, 218.
122. B. C. McCollum and J. D. Corbett: *Inorg. Chem.* **5** (1966), 938.
123. G. Natta, P. Corrandini, I. W. Bassi, and L. Porri: *Atti Accad. Nazl. Lincei Rend. Classe Sci. Fis. Mat. Mat.* **24** (1958), 121.
124. J. Lewis, D. J. Machin, I. E. Newnham, and R. S. Nyholur: *J. Chem. Soc.* (1962), 2036.
125. J. M. Sherfey: *Inorg. Synth.* **6** (1960), 57.
126. T. R. Ingraham, W. K. Downes, and P. Marier: *Inorg. Synth.* **6** (1960), 52, *Can. J. Chem.* **35** (1957), 850.
127. R. F. Rolsten and H. H. Sisler: *J. Phys. Chem.* **62** (1958), 1024.
128. R. C. Young and W. M. Leaders: *Inorg. Synth.* **2** (1946), 116.
129. R. C. Young and M. E. Smith: *Inorg. Synth.* **4** (1953), 128.
130. M. A. Oranskaya, Y. S. Lebeder, and I. L. Perfilova: *Russ. J. Inorg. Chem.* **6** (1961), 132.
131. H.-J. Seifert: *Z. anorg. allg. Chem.* **317** (1962), 123.
132. H. Bräkken: *Kgl. Norske Videnskab. Selskab. Forh.* **5**, No. 11, (1932).
133. R. J. Simme and N. W. Gregory: *J. Am. Chem. Soc.* **82** (1960), 93.
134. E. R. Epperson, S. M. Homer, K. Knox, and S. Y. Tyree: *Inorg. Synth.* **7** (1966), 163.
135. N. Wooster: *Z. Krist.* **83** (1932), 35.
136. B. Rapp Tarr: *Inorg. Synth.* **3** (1950), 191.
137. S. S. Todd and J. P. Coughlin: *J. Am. Chem. Soc.* **73** (1951), 4184.
138. S. Blairs and R. A. J. Shelton: *J. Inorg. Nucl. Chem.* **28** (1966), 1855.
139. R. O. McLaren and N. W. Gregory: *J. Phys. Chem.* **59** (1955), 184.
140. N. W. Gregory: *J. Am. Chem. Soc.* **73** (1951), 472.
141. N. W. Gregory and B. A. Thackrey: *J. Am. Chem. Soc.* **72** (1950), 3176.
142. J. A. A. Ketelaar, C. H. Macgillavry, and P. A. Renes: *Rec. Trav. Chim.* **66** (1947), 501.
143. D. H. Templeton and G. F. Carter: *J. Phys. Chem.* **58** (1954), 940.
144. H. Hecht, G. Jander, and H. Schlapmann: *Z. anorg. allg. Chem.* **254** (1947), 255.
145. A. Vavoulis, T. E. Austin, and S. Y. Tyree: *Inorg. Synth.* **6** (1960), 129.
146. G. B. Heisig, B. Fawkes, and R. Hedin: *Inorg. Synth.* **2** (1946), 193.
147. G. Natta: *Rend. Accad. Lincei* **5** (1927), 592.
148. N. Wooster: *Z. Krist.* **74** (1930), 363.
149. S. N. Anderson and F. Basolo: *Inorg. Synth.* **7** (1963), 214.
150. E. V. Stroganov and K. V. Ovchinnikov: *Vestn. Leningr. Univ.* **12**, No. 22; *Ser. Fiz. Khim.*, No. 4, (1957), 152.
151. E. Nicklaus and F. Fischer: *J. Cryst. Growth* **12** (1972), 337.
152. W. Nieuwenkamp: Thesis, University of Utrecht, 1932.
153. W. Nieuwenkamp and J. M. Bijvoet: *Z. Krist.* **81** (1932), 469.

154. J. Schoonman, G. J. Dirksen, and G. Blasse: *J. Solid State Chem.* **7** (1973), 245.
155. W. Nieuwenkamp and J. M. Bijvoet: *Z. Krist.* **82** (1932), 157.
156. S. S. Talipov: *Uzbeksk. Khim. Zh.* **7** (1963), 70.
157. I. Barin and O. Knacke: *Thermochemical Properties of Inorganic Substances*, Springer Verlag, Berlin-Heidelberg-New York, 1973.
158. O. Kubaschewski, E. Ll. Evans, and B. C. Alcock: *Metallurgical Thermochemistry*, Pergamon Press, Oxford, 4th Ed., 1965.
159. W. J. Hamer, M. S. Malmberg, and B. Rubin: *J. Electrochem. Soc.* **103** (1956), 8.
160. W. J. Hamer, M. S. Malmberg, and B. Rubin: *J. Electrochem. Soc.* **112** (1965), 750.
161. W. L. Worrell and J. Hladik: 'Thermodynamic Equilibrium Diagrams', in J. Hladik (ed.), *Physics of Electrolytes*, Vol. 2, Academic Press, London-New York, 1972, p. 747.
162. J. D. Corbett: 'Metal Halides in Low Oxidation States', in W. L. Jolly (ed.), *Preparative Inorganic Reactions*, Vol. 3, Interscience, New York-London-Sydney, 1966, p. 1.
163. 'Crystal Growth', *Disc. Faraday Soc.* **5** (1949).
164. H. E. Buckley: *Crystal Growth*, John Wiley and Sons, New York, 1951.
165. W. D. Lawson and S. Nielsen: *Preparation of Single Crystals*, Butterworths Scientific Publ., London, 1958.
166. A. Smakula: *Einkristalle, Wachstum, Herstellung, Anwendung*, Springer Verlag, Berlin, 1962.
167. *The Art and Science of Growing Crystals*, Editor J. J. Gilman, John Wiley and Sons, New York, 1963.
168. E. A. D. White: *Brit. J. Appl. Phys.* **16** (1956), 1415.
169. R. A. Laudise: *The Growth of Single Crystals*, Prentice-Hall Inc., New Jersey, 1970.
170. H. K. Henisch: *Crystal Growth in Gels*, The Pennsylvania State University Press, University Park, London 1970.
171. R. L. Parker: *Solid State Phys.* **25** (1970), 151.
172. J. W. Mullin: *Crystallization*, Butterworths Scientific Publ., London, 2nd ed., 1972.
173. F. A. Kröger: *The Chemistry of Imperfect Crystals*, 2nd revised edition, Vol. 1, *Preparation, Purification, Crystal Growth and Phase Theory*, North-Holland Publishing Co., Amsterdam-London, American Elsevier Publishing Co., Inc., New York 1973.
174. H. Schäfer: *Chemische Transport Reaktionen*, Verlag Chemie, Weinheim, 1962.
175. H. Schäfer et al.: *Z. anorg. allg. Chem.* **286** (1956), 27, 42; **290** (1957) 279; **291** (1957), 221, 294.
176. K. J. Best: *Phys. kondens. Materie* **1** (1963), 316.
177. M. R. Tubbs: *J. Phys. Chem. Solids* **29** (1968), 1191.
178. *Proceedings of the Second International Conference on Crystal Growth.* Boston, F. C. Frank, J. B. Mullins, and H. S. Peiser (eds.), *J. Crystal Growth* **3/4** (1968).
179. E. Kaldis and M. Schieber: *J. Crystal Growth* **9** (1971).
180. H. Schäfer: in P. Hartman (ed.), *Crystal Growth an Introduction*, North-Holland Series in Crystal Growth, 1973.
181. C. F. Powell, I. E. Campbell, and B. W. Gonser: *Vapor Plating*, John Wiley and Sons, New York, Chapman and Hall, London, 1955.
182. R. F. Rolsten: *Iodide Metals and Metal Iodides*, John Wiley and Sons, New York, 1962.
183. D. C. Cronemeijer: *J. Appl. Phys.* **29** (1958), 1730.
184. J. C. Marinace: *I.B.M. J. Res. Develop.* **4** (1960), 248.
185. A. Reisman, M. Berkenblit, and S. A. Alyanakyan: *J. Electrochem. Soc.* **112** (1965), 241.
186. J. C. Brice: *The Growth of Crystals from Liquids*, North-Holland Publishing Co., Amsterdam-London, 1973.
187. A. N. Holden: *Disc. Faraday Soc.* **5** (1949), 313.
188. H. K. Henisch, J. Dennis, and J. I. Hanoka: *J. Phys. Chem. Solids* **26** (1965), 493.
189. F. G. Wakin, H. K. Henisch, and H. Atwater: *J. Phys. Chem.* **42** (1965), 2619.
190. J. Dennis and H. K. Henisch: *J. Electrochem. Soc.* **112** (1965), 1240.
191. H. K. Henisch, J. I. Hanoka, and J. Dennis: *J. Electrochem. Soc.* **112** (1965), 627.
192. J. Dennis and H. K. Henisch: *J. Electrochem. Soc.* **114** (1967), 263.
193. E. S. Halberstadt and H. K. Henisch: *J. Cryst. Growth* **3/4** (1968), 363.
194. J. J. Nickel and H. K. Henisch: *J. Electrochem. Soc.* **116** (1969), 1258.
195. H. Kawamura, I. Shirotani, and H. Inokuchi: *Chem. Phys. Letters* **24** (1974), 549.
196. W. Kleber and P. Fricke: *Z. Phys. Chem. (Leipzig)* **224** (1963), 353.
197. T. Goto and M. Ueta: *J. Phys. Soc. Japan* **29** (1970), 1512.
198. S. Zerfoss, L. R. Johnson, and P. H. Egli: *Disc. Faraday Soc.* **5** (1949), 166.

199. R. M. Yu: *J. Phys. Chem. Solids* **30** (1969), 63.
200. D. K. Wright and M. R. Tubbs: *Phys. Stat. Sol.* **37** (1970), 551.
201. E. O. Wollan, W. C. Koehler, and M. K. Wilkinson: *Phys. Rev.* **110** (1958), 638.
202. C. W. Bunn and H. Emmet: *Disc. Faraday Soc.* **5** (1949), 119.
203. P. L. Jones, C. N. W. Sitting, D. E. Mason, and V. A. Williams: *J. Phys. D.* **1** (1968), 283.
204. S. Nakashima, H. Yoshida, and T. Fukumoto: *J. Phys. Soc. Japan* **31** (1971), 1847.
205. R. S. Mitchell: *Z. Krist.* **117** (1962), 309.
206. R. M. A. Lieth and J. C. J. M. Terhell: unpublished results.
207. E. N. Sickafus and D. R. Winder: *J. Appl. Phys.* **35** (1964), 2541.
208. E. N. Sickafus: *J. Appl. Phys.* **37** (1966), 2607.
209. M. R. Tubbs: *Proc. Roy. Soc.* **A280** (1964), 566.
210. R. M. A. Lieth and G. E. van Egmond: unpublished results.
211. J. F. Verwey: Thesis, University of Utrecht, 1967.
212. G. D. Currie, J. Mudar, and O. Risgin: *Appl. Opt.* **6** (1967), 1137.
213. Ch. Gähwiller and G. Harbecke: *Phys. Rev.* **185** (1969), 1141.
214. S. Nikitine: *Compt. Rend.* **251** (1960), 935.
215. Y. Kato, T. Goto, and M. Ueta: *J. Phys. Soc. Japan* **35** (1973), 625.
216. H. K. Henisch and C. Srinivasagopalan: *Solid State Commun.* **4** (1966), 415.
217. B. L. Evans: *Proc. Roy. Soc.* **A276** (1963), 136; **A289** (1966), 275.
218. B. J. Curtis and H. R. Brunner: *Mat. Res. Bull.* **9** (1974), 715.
219. A. R. Patel and R. P. Singh: *J. Cryst. Growth* **5** (1969), 70.
220. F. K. Fong and P. N. Yokom: *J. Chem. Phys.* **41** (1964), 1383.
221. C. Sandonnini: *Gazz. Chim. Ital.* **41** (1911), 144.
222. J. M. de Siebenthal, H. Bill, and L. Lacroix: *Solid State Commun.* **14** (1974), 167.
223. *Gmelins Handbuch der Anorganischen Chemie, Achte Auflage* **29** (1931), p. 142, Verlag Chemie, Berlin.
224. H. Hettig: *Z. anorg. allg. Chem.* **167** (1927), 224.
225. J. Ferguson, D. L. Wood, and K. Knox: *J. Chem. Phys.* **39** (1963), 881.
226. H. Feseveldt: *Z. Phys.* **64** (1930), 741.
227. B. Willemsen: *J. Solid State Chem.* **3** (1971), 567.
228. J. F. Verwey and J. Schoonman: *Physica* **35** (1967), 386.
229. J. Schoonman, A. F. Halff, and G. Blasse: *Solid State Commun.* **13** (1973), 677.
230. J. Schoonman: *Solid State Commun.* **13** (1973), 673.
231. J. Woods: *British Journ. Appl. Phys.* **10** (1959), 529.
232. W. H. Yim and E. J. Stofko: *J. Electrochem. Soc.* **119** (1972), 381.
233. W. van Erk: 'Magnetic and Spectroscopic Investigations on VI_2, MnI_2 and FeI_2', Thesis, State University of Groningen, 1974.
234. C. Vettier: *Phys. Rev.* **B11** (1975), 4700.
235. P. Meglino and E. Kostiner: *J. Crystal Growth* **32** (1976), 276.
236. D. P. Belotskii, V. B. Timofeev, I. N. Antipov, V. I. Vashchenko, and V. A. Bespal'chenko: *Russ. J. Phys. Chem.* **42** (1968), 740.
237. B. Tanguy, M. Pezat, C. Fontenit, and C. Fouassier: *C.R. Acad. Sci. Paris Sér. C* **277** (1973), 25.
238. V. G. Lambrecht, M. Robbins, and R. C. Sherwood: *J. Solid State Chem.* **10** (1974), 1.
239. L. H. Brixner and J. D. Bierlein: *Mat. Res. Bull.* **9** (1974), 99.
240. L. H. Brixner: *Mat. Res. Bull.* **11** (1976), 269.
241. H. P. Beck: *J. Solid State Chem.* **17** (1976), 275.
242. M. Yuste, M. Rahmani, D. Jumeau, L. Taurel, and J. Badoz: *J. Phys. C. Solid State Phys.* **6** (1973), 3167.
243. M. Yuste, S. Lefraut, J. M. Spaeth, and L. Taurel: *J. Phys. C. Solid State Phys.* **8** (1975), 3491.
244. B. Tanguy, P. Merle, M. Pezat, and C. Fouassier: *Mat. Res. Bull.* **9** (1974), 831.
245. J. L. Sommerdijk, J. M. P. J. Verstegen, and A. Bril: *J. Luminescence* **8** (1974), 502.

BIVALENT METAL HYDROXIDES

H. R. OSWALD and R. ASPER

Institute for Inorganic Chemistry, University of Zürich, Switzerland

Table of Contents
1. Introductory Survey
2. Magnesium Hydroxide, $Mg(OH)_2$
   - 2.1. General Information
   - 2.2. Compound Preparation
     - 2.2.1. By Hydration of MgO
     - 2.2.2. By Precipitation
     - 2.2.3. By Electrolysis
   - 2.3. Special Forms
   - 2.4. Properties
     - 2.4.1. Physical
     - 2.4.2. Solubility
     - 2.4.3. Thermal Decomposition
     - 2.4.4. Thermodynamic Data
3. Calcium Hydroxide, $Ca(OH)_2$
   - 3.1. General Information
   - 3.2. Compound Preparation
     - 3.2.1. By Hydrolysis
     - 3.2.2. From Solutions
     - 3.2.3. By Precipitation
     - 3.2.4. By Electrolysis
   - 3.3. Properties
     - 3.3.1. Physical
     - 3.3.2. Solubility
     - 3.3.3. Thermal Decomposition
     - 3.3.4. Thermodynamic Data
4. Manganese (II) Hydroxide, $Mn(OH)_2$
   - 4.1. General Information
   - 4.2. Compound Preparation
     - 4.2.1. From Metallic Manganese
     - 4.2.2. By Precipitation
     - 4.2.3. By Other Methods
   - 4.3. Special Forms
   - 4.4. Properties
     - 4.4.1. Physical
     - 4.4.2. Solubility
     - 4.4.3. Oxidation in Air
     - 4.4.4. Oxidation in Alkaline Aqueous Media
     - 4.4.5. Thermal Decomposition
     - 4.4.6. Thermodynamic Data
5. Iron (II) Hydroxide, $Fe(OH)_2$
   - 5.1. General Information
   - 5.2. Compound Preparation
   - 5.3. Properties
     - 5.3.1. Physical
     - 5.3.2. Electrochemical Data
     - 5.3.3. Chemical Behaviour

  5.3.4. Solubility
 5.4. Thermodynamic Data
6. Cobalt Hydroxide, $Co(OH)_2$
 6.1. General Information
 6.2. Compound Preparation
  6.2.1. By Precipitation
  6.2.2. Standard Precipitation Procedures
  6.2.3. By Electrolysis
  6.2.4. By Other Methods
  6.2.5. Preparation of Special Forms
 6.3. Properties
  6.3.1. Physical
  6.3.2. Chemical Behaviour
  6.3.3. Electrochemical Properties
  6.3.4. Thermal Decomposition
  6.3.5. Solubility
  6.3.6. Thermodynamic Data
 6.4. Cobalt(III) Oxide Hydroxide, CoOOH
7. Nickel (II) Hydroxide, $Ni(OH)_2$
 7.1. General Information
 7.2. Compound Preparation
  7.2.1. By Precipitation
  7.2.2. By Hydrolysis
  7.2.3. By Other Methods
  7.2.4. High Purity Preparations
 7.3. Preparation of Special Forms
  7.3.1. Well Crystallized Samples
  7.3.2. Gels
  7.3.3. Suspensions, Hydrosols
  7.3.4. Other Forms
 7.4. Properties
  7.4.1. Physical
  7.4.2. Chemical Behaviour
  7.4.3. Electrochemical Properties
  7.4.4. Thermal Decomposition
  7.4.5. Solubility
  7.4.6. Thermodynamic Data
 7.5. Hydroxides with Ni in Higher Valence States
8. Copper(II) Hydroxide, $Cu(OH)_2$
 8.1. General Information
 8.2. Compound Preparation
 8.3. Properties
  8.3.1. Crystal Structure
  8.3.2. Some Chemical Properties
9. Zinc Hydroxides, $Zn(OH)_2$
 9.1. General Information
 9.2. Preparation and Characterization of the $Zn(OH)_2$ Modifications
  9.2.1. Amorphous $Zn(OH)_2$
  9.2.2. α-$Zn(OH)_2$
  9.2.3. $Zn(OH)_2$ of the C6 Structure Type
  9.2.4. β_1- and β_2-$Zn(OH)_2$
  9.2.5. γ-$Zn(OH)_2$
  9.2.6. δ-'$Zn(OH)_2$'
  9.2.7. ε-$Zn(OH)_2$
 9.3. Properties
  9.3.1. Stability Relations from Energetic and Solubility Data
  9.3.2. Thermal Decomposition
10. Cadmium Hydroxide, $Cd(OH)_2$
 10.1. General Information

10.2. Compound Preparation
    10.2.1. By Hydrolysis
    10.2.2. By Precipitation
    10.2.3. By Electrolysis
10.3. Special Forms
    10.3.1. α-Cd(OH)$_2$
    10.3.2. Colloidal Cd(OH)$_2$
    10.3.3. γ-Cd(OH)$_2$
10.4. Properties
    10.4.1. Physical
    10.4.2. Solubility
    10.4.3. Thermodynamic Data
11. Conclusions
References

1. Introductory Survey

In this article, the hydroxides Me(OH)$_2$ of a number of bivalent metals, in part belonging to the first transition series, are discussed. These compounds are all more or less insoluble in water, of moderate chemical or thermal stability and have many properties in common, although there is considerable variability in the details. Therefore, an introductory Table I containing information on the properties in general, on the structural types and lattice parameters, and on some thermodynamical and solubility data has been compiled. More details may be found later under the respective sections for each compound.

The fact that all these compounds or at least one modification of each hydroxide crystallize in layered type lattices, mainly of the CdI$_2$- or C6-type (see Figure 1), is a consequence of the presence of relatively small twofold positive charged cations in close proximity to the non-spherosymmetrical and highly polarizable OH$^-$ ions. Thus, the metal ions (c) arrange in hexagonal layers, and the OH$^-$ ions (A, B) form similar layers above and below the sheet c, leading to the sequence (AcB)(AcB)(AcB)... etc. with hexagonal close packing of the anions. Every OH$^-$ forms three bonds to metal ions in its own layer and is in contact with three OH$^-$ of the adjacent layer. The metal ions reach a coordination by six OH$^-$ in equal distances,

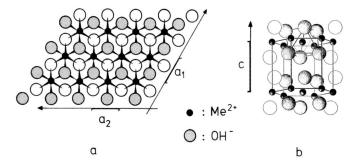

Fig. 1. Crystal structure of most hydroxides Me(OH)$_2$: CdI$_2$- (or C6-, or Mg(OH)$_2$-, or brucite-) type, hexagonal, space group D_{3d}^3-P$\bar{3}$m1 (Nr. 164). (a) layer (00.1) projected along the c-axis; (b) stacking of the layers.

TABLE I

Selected properties of bivalent metal hydroxides, Me(OH)$_2$

	Mg(OH)$_2$	Ca(OH)$_2$	Mn(OH)$_2$	Fe(OH)$_2$	Co(OH)$_2$	Ni(OH)$_2$	Cu(OH)$_2$	Zn(OH)$_2$	Cd(OH)$_2$
Electron configuration of the Me^{2+} ion	[Ne]	[Ar]	[Ar]3d^5	[Ar]3d^6	[Ar]3d^7	[Ar]3d^8	[Ar]3d^9	[Ar]3d^{10}	[Kr]4d^{10}
Colour	white	white	rose	white	blue (α) red (β)	pale green	blue	white (all forms)	white (all forms)
Natural occurence	brucite	portlandite	pyrochroite	—	—	—	—	—	—
Stability against air oxygen (25 °C)	yes	yes	no	—	—	yes	yes	yes	yes
Stability against CO$_2$ (25 °C)	no	no	no	no	no	no	no	no	no
Structure type	C6	C6	C6	C6	double layers (α) C6 (β)	C6	own type	several polymorphs; C6 (high press. form)	double layers (α) C6 (β) own type (γ)
Unit cell volume per formula unit [Å]^3	40.9$_0$	54.48	45.0$_6$	42.3$_5$	40.4$_6$ (β)	38.6$_6$	41.0$_4$	41.6$_5$ (C6) 54.4$_6$ (ϵ)	49.8$_4$ (β) 49.5 (γ)
Lattice energy [kcal mole^{-1}] (refs. cf. text)	680	617	695	715	747	757	764	748	676
Solubility product: log K_{S0}, 25 °C (ref. [33] and text)	−9.2 (act.) −10.9 (inact.)	−5.4 (inact.)	−12.8 (act.) −12.8 (inact.)	−14 (act.) −15.1 (inact.)	−14.2 (α) −14.8 (β act.) −15.7 (inact.)	−14.7 (act.) −17.2 (inact.)	−18.8 (inact.)	−15.52 (am.) −16.47 (ϵ)	−13.7 (β act.) −14.4 (β inact.)
Standard enthalpy of formation $\Delta H°_{298.2}$ [kcal mole^{-1}] (refs. cf. text)	−221.0	−235.7	−166.2	−137.2	−129.4	−128.6	−106.1	−153.5	−131.9
Standard free enthalpy of formation $\Delta G°_{298.2}$ [kcal mole^{-1}] (refs. cf. text)	−199.3	−214.8	−147.0	−117.6	−110.0	−108.3	−85.3	−132.8	−113.2

representing an octahedron which may, however, be stretched or flattened along its trigonal axis. In practice, flattened octahedra are found as far as exact crystallographic results are available, a fact which is explained by de Haan [1] as being due to a substantial increase of the Madelung energy. The protons are located on the trigonal axis, and from the manner in which they are arranged, resp. from the relatively long O—O-distances and the high wave numbers of the O—H stretching vibration (~ 3700 cm^{-1}) in the infrared spectra, it becomes evident that hydrogen bonds O—H·····O—H in the commonly accepted sense cannot be present.

Other types of crystal structures occurring are:

– α-forms: α-$Zn(OH)_2$, α-$Co(OH)_2$, and α-$Cd(OH)_2$ crystallize in characteristic double layered lattices, in which tetrahedrally as well as octahedrally coordinated metal ions are present. These more or less disordered phases are quite unstable and tend to transform into more stable ones, i.e. β-$Co(OH)_2$, ε-$Zn(OH)_2$ or $ZnO + H_2O$, and β-$Cd(OH)_2$. It is very probable that small amounts of anions other than OH^- are indispensable in stabilizing the crystal structure of the α-phases, so that these could be considered as hydroxy-salts ('basic salts') with rather a very high OH^- to salt anion ratio.

– $Cu(OH)_2$ forms its own type of lattice – a hydrogen bonded layered structure which is closely related to the structure of γ-FeOOH (lepidocrocite resp. EO_4-type). This exceptional behaviour is related to the well-known tendency of the Cu^{2+} ion to occupy coordination octahedra with a strong tetragonal (4+2)-distortion, a fact usually explained in terms of a static Jahn-Teller effect.

– $Zn(OH)_2$ can adopt the CdI_2-structure only in a modification which is obtainable by high pressure treatment and which then persists at normal pressure. Besides this modification and the above mentioned α-form, a number of other $Zn(OH)_2$-polymorphs have been characterized: $β_1$- and $β_2$-$Zn(OH)_2$ (unknown layer structure); γ-$Zn(OH)_2$ (known chain structure with tetrahedral coordination); δ-'$Zn(OH)_2$' (unknown layer structure, rather a hemihydrate $2 Zn(OH)_2 \cdot H_2O$); ε-$Zn(OH)_2$ (the most stable and best known $Zn(OH)_2$-structure, space lattice with tetrahedrally coordinated Zn^{2+}). This particular behaviour is a consequence of the property of the Zn^{2+} ion to enter octahedra, or tetrahedra, or both in one and the same structure as well – a fact which cannot be explained by an influence of the electrostatic ligand field. The trend to tetrahedral coordination is rather due to the high polarizing effect of the relatively small Zn^{2+} ion with its completely filled 3d-shell, i.e. the tetrahedral Zn—O-bonds have a considerable degree of covalency.

– $Cd(OH)_2$ forms, besides an α- and the rather stable β-form in the CdI_2-type, a phase γ with octahedral coordination around the Cd^{2+} ion, but not isotypic with any of the $Zn(OH)_2$ polymorphs.

It is of some interest to compare the unit cell volumes V_z of the different hydroxides from $Ca(OH)_2$ to $Zn(OH)_2$ as a function of the respective number of 3d-electrons of the Me^{2+} ions, as shown in Figure 2. The resulting diagram reflects well the expectations from the consideration of ligand field stabilization energies (LFSE's) for octahedrally coordinated high-spin ions:

	$3d^0$	$3d^5$	$3d^6$	$3d^7$	$3d^8$	$3d^9$	$3d^{10}$
	Ca^{2+}	Mn^{2+}	Fe^{2+}	Co^{2+}	Ni^{2+}	Cu^{2+}	Zn^{2+}
LFSE	0	0	$2\Delta_0/5$	$4\Delta_0/5$	$6\Delta_0/5$	$3\Delta_0/5$	0

(where Δ_0 is the energy difference between the e_g and the t_{2g} orbitals in the octahedral case, and neglecting the influence of configuration interaction for $3d^7$).

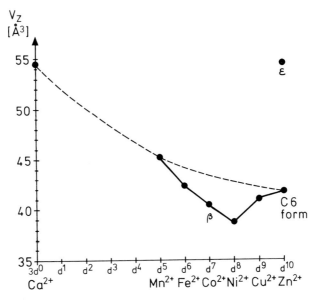

Fig. 2. Volume per formula unit $Me(OH)_2$ (V_z) as a function of the number of 3d-electrons of the Me^{2+} ions.

Without the effect of ligand field stabilization, the points for $Fe(OH)_2$, $Co(OH)_2$, $Ni(OH)_2$, and $Cu(OH)_2$ were expected to be found on a steady curve connecting V_z of $Ca(OH)_2$ over $Mn(OH)_2$ to $Zn(OH)_2$.

An analogous curve is, in principle, also obtained if the lattice energies of the hydroxides are shown as a function of the number of 3d-electrons, Figure 3, with the difference that the minimum of the curve, corresponding to the maximum in the absolute value of lattice energy, is reached here for $Cu(OH)_2$. Some particular behaviour of $Cu(OH)_2$ can be expected, however, as it crystallizes in a different structure type with distorted octahedral coordination around the Cu^{2+} ions.

Moreover, a similar picture is obtained if the solubility products, resp. log Ks_0, are plotted as a function of the number of 3d-electrons of the Me^{2+} ions. This is shown in Figure 4, where also values for some less stable hydroxides are included. Here, too, the minimal solubility is reached for $Cu(OH)_2$. If, instead, the log Ks_0-values are presented as a function of the radii of the Me^{2+} ions, the points for the C6-type hydroxides of the 3d metals fall on a straight line leading from $Ni(OH)_2$ to $Ca(OH)_2$ in Figure 5. On the other hand, $Cu(OH)_2$ and ε-$Zn(OH)_2$ (with their different structure

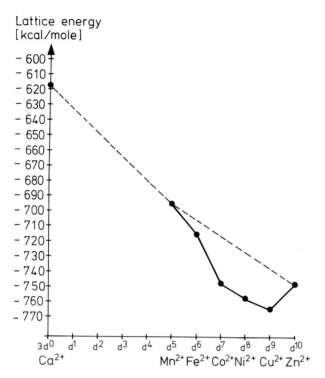

Fig. 3. Lattice energies of Me(OH)$_2$ as a function of the number of 3d-electrons of the Me^{2+} ions (values by Yatsimirskij [97].)

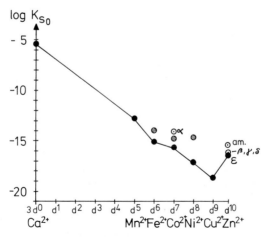

○ : unstable mod.
◉ : active form of stable mod.
● : inactive form of stable mod.

Fig. 4. Log of the solubility products (Ks_0) of Me(OH)$_2$ as a function of the number of 3d-electrons of the Me^{2+} ions (Ks_0 values by Feitknecht and Schindler [33].)

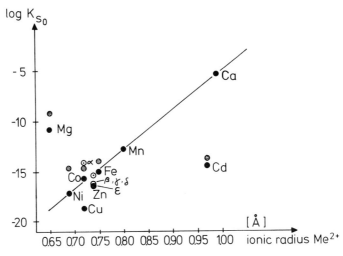

Fig. 5. Log of the solubility products (K_{S_0}) of Me(OH)$_2$ as a function of the Pauling radii of the Me^{2+} ions (K_{S_0} values by Feitknecht and Schindler [33].)

types) as well as Mg(OH)$_2$ and β-Cd(OH)$_2$ (C6-type, but metals with different electron shells) behave exceptionally. Thus, such a representation is capable of reflecting very sensitively the crystal chemical particularities of the phases under consideration.

From the viewpoints of preparative chemistry, purity and general chemical stability, the entire group of bivalent metal hydroxides is rather problematic. The compounds can in principle easily be prepared from the corresponding metal salt solutions upon sufficient raise of pH, but their usual appearances are microcrystalline to extremely highly dispersed forms, in aqueous systems frequently even colloidal (sols or gels). Due to the high specific surface and the layered lattice type, there is a marked tendency for adsorption and/or inclusion of foreign anions – which are very difficult and sometimes nearly impossible to remove by washing – and of varying amounts of loosely bonded water molecules. On the other hand, the amount of impurities in the cation sublattice is rather given by the content of foreign metal ions in the starting solutions, and may therefore be more easily controlled.

Evidently, these highly dispersed preparations show also characteristic disorder in their crystal lattice. Frequently, the only relatively sharp lines in the X-ray powder patterns are the prism reflections (hk.0), whereas the basal and pyramidal reflections (00.1) and (hk.1) are extremely broad or even missing completely (so-called 'cross-grating patterns', pointing to extremely small particle size in the direction of the trigonal axis and/or disorder, 'roughened' layer lattices or 'Arnfelt'-disorder).

Due to the small solubilities, it is relatively difficult to grow larger single crystals from solutions, except for the significantly better soluble Ca(OH)$_2$, and most other classical crystal growth methods are not applicable due to the low decomposition temperatures. Gel-growth seems promising, but little work has been done on gels which are well suited to alkaline solutions.

Once the preparations are made, they suffer from their relative chemical instability. As Table I shows, all phases considered are sensitive to carbon dioxide from the air, particularly if they are finely divided and in the presence of humidity. The formation of basic carbonates by Zn(OH)$_2$ was studied in detail by Feitknecht and Oswald [2]. Usually, these reactions are topotactic in nature. Large single crystals are much less reactive, as they are protected by a thin layer of the product formed. In cases where the metal ions can easily adopt higher valency states than $+$ II, the preparations also have to be protected from air oxygen. This holds in order of decreasing sensibility for Fe(OH)$_2$, Mn(OH)$_2$, and Co(OH)$_2$. Again, small particle size and high relative humidity (or aqueous medium) enhance the reactivity very much. For Mn(OH)$_2$, these influences were investigated in detail by Feitknecht et al. [3] resp. Oswald et al. [4]. Attack by oxygen always takes place parallel to the (00.1) planes. Thus, the rare mineral pyrochroite is always found in nature in a more or less oxidized state.

In various solutions, the hydroxides are reactive as well: acids dissolve them to form the respective salts; alkali hydroxides solve Me(OH)$_2$ under formation of hydroxo-complexes, e.g. [Me(OH)$_4$]$^{2-}$; and in salt solutions, there is a tendency to replace OH$^-$ ions by the respective anions, i.e. basic salts are formed, mainly through topotactic reactions.

Finally, also the thermal stability is rather low. In the range between less than 100 and about 200 °C, depending on which particular metal hydroxide and which polymorphic form is studied, the OH$^-$ ions condense under formation of the corresponding oxide and of water. Some general aspects regarding the thermal stability of hydroxides were presented by Freund [5].

From this general characterization of the bivalent metal hydroxides it may be easily understood that they certainly can be considered as interesting from the structural chemical viewpoint, or as reactive intermediates for the study of reactions in solid/gas, solid/liquid, and solid/solid systems; but, on the other hand, they are not of great technical significance compared to many metal oxide or sulphide phases. There are, however, exceptions like Ni(OH)$_2$ or Cd(OH)$_2$ which are both important in electrical storage cells.

If besides Me^{2+} and OH$^-$ ions other anions are included as a third component in the consideration of layered structures, then one enters the very extended field of the hydroxy- (or basic) salts. They have to be mentioned for the sake of completeness, as their structures are in the great majority of cases more or less closely related to the CdI$_2$- (or C6-) type. Unfortunately, it is impossible to treat these compounds systematically in this article. In the simplest case, some part of the OH$^-$ ions in the metal hydroxide structure is replaced by anions like Cl$^-$, Br$^-$, NO$_3^-$ etc., which may happen statistically and lead to non-stoichiometric compounds. But at least for stable phases, the ordered replacement of a certain fraction of the OH$^-$ ions, such as e.g. $\frac{1}{2}$ or $\frac{1}{4}$, by other anions is preferred. This means the formation of new structures, normally under lowering of the symmetry to orthorhombic or monoclinic, but still of the single layered type. In more complicated cases then, intermediate layers may be present leading to double layered lattices, such as e.g. in $Zn_5(OH)_8Cl_2 \cdot H_2O$, or the layered structures may degenerate to chain structures, or there is an evolution towards three-dimensional space lattice types. The widespread idea that basic salts are essentially disordered non-stoichiometric phases deriving from hydroxides by mere replacement of some OH^- by other anions is, at least, much too crude. On the contrary, nature forms a high number of stoichiometric, well-ordered and individual crystal structures, all of them with their own characteristic properties. Through insertion of new degrees of freedom, e.g. the presence of two sorts of Me^{2+} ions and/or two sorts of foreign anions besides OH^- within one and the same structure (cationically or anionically mixed hydroxy-salts), the situation gets even more interesting, but also complicated from the structural chemical viewpoint. Many of these phases can be crystallized and are known as minerals or corrosion products of metals and alloys in aqueous, salt-containing media.

The only broader review on solid hydroxy-salts of bivalent metals, written by Feitknecht [6], is unfortunately no longer up to date. It should be worthwhile indeed to treat the field again nowadays, since much more reliable structural data are available. As examples, a few references are given below. The salts MeOHCl of bivalent metals (earlier $Me(OH)_2 \cdot MeCl_2$) with single layered lattices have been treated by Oswald and Feitknecht [7]. Similarly, Oswald and Feitknecht [8] have

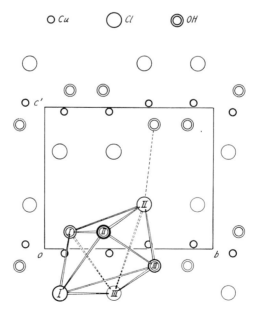

Fig. 6. Crystal structure of CuOHCl, projection along [100]$_{monocl.}$ to show the layers (001). (Iitaka et al. [9].)

reviewed the hydroxy-halides $Me_2(OH)_3X$ (earlier $3 Me(OH)_2 \cdot MeX_2$, with $X=Cl$, Br, I), crystallizing in layered structures or structures evolving towards space lattices. For examples of single-layered lattice structures see CuOHCl, Iitaka, Locchi and Oswald [9], Figure 6; and $Cu_2(OH)_3Br$, -I, Oswald, Iitaka, Locchi and Ludi [10], Figure 7.

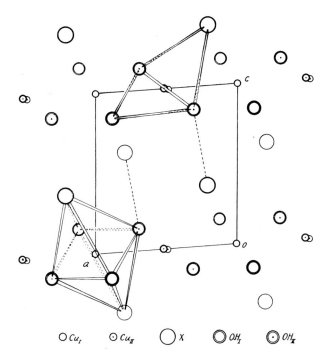

○ Cu_I　⊙ Cu_{II}　○ X　◎ OH_I　◉ OH_{II}

Fig. 7. Crystal structure of $Cu_2(OH)_3X$ ($X=Br$, I), projection along $[010]_{monocl.}$ to show the layers (001). (Oswald et al. [10].)

$Zn_5(OH)_8(NO_3)_2 \cdot 2H_2O$ (earlier $4Zn(OH)_2 \cdot Zn(NO_3)_2 \cdot 2H_2O$) has a characteristic double layer lattice (see Stählin and Oswald [11], Figure 8). For such structures a notation used by Feitknecht [6] tried to express that so-called 'main layers' of the metal hydroxide (here containing octahedrally coordinated Zn^{2+} and OH^-) alternate with 'intermediate layers' (here containing tetrahedrally coordinated Zn^{2+}, NO_3^- and water): $4Zn(OH)_2 \rightleftharpoons Zn(NO_3)_2, 2H_2O$. A similar building principle will have to be mentioned again later in connection with the assumed structure of the α-forms of some hydroxides. The old notation is, however, not fully correct, as the above compound has to be written structurally as follows: $Zn_3^{(oct.)}(OH)_8 \cdot Zn_2^{(tetr.)}(NO_3)_2(H_2O)_2$.

Finally, the structure of $CdCu_3(OH)_6(NO_3)_2 \cdot H_2O$, Oswald [12], is an example of a cationically mixed hydroxy-salt, again with a single-layered lattice.

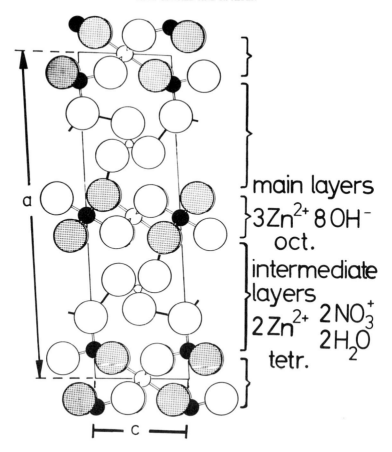

Fig. 8. Crystal structure of $Zn_5(OH)_8(NO_3)_2 \cdot 2H_2O$, projection along $[010]_{monocl.}$ to show the (100) double layered lattice characteristics. (Stählin and Oswald [11].)
Small circles: nitrogen; medium circles: zinc; large circles: OH^- or H_2O or NO_3^- oxygen.

In the following text, each section is dedicated to the hydroxides of one metal, these being arranged in the order of increasing atomic number. The further splitting up of the sections was adapted to the requirements for a clear presentation of the various phases, with special emphasis on the facts regarding preparation and chemical reactivity.

2. Magnesium Hydroxide, $Mg(OH)_2$

2.1. General information

For magnesium only one hydroxide is known, $Mg(OH)_2$, which occurs in the CdI_2- (C6- or sometimes also named brucite-) structure type. The mineral $Mg(OH)_2$ – called brucite or texalite – has given its name to the whole group of compounds with the same structure. Due to the fact that $Mg(OH)_2$ – just like the closely related $Ca(OH)_2$ – is sensitive to water and carbon dioxide, its mineralic occurrences are not frequent, and in particular well-shaped transparent crystals with a typical plate-like habit are rare. The fibrous variety of $Mg(OH)_2$ – also having the brucite structure – is called nemalite or nematolite when in mineral form.

2.2. Compound preparation

During synthesis, the sensitivity to CO_2 and H_2O should be taken into account, and the preparation of waterfree samples requires a carefully conducted drying procedure [13]. The aging of freshly prepared samples of $Mg(OH)_2$ will be discussed in the section on special forms.

2.2.1. By Hydration of MgO

Extensive literature exists – for the most part of older date – on the preparation of $Mg(OH)_2$ via hydrolysis of MgO [14, 15]. The rate of conversion of the oxide into the hydroxide is very dependent on the particle size and inner morphological structure of the oxide; its history, including the way it has been prepared, strongly affects the conversion into $Mg(OH)_2$. In case $(NH_4)_2Mg(CO_3)_2$ is used for the preparation of the oxide instead of $Mg(NO_3)_2$, the rate of formation is increased six times [16].

It was recognized quite early that prior to the hydroxide formation, water adsorption on the oxide takes place. Detailed investigations of this multiple-step conversion process were carried out and reported by Feitknecht and his co-workers [17]. Their experiments revealed the following stages:

(i) physisorption and chemisorption of H_2O on MgO;
(ii) diffusion of Mg^{2+} and OH^- in the adsorbed water film;
(iii) nucleation of $Mg(OH)_2$;
(iv) crystal growth of $Mg(OH)_2$.

In this process, epitaxial growth of $Mg(OH)_2$ on MgO was also observed [18]. The following procedure leads to the formation of microcrystalline $Mg(OH)_2$: pyrolysis of magnesium oxalate – at 600 °C – to MgO, followed by the addition of 10 ml H_2O to about 0.1 g MgO. The hydrolysis is finished in 24 h [19]. A faster hydration occurs by treatment with water vapour near 100 °C. Crystals with sizes up to 2 mm are obtained when MgO is treated with a NaOH solution at 400 °C [20].

2.2.2. By Precipitation

Solutions of magnesium salts like $MgCl_2$, $MgSO_4$ and $Mg(NO_3)_2$ are used in precipitation processes with the aid of hydroxide ions. Bases used are aqueous NaOH, KOH, K_2CO_3 and NH_3. When $Mg(NO_3)_2$ is used, the formation of basic nitrates can have a disturbing effect [21]. Furthermore, in following these preparative procedures, anion or cation adsorption – from the salt or base respectively – on the small particles of the precipitate can lead to a contaminated end product. Citric acid addition [22] or ultrasonic precipitation [23] enhances the formation of larger $Mg(OH)_2$ particles.

Brauer [24] has described the preparation of microcrystalline $Mg(OH)_2$ from $MgCl_2$ with the aid of NH_3. Likewise the growth of crystals – a few tenths of a mm in length – as a result of the reaction of KOH on $MgCl_2$ is presented in [24]. Diffusion processes through a ceramic membrane between a Mg^{2+} salt solution and a NaOH or KOH solution result in the growth of crystals up to 1 mm; these are discussed in [25].

2.2.3. By Electrolysis

For a report on the high current density electrolytic preparation of $Mg(OH)_2$, utilizing a $MgCl_2$ solution and yielding a very pure product, the reader is referred to [26].

2.3. SPECIAL FORMS

During the precipitation procedure, so-called 'active' material precipitates first. This substance, although of very small particle size, is always of microcrystalline character and shows, in particular, no expansion of the lattice. In contact with the mother liquor at ordinary temperature, the aging process is completed in the order of ten days and results in a free energy decrease of up to 2 kcal mole^{-1}. This value holds for the comparison between the very first, not yet isolated active product and the inactive hydroxide, whereas for isolated $Mg(OH)_2$ preparations differences in energy of up to 850 cal mole^{-1} have been found [14]. During the aging process an increase in particle size and a decrease in lattice defects take place. As an example, plate-like crystals of 300 Å in length and 100 Å thick in average have grown after aging to about 1300 Å × 500 Å [27]. The different sorts of aging reactions, as observed for the various $Me(OH)_2$ compounds, were the subject of Feitknecht's investigations [28]. Crystal growth during the aging of $Mg(OH)_2$ is further discussed in detail in [29].

$Mg(OH)_2$ can also be obtained in various sol- or gel-like forms [14]; in principle, even in this state the colloidal particles still exhibit a crystalline structure.

2.4. PROPERTIES

2.4.1. Physical

Reliable lattice dimensions are given by Swanson *et al.* [30]; see also de Haan [1]:

$a = 3.147$ Å $\quad\quad c = 4.769$ Å $\quad\quad c/a = 1.515$
$Z = 1$ $\quad\quad V = 40.9_0$ Å^3 $\quad\quad D_x = 2.36_7$ g cm^{-3}
free parameter $z = 0.22$ $\quad\quad D_{exp.} = 2.36$ g cm^{-3} [14]

For the lattice energy, Finch and Gardner [31] give besides 686 kcal mole^{-1} a value of 680 kcal mole^{-1}, which was determined following the approach of Kapustinkij-Yatsimirskij.

A number of other physical data on $Mg(OH)_2$ can be found in [14].

2.4.2. Solubility

The data in Table II present the solubility of $Mg(OH)_2$ in water as a function of temperature. The decrease in solubility with increasing temperature indicates that $Mg(OH)_2$ – like $Ca(OH)_2$ – has a positive heat of solution [32]. The solubility of the hydroxide under consideration is strongly dependent on the state of aging of the substance and – in case the hydroxide has been prepared by hydrolysis of the oxide – it is also dependent on the procedure followed in MgO preparation. The reported values of aged $Mg(OH)_2$ and 'active' $Mg(OH)_2$ lie in the range 1×10^{-3} to 4×10^{-3} g per 100 ml H_2O at 18 °C.

TABLE II
Solubility in water as a function of temperature for $Mg(OH)_2$

Compound	Solubility [g per 100 ml]	Temperature [°C]	Ref.
$Mg(OH)_2$	9.8×10^{-4} 6.9×10^{-4} 4.3×10^{-4} 2.2×10^{-4}	35 72 110 150	Travers [32]

For the solubility product, Feitknecht and Schindler [33] recommend, based on reliable and coinciding determinations of Gjaldbaek [34], Feitknecht, and Näsänen [35], to use the values of $\log Ks_0 = -9.2$ for active and $\log Ks_0 = -10.9$ for inactive $Mg(OH)_2$ (25 °C, $J=0$).

2.4.3. Thermal Decomposition

Like the solubility, the decomposition pressure of $Mg(OH)_2$ is extremely dependent on the state of aging, as the water vapour pressure values presented in Table III show [36]. In the literature there exists a variety of temperature values for the decomposition of the hydroxide. Gill [37] gives 375–380 °C, Richardson [38] 400–450 °C.

TABLE III
Water vapour pressure over $Mg(OH)_2$ as a function of temperature

Compound	Water vapour pressure [torr]		Ref.
'active' $Mg(OH)_2$	1.5×10^{-4} 1875	(20 °C) (350 °C)	
aged $Mg(OH)_2$	1.4×10^{-6} 209	(20 °C) (350 °C)	Fricke and Lüke [36]

Kurnakow et al. [39] report for the pure material 410 °C, and for contaminated material 480 °C, while Gregg and Packer [40] give 375–400 °C for the pure and 425–450 °C for the contaminated hydroxide.

According to Hackspill et al. [41] the starting temperature lies at 350 °C and for the release of the last 10% of water at 700 °C. Finally Berg et al. [42] report a start at 400 °C and 700 °C for the last 10%.

Electron microscopical investigations carried out by Goodman [43] revealed that $Mg(OH)_2$ loses water by a two-step mechanism, and details about this process – compared to that of $Ca(OH)_2$ – were later explained by Freund [5], who gives the following sequence: H_2O is built up from two OH^- ions in the first step, while in the second step desorption of this H_2O takes place. Evidence for the tunneling mechanism of the proton – which enables H_2O formation from the OH^- ions – was found by Freund and Naegerl [44, 45] in their studies on $Mg(OH)_2$ and $Mg(OD)_2$.

2.4.4. Thermodynamic Data

Data on $Mg(OH)_2$, unfortunately enough, are scarce. For the standard enthalpy of formation from the elements, ΔH^0, Thomsen [46] determined -217.3 kcal mole^{-1}. From specific heat measurements cited in [14], a standard entropy S^0 cal mole^{-1} K^{-1} (25 °C, 1 atm) of 15.1 was calculated. Latimer ([105] p. 316) gives $\Delta H^0 = -221.0$ kcal mole^{-1}, $\Delta G^0 = -199.3$ kcal mole^{-1} and $S^0 = 15.09$. In case $Mg(OH)_2$ is prepared by the reaction of liquid H_2O on MgO, the heat of reaction is dependent on the preparation procedure followed in preparing MgO, and values between -8.19 and -11.77 kcal mole^{-1} were presented by Roth and Chall [47] resp. Fricke and Lüke [36].

3. Calcium Hydroxide, $Ca(OH)_2$

3.1. General information

The only known hydroxide of calcium is $Ca(OH)_2$, and no stable hydrates are known.

In many commonly used preparative methods, the resulting hydroxide is a microcrystalline or even amorphous product, which contains strongly adsorbed water. Microcrystalline samples also quite easily adsorb water and carbon dioxide from the air. On the other hand, well crystallized $Ca(OH)_2$ is – especially in dry air – quite stable. Sometimes the hydroxide is found in mineral form (so-called portlandite), and in both this naturally occurring material, as well as in the substance synthesized in the laboratory, the colourless, easy-to-cleave crystals with their plate-like habit show the layered brucite type structure.

Technical applications of the hydroxide have been known since early times. Preparation proceeds by thermal decomposition ('burning') of $CaCO_3$ to CaO which is subsequently hydrated to amorphous $Ca(OH)_2$. Due to its low solubility in water, a suspension – the so-called 'lime-milk' – or a gel-like solid is formed, yielding the 'lime-mortar' by admixture of sand. In this application as plaster or inexpensive white lime-wash for walls, the conversion of $Ca(OH)_2$ into $CaCO_3$ proceeds slowly by way of CO_2 uptake from the air and H_2O desorption. Further applications of $Ca(OH)_2$ are found in the sugar and in the tannery industries. Its application in medicine as an antidote against acid poisoning is based on its reactivity towards diluted acids.

3.2. Compound preparation

All preparative methods have in common that a rapid formation of $Ca(OH)_2$ leads to a microcrystalline or even amorphous product. Only in a slow growth process can one expect to obtain $Ca(OH)_2$ which is not X-ray amorphous. In most cases the crystalline product is plate-like (00.1) in habit. Sometimes, however, thin hexagonal prismatic columns [00.1] are observed. An amorphous sample can – by aging in a saturated solution – attain a crystalline form.

3.2.1. By Hydrolysis

In a way analogous to the formation from the oxide, $Ca(OH)_2$ can be prepared by hydrolysis from many calcium compounds, for example CaH_2, Ca_3N_2, CaC_2 [48]. Further reactions are those between $CaCO_3$ and 35% KOH [49, 50] and between $CaSO_4$ and NaOH [51] which yields a crystalline product.

3.2.2. From Solutions

Careful crystallization from a saturated aqueous $Ca(OH)_2$ solution yields a good crystalline product. Well-known techniques are evaporation of the solution over concentrated H_2SO_4 [52], slow heating of a saturated solution which results in a decrease of the solubility [53], and vacuum evaporation of a saturated solution [54].

3.2.3. By Precipitation

Precipitation of the hydroxide by increasing the OH^- concentration – achieved by addition of concentrated KOH to a saturated $Ca(OH)_2$ solution – leads only under special conditions to a crystalline product [48]. For further discussions about the precipitation of the hydroxide by KOH or NaOH addition to Ca^{2+}-solutions, the reader is referred to [55] and [56].

Crystals having dimensions on the order of centimeters are obtained when two containers filled with a $CaCl_2$ and a NaOH solution respectively – kept in a larger container – are carefully flooded with water. In some weeks' time a good product has grown via the slow diffusion process [57].

3.2.4. By Electrolysis

In old literature dating back to 1808, a report describing the production of calcium hydroxide single crystals by electrolysis of a $CaCl_2$-solution exists [58]. A modern sophisticated modification based on the same principle appeared in 1967 [59].

3.3. Properties

3.3.1. Physical

Structurally, no hydroxide of the C6-group has been more investigated than $Ca(OH)_2$.

Due to the availability of single crystals of sufficient size and perfection, detailed structure analyses with neutrons by Busing and Levy [60] and with X-rays by Petch [61] became possible. Both studies support the hydrogen positions predicted as early as 1935 by Bernal and Megaw.

There are numerous accurate determinations of the lattice constants available, among which those determined by Bunn et al. [62] seem particularly reliable:

$a = 3.5844 \pm 0.0003$ Å $c = 4.8962 \pm 0.0007$ Å $c/a = 1.3660$
$Z = 1$ $V = 54.48$ Å^3 $D_x = 2.258$ g cm^{-3}
   $D_{exp.} = 2.241$ g cm^{-3} [48]

free parameter $z_0 = 0.2330 \pm 0.0004$ [61]
   0.2341 ± 0.0003 [60]
free parameter $z_H = 0.395 \pm 0.008$ [61]
   0.4248 ± 0.0006 [60]

The lattice energy is given by Finch and Gardner [31] with 599 kcal mole^{-1}, and, using an expression by Kapustinkij-Yatsimirskij, with 617 kcal mole^{-1}.

For more physical data on Ca(OH)$_2$ see [48].

3.3.2. Solubility

Here, as in the case discussed for Mg(OH)$_2$, a positive heat of solution exists which has been found to be on the order of 2.8 kcal mole^{-1} [63]. Solubility as a function of temperature [64] is presented in Table IV. The differences observed in the presented values can be explained by the dependence of the solubility on particle size, and this effect seems to be considerable. Depending on the preparation technique utilized, microcrystalline to amorphous samples are formed. The three different values for three different kinds of Ca(OH)$_2$ – all at 20 °C – as presented in Table IV, illustrate this size effect clearly [65].

TABLE IV
Solubility in water as a function of temperature and of crystallinity for Ca(OH)$_2$

Compound	Solubility [g per 100 ml]	Temperature [°C]	Ref.
Ca(OH)$_2$	0.130	0	
	0.122	15	
	0.109	30	Bassett
	0.096	45	[64]
	0.082	60	
	0.059	90	
crystalline Ca(OH)$_2$	0.121	20	Nacken
freshly slaked Ca(OH)$_2$	0.133	20	and
wet-ground Ca(OH)$_2$	0.150	20	Mosebach [65]

The solubility products found in older literature [66] based on electrometric titration measurements are on the order of 6×10^{-6} at 20 °C, while more recent work tends to somewhat lower values. Feitknecht and Schindler [33] recommend for a crystalline $Ca(OH)_2$ to use log $Ks_0 = -5.43$, corresponding to $Ks_0 = 3.72 \times 10^{-6}$ (25 °C, $J=0$).

3.3.3. Thermal Decomposition

The dehydration of $Ca(OH)_2$ has long been the subject of investigations, and various temperatures – corresponding with a dissociation pressure $p_{H_2O} = 760$ torr – have been reported in older literature, i.e. 450 °C, 513 °C, 547 °C and 560 °C [48]. More recent investigations, carried out isothermally at 260°, 300°, 350° and 380 °C in a vacuum, have shown a two-step decomposition mechanism. After Mikhail and his co-workers [67], the movement of the phase boundary $Ca(OH)_2/CaO$ into the crystals acts as a rate-determining step, with an activation energy of 14.3 kcal mole^{-1}. The fact that in the same year (1966) Dave and Chopra [68] published the extremely different value of 45 kcal mole^{-1} for the activation energy of the $Ca(OH)_2$-decomposition illustrates that kinetic parameters for solids have to be considered with great care and their significance needs further discussion – a fact which holds even today as well.

In his review on thermal decomposition of hydroxides presented in 1968, Freund [5] gives a two-step mechanism similar to that described for $Mg(OH)_2$: formation of H_2O from two OH^- ions, followed by the desorption of the water molecules.

X-ray single crystal investigations have shown that thermal decomposition is a topotactic process. Compared to a similar process in air, proceeding in a vacuum gives a more perfect orientational relationship between the initial crystal and the product crystallites. Furthermore, the shift of the $Ca(OH)_2/CaO$ phase boundary into the volume of the crystal can be traced [69]. The presence of a two-step mechanism was recently confirmed with the aid of electron microscopy by Krylow and his co-workers [70].

3.3.4. Thermodynamic Data

Values for the enthalpy and free enthalpy of formation – under standard conditions – were reported by Rossini [71], Hatton [72] and Kelley [73]:

$\Delta H°$ [kcal mole^{-1}]	$\Delta G°$ [kcal mole^{-1}]	Ref.
-235.8	-214.3	[71]
-235.7	-214.8	[72]

Entropy values calculated from specific heat data in the range 21–86 K [73] and 19–33 K [72] respectively result in the values $S^0_{298} = 17.4 \pm 1.0$ and $S^0_{298} = 19.93 \pm 0.10$ cal mole^{-1} K^{-1} respectively.

Using the additivity rule for ionic crystals, a value for S^0_{298} is found which lies between the above-mentioned values, namely $S^0_{298} = 18.2$ cal mole^{-1} K^{-1} [74]. The heat of reaction for the formation of $Ca(OH)_2$ from CaO and liquid H_2O at 298 K was determined from the heat of solution of CaO and $Ca(OH)_2$ in diluted acids and from the heat of hydration of CaO in a saturated $Ca(OH)_2$-solution. Taylor and Wells [75] reported a value of -15.5 kcal mole^{-1}.

4. Manganese(II) Hydroxide, $Mn(OH)_2$

4.1. *General Information*

A number of hydroxides, oxide hydroxides and oxide hydrates of manganese with oxidation states +II, +III and +IV are known and listed, e.g. in [76]. Whereas many of these compounds are not very well-defined phases, the sole compound of Mn(II) is $Mn(OH)_2$ and represents a definite hydroxide, its lattice being built up from Mn^{2+} and OH^- ions.

The solubility is very slight in water but good in acids, and in strong bases $Mn(OH)_2$ may be solved to some extent at higher concentrations and temperatures only. The compound is thus a strong base with some weak amphoteric character. Although +II represents the most stable valence state of manganese in general, $Mn(OH)_2$ is, in accordance with the redox potentials, readily oxidized in alkaline media, thereby changing its colour from white to brown. Its preparation in pure state by precipitation with alkalies from aqueous Mn^{2+}-solutions is therefore attainable only by careful exclusion of air oxygen during all operations. The product is described as a flocculent, 'amorphous' or very finely crystalline precipitate, with a tendency to rapid aging, of a white to light gray appearance, sometimes exhibiting a pearly luster due to its layered structure. Recrystallization from hot concentrated alkali solution leads to better crystallized material, which, in a very pure state, is more stable. When sufficiently protected from the influence of humidity and air, crystalline $Mn(OH)_2$ samples may be preserved over very long periods. The compound is indeed even found – although very rarely – in nature as the mineral pyrochroite. The transition into higher manganese oxide compounds occurs by way of a topochemical reaction over β-MnOOH (as mineral: feitknechtite).

$Mn(OH)_2$ and $Cu(OH)_2$ or $Co(OH)_2$ precipitation mixtures have been used for the preparation of thermal resistors (MMT or KMT type); see [77]. $Mn(OH)_2$ is also used for desulfurization of heavy oil fractions (see [78]), and was further discussed as a raw material for driers and varnish, and as a catalyst for the synthesis of methane from H_2 and CO. It is also catalytically active in the decomposition of hypochlorite, in the oxidation of As_2O_3 and of Cr(III) to Cr(VI).

4.2. Compound preparation

4.2.1. *From Metallic Manganese*

If Mn is boiled under exclusion of air in sodium hydroxide solutions, it is slowly transformed into $Mn(OH)_2$. The reaction goes to completion given sufficient time, excluding an amount of product which remains in solution as hydroxo complexes [79].

Klingsberg and Roy [80] reacted manganese powder e.g. at 350 or 360 °C and a water vapour pressure of the order of 15 000 to 2000 psi (1020.7 to 136.1 atm) in a silver-plated autoclave for 24 or 48 h. The product was a white powder, built up from microscopic idiomorphous hexagonal crystals. During the period of cooling down to ambient temperature, the water vapour pressure was maintained to prevent a partial decomposition to MnO.

$Mn(OH)_2$ was also observed electrochemically in the anodic dissolution process of manganese [81].

4.2.2. *By Precipitation*

In order to make the precipitation from diluted $MnCl_2$ solutions with alkali hydroxide under hydrogen as complete as possible, a pH value of 9.5 at 25 °C is necessary, whereas between pH 8.2 and 8.5, the reaction remains incomplete [82]. For 0.024 M $MnCl_2$ solution and 0.09 M NaOH, e.g. a beginning of the precipitation was found with pH 8.41 at 18 °C [83]; cf. also [84]. Similarly, the formation of $Mn(OH)_2$ from diluted $MnSO_4$ solutions has been studied in several older papers (for details see [85]). Work by Heubel [86] has shown that the addition of NaOH to $MnSO_4$ solutions (between 2 and 0.03 M) at 20 °C resulted in a product which was never completely free from sulphate ions, whereas on the other hand in alkaline media, i.e. by addition of $MnSO_4$ solution to NaOH, no formation of basic manganese(II) sulphates could be detected. If $Mn(OH)_2$ is precipitated by addition of ammonia solutions, the reaction does not go to completion or is, in the presence of a sufficient quantity of ammonium ions, even completely prevented due to complex formation. For further details regarding the reaction of NH_3 solutions with diluted $MnSO_4$, $Mn(NO_3)_2$ and $Mn(ClO_4)_2$ solutions the reader is referred to the literature; see [85].

The standard procedures for the preparation of crystallized $Mn(OH)_2$ – 'synthetic pyrochroite' – operate under redissolution of the precipitate preliminarily obtained from concentrated KOH and $MnCl_2$ solutions, which is achieved by a longer heating process under the mother liquor up to 190–200 °C. When the clear solution is cooled slowly, a modest yield of small pure $Mn(OH)_2$ crystals is obtained. All steps are carried out under a flow of purified H_2. Detailed prescriptions, based in part on older literature, are given e.g. by Brauer [87] or Berg and Kovyrzina [88]. Larger amounts of a microcrystallized $Mn(OH)_2$ have been prepared in a 1 l silver vessel, employing hydroxylammonium chloride as a reducing agent in the different steps. Subsequent recrystallization from 40 to 50% NaOH leads to well-formed, thin hexagonal plates; see Scholder and Kolb [79]. Similar products have been obtained by hydrothermal synthesis, e.g. at 230 °C and 300 atm during 24 h by Christensen and Ollivier [89]; see also [90].

For available literature on the production of $Mn(OH)_2$ by precipitation with alkali solution on a technological base, cf. the citations in [91].

4.2.3. *By Other Methods*

A number of other procedures have been described, but are not likely to have advantages over the mentioned ones, as e.g. hydrothermal hydrolysis of aqueous

MnCl$_2$ solutions under hydrogen (380 °C, 350 atm), hydrolytic decomposition of Mn(III) salts leading to Mn(II) and Mn(IV), reduction of MnO$_2$ by atomic hydrogen or electrolytically, cathodic deposition of Mn(OH)$_2$ from Mn(II) salt solutions.

Mn(OH)$_2$ electrodes may be prepared by cathodic deposition on Pt from 0.1 N Mn(NO$_3$)$_2$ solution at pH $\cong$ 3.8 or 7, $j = 1$ mA m^{-2} [92].

4.3. SPECIAL FORMS

Some older work reported on the preparation of various colloidal forms of Mn(OH)$_2$. Gels or sols are obtainable in principle by the precipitation of Mn^{2+} salts with alkali as described before, if suitable substances like proteins (gelatine) are added to the system. Similarly, rhythmic precipitation leading to 'Liesegang rings' has been studied; cf. [93].

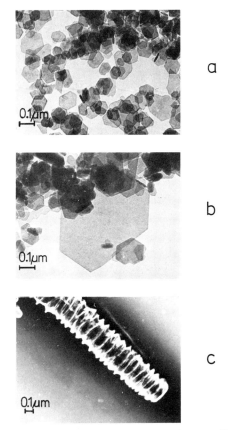

Fig. 9. Morphology of microcrystalline Mn(OH)$_2$. (a): precipitated from Mn^{2+} solution at 20 °C; (b): precipitated from Mn^{2+} solution at 85 °C; (c): preparation according to Brauer [87], rare prismatic shape [00.1].
(Transmission electron microscopy, (a) and (b) direct preparation, (c) replica technique; Oswald and Feitknecht [4].)

Zinc oxide in aqueous suspension containing Mn^{2+} ions is coated by $Mn(OH)_2$ under the influence of irradiation with $\lambda < 5100$ Å. The reaction $Mn^{2+} + Zn(OH)_2 \rightarrow Mn(OH)_2 + Zn^{2+}$ requires an activation energy of 3.3 kcal mole^{-1} (Elmore and Tanner [94].

4.4. Properties

'Synthetic pyrochroite' forms regular flattened hexagonal prisms to very thin hexagonal plates (00.1). The lower the temperature of preparation, the more the prismatic habit is developed. As Figure 9 demonstrates by electron microscopy, also highly dispersed powder samples precipitated from diluted Mn^{2+} solution at 20 or 85 °C consist of tiny hexagonal platelets (00.1) with diameters in the range of 1000 to a few 1000 Å.

The aging properties of freshly precipitated $Mn(OH)_2$ were studied very carefully by Feitknecht and his co-workers; see e.g. [28] and earlier papers. Compared to $Ni(OH)_2$ or $Fe(OH)_2$, $Mn(OH)_2$ shows a much higher aging velocity, about comparable to that of $Cd(OH)_2$.

In chemical behaviour, the most important aspects are the solubility determination in different media, the already mentioned easy oxidizability, studied on solid $Mn(OH)_2$ in gas atmosphere as well as in aqueous suspension, and the thermal decomposition of solid $Mn(OH)_2$.

4.4.1. Physical

A determination of the lattice constants by Oswald [95] and confirmed by de Haan [1] resulted in

$a = 3.316 \pm 0.005$ Å $c = 4.732 \pm 0.008$ Å $c/a = 1.427$
$Z = 1$ $V = 45.0_6$ Å^3 $D_x = 3.27_7$ g cm^{-3}
                                                   $D_{exp.} = 3.258$ g cm^{-3} [76]

free parameter $z_{OH} = 0.222$ [1].

An older value for the lattice energy by Yatsimirskij [96], 679 kcal mole^{-1}, was modified by the same author [97] to 695 kcal mole^{-1} later.

For further details on physical properties, see [76].

4.4.2. Solubility

Figure 10 illustrates the general solubility behaviour of $Mn(OH)_2$. Only scarcely soluble in pure water, it is readily dissolved in HCl (part a), but much less in NaOH, as long as the conditions are not extreme (part b). When high base concentration and temperature are applied, the amphoteric character becomes more evident (part c).

Determination of the solubility in pure water has led to considerably different results following the various methods used. Feitknecht and Schindler [33] state that in the case of $Mn(OH)_2$ fresh precipitates and aged materials have practically the same solubilities. This is consistent with the fact that well-formed crystallites as shown in Figure 9 are readily obtained at 20 °C. The authors recommend to use the value log $Ks_0 = -12.8$ which was originally determined by Fox et al. [98] for

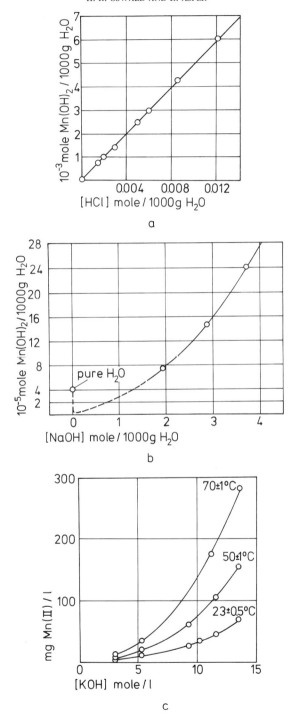

Fig. 10. Solubility of Mn(OH)$_2$ in strong acid (a) and basic (b) media at 25 °C. In (c) the solubility in concentrated KOH as a function of temperature is shown. (According to GMELIN [76].)

25 °C and $J=0$. An evaluation of some more recent determinations led Zavodnov and Fesenko [99] to the somewhat differing result of log $Ks_0 = -12.54$ at 25 °C.

In acids, the dissolution of $Mn(OH)_2$ yields Mn^{2+} ions and water. The solubility in 25% ammonia is—with 19 mg $Mn^{2+}l^{-1}$ at 20 °C—quite small, but increases significantly in the presence of other anions besides OH^-, leading to the formation of the $[Mn(NH_3)_6]^{2+}$ complex cation. In strong aqueous bases, the dissolution follows the reaction

$$Mn(OH)_2 \text{ (solid)} + OH^- \to HMnO_2^- + H_2O,$$

with an equilibrium constant of about 1×10^{-5} at 25 °C, according to Fox et al. [98]. From boiling 50% NaOH saturated with $Mn(OH)_2$, pure sodiumhydroxomanganate-(II), $Na_2Mn(OH)_4$, is obtainable between 120 and 75 °C (Scholder and Kolb [79]).

4.4.3. Oxidation in Air

The oxidation velocity depends very much on crystallinity and purity. Well-crystallized products free from superficially adhering traces of alkalies and amorphous admixtures are stable, i.e. remain white in an air-filled desiccator for weeks, if the humidity is very low, but turn dark brown within days outside the desiccator (Klingsberg and Roy [80]). The reaction behaviour of $Mn(OH)_2$ preparations with molecular oxygen dependent from the relative humidity was studied with X-rays, chemical analysis and electron microscopy by Feitknecht et al. [3], resp. Oswald and Feitknecht [100]. At a relative humidity $\leqslant 35\%$, β-MnOOH is obtained by a topotactic reaction, whereas at $\geqslant 45\%$, a mixture of β-MnOOH and Mn_3O_4 results. Between $Mn(OH)_2$ and β-MnOOH (synthetic feitknechtite), there exists according to Bricker [101] possibly a continuous transition series:

$$Mn(OH)_2 \rightleftharpoons Mn(II)_{1-x} Mn(IV)_x(OH)_{2-2x}O_{2x} \rightleftharpoons \beta\text{-MnOOH}$$

A close relationship between the crystal structures of $Mn(OH)_2$ and β-MnOOH has to exist.

4.4.4. Oxidation in Alkaline Aqueous Media

The co-called 'autoxidation' is a very complex reaction. A number of observers agree that two steps can de distinguished, the first with high and the second with low velocity, but there is considerable disagreement in interpretation. For detailed literature the reader is referred to [102]. The degree of oxidation finally obtained varies widely. Such different factors as temperature, alkalinity, Mn2+ or foreign ion content of the liquid phase, velocity of air or oxygen flow, as well as purity, age and crystallinity of the $Mn(OH)_2$ indeed influence the product(s) and the kinetics of the reaction in a most complicated manner.

4.4.5. Thermal Decomposition

The thermal decomposition follows the general reaction path $Mn(OH)_2 \to MnO + H_2O$ and has been studied mainly by thermal analytical techniques (TGA, DTA, etc.). Dehydration and subsequent reactions are strongly influenced by the crystallinity of the initial material and by the surrounding atmosphere. After the loss of physisorbed water molecules, the condensation of OH⁻ ions starts around 200 °C. In air, the MnO is oxidized to γ-Mn_2O_3 (640 °C), and later β-Mn_3O_4 is formed (960 °C). A hemihydrate $MnO \cdot 0.5\, H_2O$ formerly supposed could not be confirmed. Berg and Kovyrzina [88] investigated the isobaric dissociation of $Mn(OH)_2$ for pressures from 10 to 700 torr in nitrogen atmosphere. The resulting temperature-time curves show only one sharp endothermic peak at 166 to 200 °C, depending on pressure. For the dependence of the dissociation pressure p from the absolute temperature T a straight line $\log p - 1/T$ is obtained, and from this an equilibrium constant $K_p = 7 \times 10^{-9}$ atm at 298 K. The thermodynamic data for the reaction are:

$\Delta H_{298} = 25.86$ kcal mole⁻¹; $\quad \Delta G_{298} = 15.05$ kcal mole⁻¹
$\Delta S_{298} = 36.27$ cal mole⁻¹ K⁻¹

4.4.6. *Thermodynamic Data*

The standard enthalpy ΔH^0 and standard free enthalpy ΔG^0 in kcal mole⁻¹ for the formation of $Mn(OH)_2$ from the elements are

$\Delta H^0 = -166.2 \qquad \Delta G^0 = -147.0 \qquad (25 °C)$

(for the precipitated, 'amorphous' compound, according to [103]).
These data replace the respective older values

$\Delta H^0 = -165.8 \qquad \Delta G^0 = -145.9$

by Rossini [104].

For a crystalline $Mn(OH)_2$, there exists a value of $\Delta H^0 = 163.2$ (at 18 °C) derived from older calorimetric measurements.

Bricker [101] calculated $\Delta G^0 = -147.34 \pm 0.05$ from the solubility product Ks_0 at 25 °C determined by himself, using the ΔG^0 values – for Mn^{2+} and OH⁻ in infinitely diluted aqueous solution – by Latimer [105].

$\Delta G^0 = -146.9$ is in good agreement with the Ks_0 values given by Fox et al. [98].

The standard entropy S^0 [cal mole⁻¹ K⁻¹] of $Mn(OH)_2$ at 25 °C was derived from literature data on precipitated 'amorphous' material by Wagman et al. [103]: $S^0 = 23.7$, replacing an older value of 21.1 by Rossini.

For the formation of $Mn(OH)_2$ from MnO and liquid H_2O resp. water vapour, ΔH values of -4.2 ± 0.5 resp. -14.8 ± 0.5 kcal mole⁻¹ (at 18 °C) are given in [106].

5. Iron(II) Hydroxide, $Fe(OH)_2$

5.1. General information

In chemistry $Fe(OH)_2$ is – although unstable and not easy to prepare, at least in the pure state – well known. This dates back to the early 19th century when the white unoxidized compound was first observed by Thénard [107] in 1805 ('Thénard's white oxide'). Among others, Proust, Davy, and Berzelius considered it a hydrate of Fe_3O_4, and it was Berzelius [108] who in 1826 recognized the compound as a 'hydrate of FeO'. White $Fe(OH)_2$ is extremely sensitive to oxidation and thereby transforms, through green and greenish black intermediates, into red brown iron(III) oxidehydroxide. Descriptions of the green, partially oxidized form of $Fe(OH)_2$ are even older, dating back to de Foucroy (1785) and Scheele (1793).

The older literature on $Fe(OH)_2$ is well reviewed in *GMELINS Handbook* [109], but unfortunately only up to the year 1929.

Today, $Fe(OH)_2$ is characterized as a true hydroxide and crystalline compound having the CdI_2- (or brucite-) structure, but it is not frequently prepared or used, and newer literature on it is scarce.

5.2. COMPOUND PREPARATION

There are a number of reactions which enable the formation of more or less pure $Fe(OH)_2$, at least intermediately as e.g. during the corrosion of metallic iron under oxygen-deficient conditions, or electrolytically from alkali salt solutions with iron anodes, or during electrolysis of iron(II) salt solution on the cathode.

The best-known preparations, however, are based on the precipitation from Fe^{2+} solutions with ammonia or alkali hydroxides under extremely careful exclusion of oxygen and carbon dioxide. With gaseous ammonia under H_2-atmosphere, a pure white, fine product is obtained – according to Deiss and Schikorr [110] – which is washed with oxygen-free water. The NH_3 stream has to be stopped at the right time, otherwise the hydroxide is redissolved. In general, the precipitations with NH_3 are incomplete and, in the presence of greater amounts of NH_4^+-salts, even fully hindered, due to the fact that the solubility product of $Fe(OH)_2$ is no longer reached. Moreover, the precipitate tends to adsorb NH_3.

With an excess of NaOH or KOH solutions, more complete precipitation is reached, but it is of utmost importance to completely exclude oxygen during all operations. A standard procedure for obtaining pure $Fe(OH)_2$ was described by Rihl and Fricke [111] as a test for equipment to work under inert gas atmosphere: Into the apparatus, which was first evacuated to high vacuum and thereafter set under a continuous stream of purified N_2, a previously prepared oxygen-free solution of high purity $FeCl_2$ in concentrated ammonia is introduced through a filter. Then it is diluted with ample oxygen-free water in order to precipitate the hydroxide. A subsequent aging process during 3 h at 80 °C aims at producing a more compact product. After separation from the mother liquor, it is decanted 10 to 12 times with water until no more Cl^- ions are detectable in the water. To dry the product, the residue is frozen out and the water sublimated in high vacuum into a cooled vessel. The final drying is achieved over P_2O_5 in high vacuum. A practically white microcrystalline $Fe(OH)_2$ is obtained, which is pyrophoric when exposed to air. According to our own experience, completely dry products are much less vulnerable to attack by traces of oxygen in a dry atmosphere than in the presence of even small amounts of water vapour.

As a special form, colloidal solutions of $Fe(OH)_2$ can be obtained. Deiss and Schikorr [110] precipitated white $Fe(OH)_2$ from $FeCl_2$ solution with gaseous ammonia in a vacuum apparatus and treated the product with water till it dissolved in part to white, in transmitted light reddish opalescent $Fe(OH)_2$ sols with a content of up to

4 g l^{-1}. Intensely green coloured sols containing partially oxidized Fe(OH)$_2$ particles are obtained by electric vaporization of iron rods in a narrow cylinder filled with gelatine solution.

The white sol is not stable for more than 6 days. It is positively charged. The green sol is even less stable. At higher concentrations if flocculates even upon shaking.

5.3. Properties

5.3.1. Physical

Old unit cell dimensions by Natta and Casazza [112], $a=3.24$ Å, $c=4.47$ Å, which are sometimes referred to in books, should no longer be used. There is a reasonable agreement between the values given by Clark et al. [113]

$a=3.25$ Å , $c=4.61$ Å

and the determination given by Bernal et al. [114]

$a=3.262$ Å , $c=4.596$ Å , $c/a=1.409$

which are probably the best and which have also been taken over in the work of de Haan [1]. The further data are

$Z=1$ $V=42.3_5$ Å^3 $D_x = 3.52_3$ g cm^{-3}
$D_{exp.} = 3.4$ g cm^{-3}

free parameter z_{OH}: not known, estimated to 0.22 [1].

The lattice energy of 709 kcal mole^{-1} given by Yatsimirskij [96] was later modified by the same author [97] to 715 kcal mole^{-1}.

5.3.2. Electrochemical Data

The potential of Fe(OH)$_2$ measured under hydrogen at 20 °C against a normal hydrogen electrode is -0.740 V. Another value is -0.75 V, according to [109].

5.3.3. Chemical Behaviour

With air, Fe(OH)$_2$ undergoes rapid oxidation and is finally transformed into brown Fe(III) oxide hydroxides. The details of the occurring reactions and the structural chemistry of the green intermediate phases are difficult subjects, mainly due to the non-stoichiometric character and to the high disorder which is present in these typically laminar dispersed systems. At least some 10% of the Fe^{2+} ions may be converted to Fe^{3+} under conservation of the initial crystal lattice, except for a small shrinking of the a-axis. Variations in the c-axis are much harder to detect, considering the very broad character of the 00.l reflections as a consequence of the laminar character of the particles. The green colour is maintained up to 50% oxidation.

Feitknecht and Keller [115] have studied the oxidation by air in aqueous medium of freshly prepared Fe(OH)$_2$ suspensions with plate diameters of 500 to 1000 Å and of about 30 Å thickness. They state that the intermediate 'green ferroferrites' are quite similar to products which can be obtained by oxidation of neutral buffered Fe^{2+} salt solutions, e.g. FeCl$_2$, and by corrosion of metallic iron under certain conditions ('green rusts'). The latter compounds are basic salts with double layer lattices, as e.g.

4 $Fe^{II}(OH)_2 \cdot Fe^{III}OCl \cdot aq$. They may contain considerably more Fe(III) instead of Fe(II) than in the ideal formula given, and are related structurally to similar double layer lattice compounds, e.g. to green basic cobalt(II), (III) halides, and to oxidation products of $Ni(OH)_2$. There is evidence that also green Fe(II), (III) hydroxides (resp. oxide hydroxides) without foreign anions like Cl^- can be derived from this type of crystal lattice, but the extremely laminar dispersed products, tending to form dark green colloidal solutions, are indeed very difficult to characterize structurally.

Bernal et al. [114] studied the transformations in the iron oxide hydroxide system with special attention to topotactic principles. Very rapid oxidation of $Fe(OH)_2$ leads to δ-FeOOH (earlier δ-Fe_2O_3, see also Francombe and Rooksby [116]), and in its formation as well as in the dehydration of δ-FeOOH to α-Fe_2O_3 (hematite) the hexagonal close packed oxygen layers are maintained even in their stacking. This is in contrast with the non-topotactic production of γ-Fe_2O_3 by slow wet oxidation of $Fe(OH)_2$. Green rusts, resultant from slow oxidation of $Fe(OH)_2$ precipitates, have been recovered as hexagonal superstructures in a 9-layer and a 4-layer sequence, but they deteriorate to a γ-FeOOH which is then dehydrated to γ-Fe_2O_3.

In the dehydration of $Fe(OH)_2$ to FeO (and/or Fe_3O_4) and in the FeO–Fe_3O_4 oxidation reaction, the original hexagonal c axis is retained as one of the cubic 3-fold axes [114].

Stability against water: if the system is free from dissolved oxygen, $Fe(OH)_2$ remains stable in contact with water [110], but in the presence of substances which can take up hydrogen, water is disintegrated by $Fe(OH)_2$. Upon addition of a small amount of $PdCl_2$ to a system containing $Fe(OH)_2$ in water, $Fe(OH)_2$ is oxidized under decomposition of water. If $Fe(OH)_2$ is heated with water under pressure in the absence of molecular oxygen, it is oxidized mainly to Fe_3O_4 under development of hydrogen.

Behaviour in acids: $Fe(OH)_2$ is soluble in diluted acids as HCl, H_2SO_4, etc. under pronounced heat development, but the measured thermochemical data are all very old (cf. [109]).

Reactivity towards salts: freshly precipitated $Fe(OH)_2$ in an aqueous medium acts as a strong reducing agent and reduces e.g. solutions of alkali nitrites and nitrates in alkaline medium to NH_3. The same happens with hydroxylamine. Alkali peroxides, copper(II)-, mercury(II)-, platinum salts and numerous organic compounds are reduced as well.

5.3.4. Solubility

The values for the solubility product of $Fe(OH)_2$ given in [109] are old and not reliable. At present, the best data are probably those given by Feitknecht and Schindler [33]:

log Ks_0 for an active form $= -14$, at 25 °C (estimated from results by Oka [117])

log Ks_0 for an aged, inactive form $= -15.1$, at 25 °C, $J=0$ (determination by Leussing and Kolthoff [118]).

In ammonia solutions, Fe(OH)$_2$ is soluble if NH$_4^+$ ions are present. Due to the formation of iron(II) ammine complexes and lowering of the free OH$^-$ concentration, the solubility product of Fe(OH)$_2$ can no longer be reached.

Under excess of OH$^-$ ions, hydroxo complexes such as FeOH$^+$, Fe(OH)$_3^-$, and Fe(OH)$_4^{2-}$ are formed, referring to work by Hedström [119] or Gayer and Woontner [120]. In [33], respective equilibrium constants are listed, taken from [120]. These authors state that Fe(OH)$_2$ possesses weak acidic properties.

It is also possible to prepare solid compounds containing Fe(II)hydroxo complexes, e.g. by boiling highly dispersed metallic iron in 50% NaOH under exclusion of oxygen. On subsequent cooling, fine bluish-green crystals of Na$_4$[Fe(OH)$_6$] are obtained, and the respective Sr- or Ba-salts may be precipitated as well (Scholder *et al.* [121]).

5.4. THERMODYNAMIC DATA

The few values indicated below are taken from *JANAF Thermochemical Tables* [122].

Standard enthalpy of formation, $\Delta H^0_{298} = -137.2 \pm 0.7$ kcal mole^{-1} (based on the determination of the heat of combustion according to the chemical reaction Fe(OH)$_2$(s) + $\frac{1}{4}$ O$_2$ (g) = $\frac{1}{2}$ Fe$_2$O$_3$(s) + H$_2$O(g) by Fricke and Rihl [123], which resulted in -29.8 ± 0.65 kcal mole^{-1}).

Standard free enthalpy of formation, $\Delta G^0_{298} = -117.6$ kcal mole^{-1}.

Standard entropy S^0 [cal mole^{-1} K^{-1}] = 21 $\pm$ 2 (25 °C).

6. Cobalt Hydroxide, Co(OH)$_2$

6.1. GENERAL INFORMATION

Cobalt forms the true hydroxide Co(OH)$_2$ and the cobalt(III) aquoxide Co$_2$O$_3$·aq resp. CoOOH which may appear with varying water content. Phases with varying degrees of oxidation were also obtained, and the existence of Co$_3$O$_4$·H$_2$O and CoO$_2$·aq is supposed. Further, an unstable layer lattice compound of the composition 4 Co(OH)$_2$·CoOOH was isolated.

The main emphasis is laid here on the particularly important Co(OH)$_2$, whereas some of the above-mentioned other compounds will only be treated as products of its oxidation. From Co(OH)$_2$, a blue α-form and a pink to red coloured β-modification are known. Upon aging, fresh α-precipitates transform into the more stable β-phase. The need to distinguish between α and β is definitely not due to possible variations, e.g. in particle size and degree of disorder in the lattice only, as there exist two really different phases from the structural chemical point of view.

The extended older literature on this subject is reviewed in [124], and in the respective supplementary volume [125] original work up to about 1960 is considered.

6.2. Compound preparation

6.2.1. *By Precipitation*

$Co(OH)_2$ is obtained when Co^{2+} and OH^- ions are brought together, e.g. through precipitation from dissolved Co^{2+} salts with alkaline hydroxide solutions in equivalent amounts or with a small excess of OH^-. When the hydroxide is added to the salt solution, basic salts are formed first which are then transformed into the hydroxide through rapid reversible topochemical reactions, see e.g. Feitknecht [126] and Heubel [86]. According to Feitknecht and Bédert [127], excess alkali acts in favour of the formation of β-$Co(OH)_2$.

Through precipitation in the presence of K_2SO_4, albumin or gelatine, α-$Co(OH)_2$ is stabilized for a short time only, whereas by addition of higher alcohols or sugars, α can be stabilized for several months; cf. Weiser and Milligan [128]. According to Feitknecht [126, 129, 130] these additives get adsorbed and inhibit the nucleation of the β-phase. They may be added shortly after the precipitation has taken place as well. Preserved and non-preserved α-phases are not distinguishable by X-rays; cf. Feitknecht [131]. A slight oxidation of the Co(II) during the formation of $Co(OH)_2$ also has a stabilizing action upon α [127], but to obtain pure preparations, it is of course necessary to work under careful exclusion of O_2 as well as CO_2, e.g. using a N_2 or H_2 atmosphere.

A medium from which preparation of $Co(OH)_2$ is frequently started is $Co(NO_3)_2$. If its saturated solution is boiled under reflux with an excess of NaOH solution in an N_2 atmosphere, β-$Co(OH)_2$ in the form of red hexagonal platelets is obtained upon cooling (Scholder and Weber [132]). Other authors work with $CoCl_2$ or $CoSO_4$ as well, but there is a general danger that anions might be included into the precipitate. If diluted $CoSO_4$ solution (5%) is shaken with organic bases like piperidine, ethanolamine etc. in equal concentration, blue α-$Co(OH)_2$ is obtained, after MacKenzie and Edson [133].

The reaction between $Co(NO_3)_2$ solution and NaOH was followed with potentiometry and conductometry by Oka [134]. Näsänen [135] worked potentiometrically by addition of 0.005 N $CoCl_2$ to alkali hydroxide, and Heubel [86] conductometrically by addition of 0.1 N resp. 0.01 N $CoSO_4$ to 0.1 N resp. 0.01 N NaOH.

The precipitates tend to become gel-like and show, in part, peptisation on washing. When dried at normal pressure, excess water is retained.

Upon addition of sub-stoichiometric amounts of alkali hydroxide to Co^{2+} salt solutions, hydroxy-salts are formed. The appropriate limits have been studied by Feitknecht and Fischer [136]: in < 0.35 M $Co(NO_3)_2$, $CoCl_2$, $CoBr_2$, and in $< \sim 0.01$ M $CoSO_4$, basic salts are no longer stable. At 100 °C, these limits are at slightly higher concentrations. The equilibria are reached only within months at room temperature, but at 100 °C already within a few days or less. In presence of very high amounts of excess alkali hydroxide, $Co(OH)_2$ behaves amphoterically and is dissolved under formation of hydroxo complexes. According to Scholder and Weber [132], a $Na_2[Co(OH)_4]$ precipitate, which forms upon the boiling of saturated $Co(NO_3)_2$ solution with concentrated NaOH and subsequent cooling under N_2, begins to contain $Co(OH)_2$ besides the hydroxo complex if the salt to alkali hydroxide mole ratio falls below 1 to 4.

6.2.2. Standard Precipitation Procedures

In [137], the following procedures are given:

— *Rose β-Co(OH)$_2$:* to 40 g KOH dissolved in 500 g H_2O, a solution of 40 g $Co(NO_3)_2 \cdot 6H_2O$ in 1000 g H_2O is added drop by drop under vigorous shaking. Both solutions have to be at 0 °C. The blue colour of the precipitate soon turns to rose. The solid is washed by decantation with CO_2- and O_2-free water until no more K^+ and NO_3^- are detectable, then filtered and dried in a vacuum over 50% H_2SO_4. All operations have to be executed in an atmosphere free from CO_2 and O_2.

— *Blue α-Co(OH)$_2$:* a 0.1 M Co^{2+} salt solution which contains 1% of glucose is precipitated with a small excess of NaOH. The precipitate is washed under exclusion of air, first with an ethanol-water mixture, then with acetone-water, and finally with pure acetone. The product is very sensitive to oxidation. For small quantities, a process with immediate freezing of the suspension and subsequent treatment with 25% ethanol is recommended by Feitknecht [126].

6.2.3. By Electrolysis

Nicol [138] describes the cathodic production of bluish-green $Co(OH)_2$ from 0.0002 N $CoSO_4$ solution on Pt. At 0.005 N, some SO_4^{2-} ions are built in, and at > 0.1 N $CoSO_4$, metallic cobalt is deposited on the cathode.

6.2.4. By Other Methods

Treatment of $[Co(NH_3)_6]Cl_2$ in water under H_2-pressure at temperatures above 150 °C has led to $Co(OH)_2$.

Heubel [86] as well as Feitknecht and Fischer [136] described the production of $Co(OH)_2$ upon treatment of basic cobalt salts with stoichiometric amounts of NaOH solution or also with water.

According to Scholder and Weber [132], concentrated $Na_2[Co(OH)_4]$ solutions yield dark rose β-Co(OH)$_2$ upon treatment with small amounts of water.

6.2.5. Preparation of Special Forms

There is practically no literature on the preparation of large single crystals or especially pure samples. Only de Haan [139, 140] has developed a simple interdiffusion method wich enabled him to grow crystals of β-Co(OH)$_2$ with dimensions ranging from a few tenths of a mm to a few mm, starting from a 0.5 M $Co(NO_3)_2$-solution and 3 M ammonia. Nothing is said about the purity of these preparations.

On the other hand, information on the preparation of particularly highly dispersed forms is much more readily obtained. Rhytmic precipitation of '$Co(OH)_2$'-layers from $Co(NO_3)_2$ in 10% gelatine solution upon addition of 0.2 M ammonia is e.g. described by Moeller [141], but it seems doubtful whether the product consisting of 'green rings' is not a basic salt instead. Feitknecht and Studer [142] investigated by electron microscopy Co(OH)$_2$ sols with a particle size of ~170 nm which were obtained by peptisation of highly dispersed precipitates during the process of repeated washing operations or by the influence of ultrasonic waves. More recent work on the influence of ultrasonic waves on Co(OH)$_2$-sol has been done by Prakash and Pandey [143]. The fact that an α-Co(OH)$_2$ sol is blue in transmitted and green in reflected light was explained with the Tyndall effect by Weiser and Milligan [128].

If anhydrous cobalt ethylate is poured into water, a green gel is obtained, and if the ethylate was mixed with little water before, thixotropic gels are formed according to Kandelaki and Setašvili [144]. Gels formed from Co^{2+} salts and NaOH may be negatively or positively charged, depending upon their preparation conditions, but when washed, Žukov and Pigarera [145] found that all gels are positively charged.

6.3. Properties

6.3.1. *Physical*

Crystallographic data: α-Co(OH)$_2$ is laminar dispersed and very poorly crystallized. X-ray powder photographs show (hk0)-reflexions only. Feitknecht [126, 129, 130, 131] proposed a disordered double layer lattice structure for α-Co(OH)$_2$, built up from hexagonal Co(OH)$_2$ layers with Co^{2+} ions in octahedral coordination similar to those in β-Co(OH)$_2$. These layers are arranged parallel to each other, but their distance is not constant. On four layers of ordered material, there follows one intermediate layer of disordered Co(OH)$_2$ with Co^{2+} ions in tetrahedral environment. For this structure, the formula 4 Co(OH)$_2$ ⇌ Co(OH)$_2$* is proposed. Similar double layer arrangements are better known from hydroxy-salt structures, and it is indeed uncertain, how much foreign anions besides OH^- contribute to the stability of the α-Co(OH)$_2$ lattice. Approximate unit cell dimensions are $a = 3.09$ Å, $c \approx 8$ Å.

β-Co(OH)$_2$ crystallizes as hexagonal platelets or sometimes prismatic in the CdI_2-(C6-) lattice type (Feitknecht [126, 129, 130]). Lotmar and Feitknecht [146] give the following lattice parameters:

$a = 3.173 \pm 0.002$ Å; $c = 4.640 \pm 0.004$ Å; $c/a = 1.462$
$Z = 1$, $V = 40.4_6$ Å^3, $D_x = 3.81_4$ g cm^{-3}
$D_{exp.} = 3.60$ g cm^{-3}

free parameter $z_{OH} = 0.220$.

These data are in good agreement with those contained in the work by de Haan [1], where also results of structural calculations based on single crystal X-ray measurements not published in detail are referred to.

Lattice energy: is only known for β-Co(OH)$_2$. A calculated value by means of the Born-Haber cycle is 724 kcal mole^{-1} [125].

Yatsimirskij [96] published under the assumption of equal contributions of covalent and ionic bonding a value of 719 kcal mole^{-1} first, which was later modified by the

---

* For an explanation see Section 1.

same author [97] to 747 kcal mole^{-1}. Although it appears somewhat high, this result was used for Table I in the present article.

Colour: the blue colour of α-Co(OH)$_2$ is caused by the more covalently bonded, tetrahedrally coordinated Co^{2+} ions in the intermediate layers, whereas the rose to red colour of β-Co(OH)$_2$ is a consequence of Co^{2+} ions in octahedral coordination (Feitknecht [131]). If Co^{2+} salts are in excess during the precipitation, the resulting α-form may appear bluish-green or green instead of blue, a fact which favours the assumption of at least partial formation of basic salts, which are in fact green [126]. Also partial oxidation can influence the colour of α-Co(OH)$_2$ towards green [127, 129, 130, 131], whereas partially oxidized samples of β-Co(OH)$_2$ appear in more or less brownish tones [86].

Magnetic properties: the specific susceptibility χ in 10^{-6} cm^3 g^{-1} and its temperature dependence are according to Selwood et al. [147]

t[°C]	+25	−44	−89	−146
χ	141	189	237	320

Additional information on physical properties may be taken from [124, 125].

6.3.2. Chemical Behaviour

Aging: the transformation of α-Co(OH)$_2$ to the β-type upon aging under the mother liquor has already been mentioned. The very diffuse X-ray powder patterns of products obtained at lower temperatures are turning increasingly sharp, whereas products precipitated at 90 °C show sharp lines from the beginning. A more drastic procedure is to treat a Co(OH)$_2$ gel obtained from CoCl$_2$ solution with a small excess of NaOH for 1 h at 190 °C in an autoclave; see Katsurai [148].

Oxidation of Co(OH)$_2$-suspensions: blue α-Co(OH)$_2$ is oxidized under the mother liquor first to green and then to brown products. As these reactions happen – depending upon temperature – in concurrence with the transformation to the β-phase and the oxidation of the latter, such systems are not easy to describe; see Feitknecht and Bédert [127].

The end product of the oxydation by O$_2$ is the brown or black, rhombohedral CoOOH. β-Co(OH)$_2$ yields this phase directly, whereas with α-Co(OH)$_2$, a green intermediate phase with an oxidation degree of about 20% is formed, the nature of which can be understood from the assumed double layer lattice structure of α-Co(OH)$_2$. It may be written 4 Co(OH)$_2$ $\rightleftharpoons$ CoOOH* and is formed in a topochemical single phase reaction by formation of CoOOH in the disordered intermediate layer. Thereby, the varying layer distance of the initial material becomes constant and the Co-Co distances within the layers smaller. According to Figure 11 unprotected α-Co(OH)$_2$ is much more readily oxidized than a sample protected by means of glucose. The detailed kinetic investigation by Feitknecht and Bédert [149] resulted in the relatively small activation energy of 10 to 15 kcal mole^{-1}, explained by autoxidation with activation of O$_2$ by adsorption.

---

* For an explanation see Section 1.

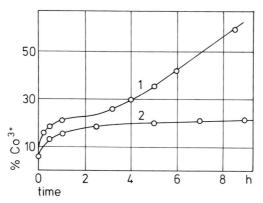

Fig. 11. Oxygen uptake by unprotected (1) and protected (2) α-Co(OH)$_2$. (Suspended in the mother liquor, O$_2$, 0 °C, according to Feitknecht and Bédert [127, 149].)

In [150], the oxidation of Co(OH)$_2$ suspensions with H$_2$O$_2$, alkaline peroxides, NaOCl, NaOBr, I$_2$+KOH, peroxosulphates, and KMnO$_4$ is reviewed, based mainly on the work of Besson [151]. As main products, CoOOH and CoO$_2 \cdot$ aq are formed.

Reduction of Co(OH)$_2$: Co(OH)$_2$ pressed to grains and reduced in streaming H$_2$ at 450 °C is reported to yield metallic Co with good catalytic activity [152].

With Co^{2+} salt solutions, α-Co(OH)$_2$ readily forms the corresponding basic salts in a finely divided state. β-Co(OH)$_2$ reacts much slower.

6.3.3. Electrochemical Properties

A Co(OH)$_2$ anode freshly deposited on a cobalt-plated nickel electrode was investigated in 6N KOH against a Hg electrode. Upon charging, the anode is oxidized and turns black, whereas the cathode takes up 0.5 to 1.5% K. Curve 1 in Figure 12

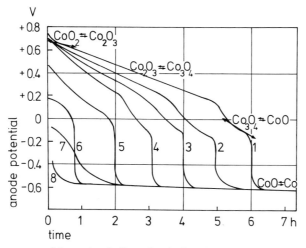

Fig. 12. Discharge curves of electrochemically or chemically oxidized Co(OH)$_2$ and Co anodes. 1: electrochemically oxidized; 2: ox. with Na$_2$S$_2$O$_8$; 3: with KMnO$_4$; 4: with NaOCl; 5: with NaOI; 6: with H$_2$O$_2$; 7: with air; 8: non exidized Co for comparison.
(Besson [151] and GMELIN [125].)

shows the dependence of the discharge potential on time. The steps appearing in the curve point to the equilibria

$$2 \text{CoO}_2 + 2\text{H}^+ + 2e \rightleftharpoons \text{Co}_2\text{O}_3 + \text{H}_2\text{O}; \quad E_h = +0.65 \text{ V} \quad (1)$$
(or CoOOH rather)
$$\text{Co}_3\text{O}_4 + 2\text{H}^+ + 2e \rightleftharpoons 3 \text{CoO} + \text{H}_2\text{O}; \quad E_h = -0.05 \text{ V} \quad (2)$$
$$\text{CoO} + 2\text{H}^+ + 2e \rightleftharpoons \text{Co} + \text{H}_2\text{O}; \quad E_h = -0.60 \text{ V} \quad (3)$$

Similar curves were in principle obtained with Co deposited on a Pt wire and oxidized in an analogous way. Curves 2 to 7 in Figure 12 show the discharge reaction of Co(OH)_2 or Co which were oxidized by different chemical media. Besson [151] concluded that Co(OH)_2 may be anodically oxidized up to $\text{CoO}_2 \cdot \text{aq}$.

6.3.4. Thermal Decomposition

Upon heating under exclusion of molecular oxygen, Co(OH)_2 decomposes to CoO and H_2O. Figlarz and Vincent [153] have studied the detailed dehydration mechanism of $\beta\text{-Co(OH)}_2$ in an Ar atmosphere by electron and X-ray diffraction, B.E.T. adsorption isotherms, and thermogravimetry. The samples used were fine powders consisting of hexagonal plates (00.1) with an average diameter of 430 Å and an average thickness of 165 Å. The dehydration takes place topotactically under conservation of the external habit of the initial crystallites, and the dehydration temperature as well as the crystallinity of the CoO formed depend strongly upon the detailed reaction conditions. In spite of the fact that the dehydration takes place at temperatures lower than $\sim 200\,°\text{C}$, the oxide obtained shows a characteristic tendency to retain small amounts of water up to more than $300\,°\text{C}$. For $\alpha\text{-Co(OH)}_2$, dehydration is reported to take place already between 150 and $170\,°\text{C}$.

Pistorius [154] followed the dehydration temperature of Co(OH)_2 up to a pressure of 100 kbar. At $p_{\text{H}_2\text{O}}$ of 80 kbar, Co(OH)_2 is stable to $320 \pm 10\,°\text{C}$. The invariant point $\text{Co(OH)}_2/\text{CoO} + \text{ice VII}/\text{CoO} + \text{liquid}$ lies at 80 ± 5 kbar and $320 \pm 10\,°\text{C}$.

When heated in air, Co(OH)_2 takes up oxygen and liberates water. Around 100 to $110\,°\text{C}$ the product obtained is CoOOH, whereas at higher temperatures – values in the literature range between 150 and $310\,°\text{C}$ (see e.g. [155]) – Co_3O_4 is formed.

6.3.5. Solubility

The amphoteric behaviour of Co(OH)_2 is expressed by formation of violet-blue solutions containing $[\text{Co(OH)}_4]^{2-}$ ions resp. wine red crystals of $\text{Na}_2[\text{Co(OH)}_4]$ in the presence of concentrated alkali hydroxides.

In ammonia with access to oxygen, Co(OH)_2 is dissolved under formation of the Co(III) ammine complex.

Recommended values for the solubility products of different Co(OH)_2 samples are listed in [33]:

	log K_{s_0}	Temperature
$\alpha\text{-Co(OH)}_2$, blue	-14.2	room temp.
$\beta\text{-Co(OH)}_2$, active	-14.8	room temp.
$\beta\text{-Co(OH)}_2$, inactive	-15.7	$25\,°\text{C}$

The data were determined by Feitknecht and Hartmann [156]. For the inactive β-Co(OH)$_2$, Gayer and Garrett [157] found a similar value.

6.3.6. *Thermodynamic Data*

The heat of solution and heat of precipitation of β-Co(OH)$_2$ were determined calorimetrically by Gedansky *et al.* [158]. Combination of these data with appropriate heats of formation yield for the standard enthalpy of formation of β-Co(OH)$_2$

$$\Delta H^0_{298} = -129.4 \text{ kcal mole}^{-1}$$

This value was combined with the standard free enthalpy of formation $\Delta G^0_{298} = -110.0$ kcal mole^{-1} for β-Co(OH)$_2$ from solubility data to yield $S^0_{298} = 22.3$ cal mole^{-1} K^{-1}.

6.4. Cobalt(III) oxide hydroxide, CoOOH

As this compound is the dominating product of the oxidation of Co(OH)$_2$ and is a typical layer lattice compound as well, it shall be briefly treated here. In the original literature, several formulas were used, besides CoOOH e.g. cobalt(III) aquoxide, Co$_2$O$_3$·aq., Co$_2$O$_3$·H$_2$O, Co(OH)$_3$. Freshly precipitated samples often contain about 3 moles of water per mole of Co$_2$O$_3$, from which two can be continuously liberated upon heating or aging, whereas the third is bonded much stronger. Thus from the viewpoint of dehydration the existence of a 'monohydrate' is confirmed, but structurally, the true formula has to be written CoOOH.

This phase is prepared by oxidation of isolated Co(OH)$_2$ with air, oxygen, or by reaction of Co(OH)$_2$ in aqueous suspension with H$_2$O$_2$, NaOCl, etc. For details, see e.g. [159]. CoOOH can of course also be prepared directly starting from Co^{2+} salt solutions and bases in the presence of oxidizing agents [151]. Another method is the decomposition of Co(III) compounds, e.g. [Co(NH$_3$)$_6$]Cl$_3$ or K$_3$[Co(OH)$_6$], by boiling or high pressure treatment in water with or without addition of bases, and finally, there are also electrochemical preparations. Sol- and gel-like forms of highly dispersed CoOOH are known.

CoOOH crystallizes in a rhombohedral unit cell, with the most probable space group D^5_{3d}-R$\bar{3}$m. Its lattice constants are:

$a = 2.849 \pm 0.001$ Å; $c = 13.130 \pm 0.005$ Å; $c/a = 4.609$; $Z = 3$

for the hexagonal and

$a = 4.676 \pm 0.002$ Å; $\alpha = 35°28'30'' \pm 30''$; $Z = 1$

for the rhombohedral cell.

The structure is typically layered, with sheets of [CoO$_6$]-octahedra parallel to (00.1). There are OH$^-$ besides of O^{2-} ions in the structure, but according to Kondrašev and Fedorova [160], there is a very short symmetrical hydrogen bond (—O—H····O^{2-}) which links oxygen atoms of adjacent layers. Thus, the structure is the same as that of NaHF$_2$. It must be mentioned, however, that the structure determination referred to cannot be considered a very accurate one. The particularly important O—O—distance of 2.36 Å given by the authors appears extremely short indeed. For Co—O, 1.94 Å are mentioned.

CoOOH is thermally stable up to about 200 °C. Between 240 and 300 °C, it is irreversibly transformed into Co_3O_4, water and oxygen [155]. Upon careful dehydration of CoOOH, the composition Co_2O_3 was approximately reached, but the proof for the existence of such a phase is very problematic.

CoOOH is, following powder X-ray data, identical with the mineral stainierite, and probably with some other ill-defined minerals listed in [161].

7. Nickel(II) Hydroxide, $Ni(OH)_2$

7.1. GENERAL INFORMATION

In the system Ni-O-H, a number of different phases are known: the true hydroxide $Ni(OH)_2$, the compound $4\,Ni(OH)_2 \cdot NiOOH$ which can appear with varying water content, and further $Ni_3O_2(OH)_4$, α-NiOOH, β-NiOOH, γ-NiOOH. The existence of $NiO_2 \cdot aq$ and some other compounds is discussed. Among these phases $Ni(OH)_2$ is the best defined and characterized compound. All the other ones contain nickel in the higher valence states +III and/or +IV, tend to variations in stoichiometry, and are more or less poorly crystallized. The interest in these higher valent nickel-containing substances is mainly based on their importance for nickel-iron or nickel-cadmium storage batteries [162, 163] which employ strong alkaline electrolytes. These work according to the following simplified principle:

$$Fe + 2\,NiOOH + 2\,H_2O \underset{\text{charge}}{\overset{\text{discharge}}{\rightleftharpoons}} Fe(OH)_2 + 2\,Ni(OH)_2 \;(\sim 1.3\,V)$$

The black, so-called 'peroxide' $NiO_2 \cdot nH_2O$ is unstable, being readily reduced by water, but can be used as an oxidizing agent for organic compounds [164, 165].

In this context, only the true hydroxide $Ni(OH)_2$ is discussed in detail, because

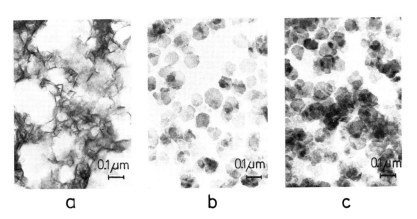

Fig. 13. Typical $Ni(OH)_2$ particles form fresh and aged suspensions, by reaction of 0.5 N $NiCl_2$ and 0.5 N NaOH. (a) freshly prepared at room temperature; (b) aged in the mother liquor, 70 °C, 30 d; (c) as (b), 60 d.
(Transmission electron microscopy; Gehrer, Portmann and Oswald [166].)

most of the higher oxidized phases are structurally related to its crystal lattice. This compound, often described as a gel-like or 'amorphous' phase, is in reality always microcrystalline, although with a more or less disordered lattice. The extraordinarily small tendency of $Ni(OH)_2$ to form larger, well-shaped crystallites is illustrated by Figure 13, where electron microscopical pictures [166] of fresh and aged precipitates are compared. After 30 days at 70 °C under the mother liquor, extremely thin irrigularly contoured platelets of a few 100 Å in diameter only are visible. Upon doubling the period of aging, no apparent increase in crystallinity occurs. In comparison, $Mn(OH)_2$ (Figure 9) crystallizes much more readily. The difference in solubilities cannot serve as a sufficient explanation, which must be sought instead in a specific kinetic hindrance of $Ni(OH)_2$ (and other Ni(II) compounds) recrystallization in aqueous media.

Only one modification is known, crystallizing in the CdI_2–(C_6–or brucite-) lattice type, and no hydrates of a definite composition exist.

7.2. Compound preparation

7.2.1. *By Precipitation*

The most common way to obtain $Ni(OH)_2$ is through precipitation of aqueous $NiCl_2$, $Ni(NO_3)_2$ or $NiSO_4$ solutions with a slight excess of NaOH- or KOH-solutions for example following Glemser [167]: 25 g KOH are dissolved in 250 ml carbon dioxide free water and added dropwise under vigorous stirring to a warm solution (35 °C) of 60 g $Ni(NO_3)_2·6H_2O$ in 250 ml water. The precipitate is repeatedly decanted with either 5 l of warm water free from carbon dioxide, then once with 5 l of similar water to which a small portion of ammonia was added, and then again with the same portions of warm carbon dioxide free water, until neither K^+ nor NO_3^- ions can be detected in the liquid as well as in the residue. After filtration, the latter is dried in vacuum over concentrated sulfuric acid. The product contains up to one mole of adsorbed water which is lost zeolitically through heating to 200 °C. Precipitation, decantation and filtration have to be carried out under strict exclusion of carbon dioxide from the air. The author recommends $Ni(NO_3)·6NH_3$ as a starting material as well, whereas $NiCl_2·6H_2O$ and $NiSO_4·7H_2O$ are less recommended, the anions Cl^- and SO_4^{2-} being more strongly adsorbed by the product, which forms an apple green microcrystalline powder. The influence of a number of variables on the precipitation of $Ni(OH)_2$ from aqueous nickel salt solutions with strong bases has been extensively studied in older literature, which is collected up to 1964 in [168]. The conditions varied are for example:

- Nature and concentration of the anion.
- Ni^{2+} concentration of the educt solution.
- Excess of strong alkali.
- Sequence of mixing (Ni^{2+} added to the alkali solution or vice versa).
- Velocity of precipitation, intensity of stirring.
- Temperature of precipitation and subsequent aging.

Adsorption effects are not the only reason that Ni(OH)$_2$ precipitates can contain considerable amounts of foreign anions; but there occur also various nickel hydroxy-salt phases. These are the primary precipitation products, especially if the alkali hydroxide is added to the nickel salt solution, and only at sufficiently elevated pH-values are they transformed into the hydroxide. This transformation of metastable nickel hydroxy-salts into hydroxide is accelerated by heating. Figure 14 shows the composition of precipitates in [OH]/[Ni] as a function of the pH of the reaction medium for room temperature, and Figure 15 contains [OH]/[Ni] ratios for precipitates from NiCl$_2$ solution as a function of pH, temperature, time and chloride ion concentration (after Singley and Carriel [169]).

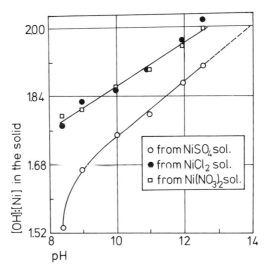

Fig. 14. Composition of precipitates from Ni^{2+} solutions as a function af pH and reaction medium, at room temperature.
(According to Singley and Carriel [169] and GMELIN [168].)

An anion which shows a particularly small tendency to provoke hydroxy-salt formation is ClO$_4^-$; see Feitknecht [28]. There is a great deal of valuable older work by Feitknecht and his co-workers on the composition and disordered structures of primarily formed nickel hydroxy-salts and on their transformation into Ni(OH)$_2$ under different conditions (see [170]).

It is also possible to obtain Ni(OH)$_2$ from NiCl$_2$ or Ni(NO$_3$)$_2$ solutions with an excess of ammonia solution at boiling temperature [171], whereas alkaline carbonates or alkaline earth hydroxides are not able to yield the hydroxide, but rather lead to basic carbonates or other hydroxy-salts of nickel.

7.2.2. *By Hydrolysis*

Microcrystalline Ni(OH)$_2$ may be obtained from the reaction of boiling water with

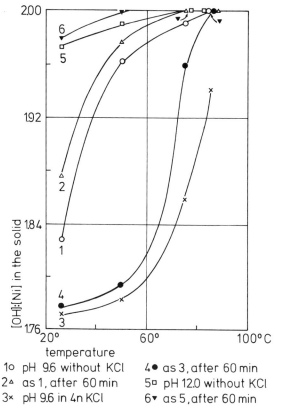

Fig. 15. [OH]/[Ni] ratio in precipitates from $NiCl_2$ solutions as a function of temperature, pH, time and Cl^- ion concentration.
(According to Singley and Carriel [169] and GMELIN [168].)

$Na_2[Ni(OH)_4]$ as well as with $Ba_2[Ni(OH)_6]$ or the analogous Sr compound (Scholder and Giesler [172]). Similarly, a number of nickel ammine salts and nickel complexes with organic ligands are hydrolyzed under formation of $Ni(OH)_2$.

7.2.3. By Other Methods

The interaction of water and metallic nickel leads to the formation of a $Ni(OH)_2$ surface layer. The same is the case if the products of the reaction between a NaOH melt and nickel (NiO, $NaNiO_2$) are decomposed hydrolytically.

It is also possible to produce $Ni(OH)_2$ electrolytically, e.g. on Ni anodes in alkaline solution. The current-density/tension curves of the anodic polarization of Ni in 80 °C, aqueous 0.1 and 5 N NaOH solutions, which are saturated with H_2, show discontinuities in the 0.1 N solution at 0.07 and 1.41 V, and in the 5 N solution at 0.18 and 1.31 V. Within these limits, $Ni(OH)_2$ is formed, whereas at higher tension, $Ni(OH)_2$ is oxidized (Volchkova et al. [173]). $Ni(OH)_2$ is frequently the final product of the reduction of higher valency nickel oxide hydroxides (compare e.g. Glemser and Einerhand [174]).

7.2.4. High Purity Preparations

Only a few authors give exact details on the purity of their $Ni(OH)_2$ products. Merlin and Teichner [175] resp. Teichner and Morrison [176] reacted NH_3 and $Ni(NO_3)_2$ solutions, boiled the mixture at ordinary or reduced pressure or treated it with water vapour. After the product was exhaustively washed and carefully dried, the following characteristics were determined: $H_2O/NiO = 1.07$; NH_3, N_2O_5: less than 0.08%, specific surface 13–20 $m^2\ g^{-1}$.

Purity data, reported in [177], of electrolytically prepared $Ni(OH)_2$ are the following: 58% Ni, 0.15–0.2% Co, 0.03–0.05% Fe, 0.015–0.025% Cu, negligible Cr, <0.002% Pb, <0.003% Zn, <0.05% Na, <0.002% S, <0.003% Cl.

'Very pure' hydroxide was made by Scholder and Giesler [172] through precipitation of aqueous $Ni(ClO_4)_2$ solution with 63% aqueous NaOH, and heating the precipitate under its mother liquor to 110 °C for several days. The microcrystals, well-suited for filtering, are washed with much water.

A $Ni(OH)_2$ of high degree of purity may be obtained from aqueous Ni^{2+} salt solutions and alkali hydroxide or strong organic bases at pH > 12.5. The Ni salt as well as the base have to be soluble in organic solvents, which are used to extract the impurities after the precipitate has been dried. As solvents, methanol, ethanol, and ethyl ether are particularly recommended [178, 179]. In the literature, electrofiltration resp. electrodialysis are also recommended for purification of $Ni(OH)_2$ (see [180]).

Following our own experience, it is difficult indeed to obtain products which are really free e.g. from adsorbed foreign anions. In a sample precipitated from $(Ni(NO_3)_2$ solution and extensively washed virtually free from NO_3^-, nitrate bands were easily detectable by IR spectroscopy. Then the sample was completely redissolved in concentrated ammonia solution, and the ammonia adsorbed by means of 1:1 H_2SO_4 in vacuum. The new precipitate, washed with ample water and dried in a vacuum, showed no more nitrate bands in the IR [181].

7.3. PREPARATION OF SPECIAL FORMS

7.3.1. Well Crystallized Samples

According to Franzen et al. [182], well crystallized samples are best produced by means of a hydrothermal treatment at 250 °C of a precipitate obtained by reaction of nickel(II)nitrate and alkali solutions. Romo [183] treated a precipitate hydrothermally for 3 days at 200 °C. More recently, Aia [184] made hydrothermal studies in the system $NiO-H_2O$. Precipitated $Ni(OH)_2$ was treated in the range of 200–400 °C and H_2O pressures from 0.2 to 2.1 kbar. At 285±5 °C, $Ni(OH)_2$ converted to NiO, with very little effect of pressure on the decomposition temperature. The best crystals of $Ni(OH)_2$ were obtained in 40% KOH solution at 200 °C and 1.0 kbar. It was not possible to convert NiO to $Ni(OH)_2$.

Coarse crystals of $Ni(OH)_2$, the habit of which corresponds to the high temperature form of $Na_2[Ni(OH)_4]$, are obtained from this compound by hydrolysis in boiling water, but not from the low temperature modification, according to Scholder and Giesler [172]. This is obviously a topotactic reaction, thus the $Ni(OH)_2$ 'crystals' might instead be built up from a huge number of highly oriented small $Ni(OH)_2$ crystallites.

7.3.2. Gels

In spite of the typically gel-like appearance of many precipitated $Ni(OH)_2$ samples – which are therefore often named gels – these products are always microcrystalline in character. The electron microscope reveals laminar disperse, plate-shaped primary crystallites, sometimes with regular hexagonal contours, as shown e.g. by Feitknecht et al. [185] and also in Figure 13. Under suitable conditions clear, transparent, thixotropic gels may be prepared. Much work has been done on the dehydration and aging behaviour of such systems, mainly by Feitknecht and his school. Mixed hydroxide gels are also known, e.g. from Ni(II) and Mg(II), or Ni(II) and Fe(III), or from the combination of positively charged SiO_2 – and negatively charged $Ni(OH)_2$ – sols.

7.3.3. Suspensions, Hydrosols

Hydrosols of $Ni(OH)_2$ are common as intermediate stages during the precipitation of nickel salt solutions with alkalies, but usually the sols are flocculated rapidly by the strong electrolytes present in the systems. Under suitable conditions, relatively stable (6–8 weeks) sols with up to 5g $Ni(OH)_2 l^{-1}$ can be obtained, as described e.g. by Feitknecht et al. [185], who also followed the coagulation and peptisation reactions in such systems (see e.g. [186]).

$Ni(OH)_2$ sols may be positively or negatively charged, depending on their preparation. By addition of substances like glycerol, sugars, and starch, protected sols were prepared und purified by dialysis.

The clear sols contain hexagonal crystalline plates of 16–20 Å thickness and a few 100 Å in diameter (see Feitknecht et al. [185]). Their aging process is characterized by the very low velocity of particle growth, depending on a number of experimental conditions. More details about the colloid chemical behaviour of $Ni(OH)_2$ can be found in the literature listed under [187].

7.3.4. Other Forms

By coagulation of particles in hydrosols, 60 to 100 Å thick $Ni(OH)_2$ layers can form on the surface and be isolated with a wire loop. Such layers are generated still faster at the phase boundary of the hydrosol with benzene, or toluene, or o-xylene added.

A special form of $Ni(OH)_2$ with a turbostratic structure was found and characterized by Le Bihan et al. [188]. It is formed when the washing of a precipitated hydroxide – in contrast to simple decantation – is speeded up by repeated centrifuging. X-ray diffractometry and IR spectroscopy show that the structure of the turbostratic compound consists of a stacking of two dimensional parallel $Ni(OH)_2$ leaflets, which are disoriented one from the other; the leaflets are separated by H_2O molecules involved in hydrogen bonding with the leaflet's hydroxide groups. In water, the compound crystallizes slowly by a diphasic mechanism (see Le Bihan and Figlarz [189]).

7.4. PROPERTIES

7.4.1. Physical

The unit cell dimensions of the only crystalline form of $Ni(OH)_2$ – so-called 'α-forms' in analogy to α-$Co(OH)_2$ or α-$Zn(OH)_2$ could not be confirmed – were determined by Lotmar and Feitknecht [146].

$a = 3.117 \pm 0.002$ Å; $\quad c = 4.595$ Å; $\quad c/a = 1.474$
$Z = 1$ $\quad\quad\quad\quad\quad\quad V = 38.6_6$ Å^3, $\quad D_x = 3.98_2$ g cm^{-3}
$\quad\quad\quad\quad\quad\quad\quad D_{exp.} = 3.56 - 3.60$ g cm^{-3} [168]
$\quad\quad\quad\quad\quad\quad\quad\quad\quad$ (too low)

free parameter $z_{OH} = 0.215$ [1].

The values for a and c agree quite well with those listed by de Haan [1], who refers to structural calculations for the determination of z_{OH} which are not published in detail. Older unit cell data occasionally mentioned in [190] are inaccurate and should be discarded.

Lattice energy: a value of 731 kcal mole^{-1} – derived by Yatsimirskij [96] from the Born-Haber process and a formula of his own, under the assumption of a 1:1 ionic and covalent character of the bond – seems too low. Later, the same author [97] indicated 757 kcal mole^{-1}, with a contribution of the crystal field of 32 kcal mole^{-1}.

Diffusion: the exchange of Ni between solid $Ni(OH)_2$ and aqueous $NiCl_2$ solutions was investigated with ^{63}Ni. One has to distinguish between surface diffusion and the slower lattice diffusion. Wyttenbach [191] gives for the lattice diffusion coefficient $D = 2.5 \times 10^{-13}$ exp. $(-14100/RT)$ as function of temperature t:

t[°C]	25	50	76	95
D[cm^2 s^{-1}]	1.2×10^{-23}	7.4×10^{-23}	1.6×10^{-22}	1.6×10^{-21}

In a similar way, the exchange behaviour between deuterium or tritium marked $Ni(OH)_2$ samples and liquid or gaseous H_2O was followed in detail on well-characterized preparations [191, 192, 193].

Magnetic properties: $Ni(OH)_2$ is paramagnetic. Below 90 K, the specific susceptibility χ is somewhat dependent on the field strength, leading to the assumption of ferromagnetism [194].

For additional information on physical properties, [168] may be consulted.

7.4.2. Chemical Behaviour

Reactions with gases and non-metallic compounds: pure $Ni(OH)_2$ is reduced to Ni metal in streaming molecular hydrogen at temperatures between 200 and 300 °C. Due to the intermediate formation of NiO, the reaction is slower at 250 °C than at 225 °C.

In air or molecular oxygen, traces of higher valency Ni hydroxides are formed superficially on freshly prepared, humid $Ni(OH)_2$. Ozone leads to rapid oxidation with a colour change to black. In presence of SO_2, $Ni(OH)_2$ undergoes an induced oxidation by air oxygen. When heated with water under pressure, no hydrates of $Ni(OH)_2$ are formed. Carbon dioxide, even in slight amounts from the air, reacts readily with $Ni(OH)_2$ – particularly if in a humid state – under formation of superficial basic nickel carbonates.

Behaviour towards acids and bases: $Ni(OH)_2$ is easily dissolved in various acids such as HF, HCl, H_2SO_4, HNO_3, diluted acetic acid.

The solubility in strong bases is small, e.g. in NaOH at 25 °C, concentration and solubility values in mole/1000 g H_2O are as follows:

Conc. of NaOH	1.6×10^{-3}	1.0×10^{-2}	1.1×10^{-1}	1.0	8 to 15
Conc. of $Ni(OH)_2$	1×10^{-7}	4×10^{-7}	2×10^{-6}	6×10^{-6}	$\sim 6 \times 10^{-6}$

(according to Gayer and Garrett [195]).

In 61.5 to 76.8% NaOH, $Ni(OH)_2$ reacts to $Na_2[Ni(OH)_4]$ [172]. The reaction takes place in 77 to 73% NaOH at temperatures up to 170 °C, in 70 to 66% NaOH up to 110 °C. Between 140 and 170 °C, the high temperature modification of $Na_2[Ni(OH)_4]$ is formed.

In cool, concentrated NaOH which contains 0.5 mole of brenzcatechol per mole $Ni(OH)_2$, the latter is dissolved under formation of a complex brenzcatechol-hydroxoanion [172].

Aqueous NH_3 solution dissolves $Ni(OH)_2$ under the appearance of a blue-violet colour. The solubility depends on the purity of the nickel hydroxide and the ammonia. Typical values are:

Conc. of NH_3 [mole l^{-1}]	0.28	1.13	2.53	4.88	10.20
Conc. of $Ni(OH)_2$ [g l^{-1}]	0.23	0.84	5.21	11.81	29.57

(according to Paris [196]).

Upon addition of NH_4^+ salts, the solubility is raised.

Salt solutions: several non-oxidizing salt solutions dissolve $Ni(OH)_2$ under formation of complexes, e.g. hot neutral aqueous NH_4F, from which light yellow $(NH_4)_2[NiF_4] \cdot 2H_2O$ may be crystallized.

In oxidizing solutions such as hypochlorites, chlorites, etc. $Ni(OH)_2$ is oxidized. During these reactions, the crystal structure of $Ni(OH)_2$ is preserved at least in the beginning, Ni^{2+} ions being replaced by Ni^{3+}, and OH^- by O^{2-} in statistical distribution. With higher degrees of oxidation, the lattice changes more drastically under formation of a $Ni_3O_2(OH)_4$ structure with disordered and distorted layers (see e.g. Feitknecht et al. [197]). It is not possible to discuss here in detail the rather complex circumstances of the topochemical or topotactic evolution of the higher valency hydrated nickel oxides.

7.4.3. Electrochemical Properties

Due to their importance for the Ni-Cd or Ni-Fe storage batteries, the electrochemistry of $Ni(OH)_2$ electrodes has been extensively studied. The literature is reviewed in [198].

The potential of a cell Ni/Ni(OH)$_2$/0.1 N NaOH relative to the normal hydrogen electrode is given to $E_h = -0.60$ V at 25 ± 0.01 °C, under N_2.

Freshly prepared Ni(OH)$_2$ layers on Ni undergo fast anodic oxidation, first to β-NiOOH and then to higher, water-containing oxides. Aged Ni(OH)$_2$ is oxidized less readily, see e.g. Briggs et al. [199].

7.4.4. Thermal Decomposition

On heating, one molecule of water is lost and NiO formed. In principle, the thermal decomposition of Ni(OH)$_2$ is doubtless a topotactic reaction. There is considerable variation in the literature regarding its temperature range, mainly due to differences in particle size and crystallinity of the initial samples and to varying experimental conditions like heating rate, atmosphere etc. Some results are as follows:
- between 250 and 300 °C, the dehydration is nearly complete; at 950 °C, the resulting NiO is completely water free
- decomposition temperature of Ni(OH)$_2$ precipitated from Ni(NO$_3$)$_2$ with NaOH, in air, in an open system: 255 ± 5 °C.

The thermogravimetric dehydration of Ni(OH)$_2$ was studied by Cabannes-Ott [171]. In such investigations, great care has to be taken to use samples really free from impurities such as e.g. CO_2, Cl^- etc., for otherwise the virtual 'dehydration temperatures' get too high. Nevertheless there is some evidence that small amounts of water are retained by the nickel oxide formed up to rather high temperatures, cf. e.g. Richardson and Milligan [194]: at 200 °C in streaming N_2, Ni(OH)$_2$ is still stable, whereas at 250 °C, it has mainly transformed into NiO. The values for 'excess oxygen' (OH$^-$?) in the product are:

t[°C]	250	300	350	500	700	1300
excess O/NiO[wt.-%]	0.36	0.37	0.32	0.15	0.03	0.00

Klingsberg and Roy [80] give as equilibrium temperature for the dehydration under 1500 psi (~ 105 atm) water vapour pressure 300 °C, and Pistorius [154] studied the decomposition for H$_2$O pressures up to > 100 kbar, indicating an invariant point at 103 ± 6 kbar and 347 ± 10 °C. At a pressure of 10^{-6} torr, on the other hand, yellowish green NiO with 0.16 wt.-% H$_2$O and high specific surface is obtained already at 200 °C.

As reaction enthalpy for Ni(OH)$_2 \rightarrow$ NiO + H$_2$O at 250–300 °C, Romo [183] gives a value of 12.0 ± 1.5 kcal mole^{-1}.

7.4.5. Solubility

The remarkable differences in the solubilities given for Ni(OH)$_2$ are, as pointed out by Feitknecht and Hartmann [156], a consequence of the presence of different active states of Ni(OH)$_2$ and of the use of less-suited determination methods. The solubility products listed below hold for solid Ni(OH)$_2$ in equilibrium with Ni^{2+} and 2 OH$^-$, in pure water.

$Ks_0 = 1.9 \times 10^{-15}$ (room temp.) for a most active form of fresh precipitate from NiCl$_2$ solution, Feitknecht and Hartmann [156].

$Ks_0 = 7.5 \times 10^{-18}$ (25 °C) for the above-mentioned product, aged at 110 °C, by the same authors. This value agrees well with $Ks_0 = 6.5 \times 10^{-18}$ (25 °C) determined for another aged preparation by Gayer and Garrett [195].

Between these extremes, there are e.g. results by Näsänen [135] (6.2×10^{-16}, 25 °C, potentiometric titration of alkali hydroxide with aqueous $NiCl_2$), and by Latimer [200] (1.6×10^{-16}, calculated from thermodynamical data about $Ni(OH)_2$).

As a consequence, Feitknecht and Schindler [33] included in their list of recommended values

$\log Ks_0 = -14.7$ (active $Ni(OH)_2$, room temp., $J=0$)
$\log Ks_0 = -17.2$ (inactive $Ni(OH)_2$, 25 °C, $J=0$)

There is another compilation of various Ks_0 values of $Ni(OH)_2$ by Sillén and Martell [201].

7.4.6. Thermodynamic Data

The available data for the standard enthalpy of formation, $\Delta H°$, and for the standard free enthalpy of formation, $\Delta G°$, of $Ni(OH)_2$ from the elements are not very convincing and have to be considered with care. The units for $\Delta H°$ and $\Delta G°$ are kcal mole^{-1}, for the entropy change $\Delta S°$ cal mole^{-1} K^{-1}; all values at 298 K:

$\Delta H°$	$\Delta G°$	$\Delta S°$	Remarks, Source
−128.6	−108.3	−	from old heats of neutralization, calc. by Rossini et al. [202]
−129.1	−112.3 −112.8	−56.6	from own e.m.f.-measurements and older thermochemical data, calc. by Murata [203]
−	−109.2	−	for H_2 pressure of 1 torr, from solubility data, cf. Murata [203]

For 291 K, Bichowski and Rossini [204] indicate a ΔH value of -133 kcal mole^{-1}.

7.5 Hydroxides with Ni in higher valence states

These phases are the products if $Ni(OH)_2$ is oxidized in the solid state or, less commonly, are also possible starting materials to produce $Ni(OH)_2$ by reduction of solids. Although (excluding $NiO_2 \cdot aq$) they crystallize in layered, double layered or band lattices which are more or less closely related to the structure of $Ni(OH)_2$, it is not possible to treat them here in detail. They are characterized by a considerable variability of their degree of oxidation, even in crystallized preparations, and by disordered structures.

An extended number of references on these compounds may be found in [205]. In this context, attention may also be drawn to the work of Glemser and Einerhand [174]. By means of Table V a rough survey of the more or less known substances is presented. In addition, a considerable number of other, rather uncertain phases were mentioned in the literature, which may be at least in part identical with the ones listed.

TABLE V
Hydroxide or hydrated oxide phases of nickel in valence states of more than (II)

Designation	Constitutional formula	Degree of oxidation	Other formulae given in literature	For structural remarks, see
$NiO_{>1\ to\ 1.23} \cdot aq$	–	$NiO_{>1\ to\ 1.23}$	$Ni(OH)_2$ with excess of oxygen	I
$4Ni(OH)_2 \cdot NiOOH$	$4Ni(OH)_2 \cdot NiOOH$	$NiO_{>1.07\ to\ 1.22}$	$NiO_{1.07\ to\ 1.22} \cdot aq$	II
$Ni_3O_2(OH)_4$	–	$NiO_{1.33}$	$NiO_{1.33} \cdot aq$ $Ni_3O_4 \cdot 2H_2O$	III
α-NiOOH	$4NiOOH \cdot NiOOH$	$NiO_{1.5}$	$NiO_{1.5} \cdot aq$ $Ni_2O_3 \cdot H_2O$	IV
β-NiOOH	–	$NiO_{1.5}$	$NiO_{1.5} \cdot aq$ $Ni_2O_3 \cdot H_2O$ $Ni(OH)_3$	V
γ-NiOOH	$3NiOOH \cdot NiOOH$	$NiO_{1.5}$	$NiO_{1.5} \cdot aq$ $Ni_2O_3 \cdot H_2O$	VI
$NiO_2 \cdot aq$	–	NiO_2	–	VII

A few short structural remarks on the different phases can be given as follows:

I $Ni(OH)_2$ structure type, with varying amount of Ni^{3+} instead of Ni^{2+} and O^{2-} instead of OH^- ions

II 1. Hexagonal-rhombohedral, related to the C19-($CdCl_2$-)type, disordered arrangement of NiOOH in intermediate layers between the rhombohedrally stacked layers of $4 Ni(OH)_2$; similar to $4 Co(OH)_2 \cdot CoOHCl$, $4 Ni(OH)_2 \cdot NiOHBr$ etc.
2. Related to the C27-type (CdI_2, modif. II), Ni^{3+} and O^{2-} ions statistically distributed to the hexagonally arranged Ni^{2+} and OH^- sites

III Three structures are proposed:
1. layer lattice similar to $4Ni(OH)_2 \cdot NiOOH$
2. disordered layer lattice with a rhombic deformation
3. CdI_2-structure analogous to $Ni(OH)_2$, oxidized

IV Strongly disordered, related to β-NiOOH, but without X-ray reflections from the basal planes

V C19-type, disordered

VI Similar to C19-type, rhombohedral, but monoclinic or rhombic unit cells have also been proposed [206]

VII Amorphous to X-rays.

8. Copper(II) Hydroxide, $Cu(OH)_2$

8.1. General information

Copper(II) does not form a hydroxide with the CdI_2- (or C6-) structure like most of the bivalent 3d transition metals, although these cations are of comparable ionic radii. This fact is related to the well-known structural chemical particularity of Cu^{2+} to adopt a strongly distorted coordination octahedron with four ligands at short and two more at much longer distances, usually interpreted as a consequence of the Jahn-Teller theorem. But as the crystal structure of $Cu(OH)_2$ belongs to the layered type as well, some details about this compound are given here.

8.2. Compound preparation

Copper(II) hydroxide has to be prepared with some care, as in aqueous media $Cu(OH)_2$ is only metastable and tends to decompose. Feitknecht [207] gives for $\Delta G°_{298.2}$ of $Cu(OH)_2 -85.41$ kcal mole^{-1}, and for the reaction $Cu(OH)_2(s) \rightarrow CuO(s) + H_2O$ (l) at room temperature $\Delta G = -1.62$ kcal mole^{-1}.

Ample preparative literature on $Cu(OH)_2$ is assembled in *GMELINS Handbook* [208]. The compound is normally precipitated from copper(II) salt or copper ammine complex solutions with alkaline hydroxides under exclusion of CO_2. Substoichiometric amounts of OH^- lead to formation of hydroxy-salts, and a huge excess of OH^- has to be avoided as well because of the tendency of $Cu(OH)_2$ to re-dissolve under formation of deep blue, soluble hydroxocuprate complexes, from which e.g. $Na_2[Cu(OH)_4]$ may be isolated (see Scholder [209] or Feitknecht and Lenel [210]).

The $Cu(OH)_2$ samples are microcrystalline powders, which are difficult to produce completely free from adsorbed foreign anions, alkali ions and water. A procedure described by Oswald and Jaggi [211] follows in its first step the work of Fricke and Kubach [212]. A basic copper(II) sulphate, previously precipitated with ammonia and washed, is transformed by NaOH solution into $Cu(OH)_2$. The well-washed microcrystalline product shows slight disorder in its crystal lattice and is therefore re-dissolved in concentrated ammonia in the next step. If this solution is slowly deprived of NH_3 over H_2SO_4 (1:1) in vacuum at room temperature, a pure and well-ordered $Cu(OH)_2$ of an intense blue colour is obtained, which is built up of aggregated flat needles of about 5×10^{-2} mm in length.

Unfortunately, up till now there has been no success in growing macroscopic single crystals of $Cu(OH)_2$. In the literature, samples are sometimes described of 'crystallized' $Cu(OH)_2$, which are obtained by careful treatment of single crystals of copper hydroxy-salts, e.g. $Cu_2(OH)_3X$ (where X is an anion like NO_3^-, NO_2^-, etc.) with diluted alkaline hydroxides (cf. e.g. Labanukrom [213]). But the careful study of these 'crystals' by electron microscopical and single crystal X-ray methods has shown that they are, although sometimes transparent and of perfect shape, in fact pseudomorphs built up from a very high number of fibrous, well oriented $Cu(OH)_2$ crystallites (cf. Oswald and Brunner [214]). These systems are excellent examples for the study of topotaxy, as in the course of the reaction

$$Cu_2(OH)_3X(s) + OH^-(l) \rightarrow 2\ Cu(OH)_2(s) + X^-(l)$$

the dominant structural element of edge-linked distorted coordination octahedra around the Cu^{2+} ions is retained to a very high degree of perfection, as shown in Figure 16.

There are also electrolytic procedures for the preparation of pure $Cu(OH)_2$, mainly described in the patent literature (see e.g. [215]). This is due to the technical importance of $Cu(OH)_2$ as a fungicide in plant protection.

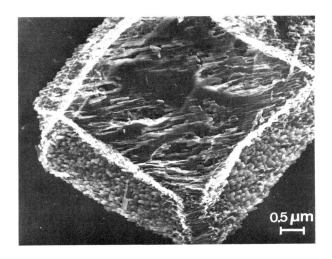

Fig. 16. 'Crystalline' $Cu(OH)_2$ by topotactic reaction of a $Cu_2(OH)_3NO_2$ crystal in 2N NaOH. (Transmission electron microscopy, replica technique, cf. Oswald and Brunner [214].)

8.3. Properties

8.3.1. *Crystal Structure*

From X-ray powder data it has been known for a long time that $Cu(OH)_2$ does not crystallize in the CdI_2-lattice, but due to the lack of single crystals, no further information became available.

An interesting detail is that in this situation Feitknecht and Maget [216] tried to calculate the a- and c-values for the unit cell of a hypothetical, not really existing $Cu(OH)_2$ phase with the CdI_2 type of lattice. This was achieved by extrapolation from accurate cell constant measurements of a series of solid solutions of $Cu(OH)_2$ in $Ni(OH)_2$ and resulted in

$$a_{Cu(OH)_2, hyp.} = 3.11 \text{ Å}, \quad c_{Cu(OH)_2, hyp.} = 4.6 \text{ Å}, \quad V = 38.5 \text{ Å}^3.$$

There exists a quite similar hydroxy-salt phase with a composition of about $Cu(OH)_{1.5}F_{0.5}$ (Oswald [217]).

From high resolution Guinier X-ray films of well-crystallized powder samples, Oswald and Jaggi [211] could later determine the unit cell of the real $Cu(OH)_2$, and subsequently the same authors calculated the complete structure based on accurate counter goniometer intensity measurements [218]. The crystal data are:

orthorhombic, $a = 2.949 \pm 0.003$ Å,   space group Cmcm-D_{2h}^{17}
$b = 10.59 \pm 0.01$ Å $Z = 4$
$c = 5.256 \pm 0.005$ Å $D_x = 3.95$ g cm^{-3}
$V = 41.0_4$ Å^3 mole^{-1} $Cu(OH)_2$ $D_{exp.} = 3.85$ g cm^{-3}

The structure which is shown in principle by Figure 17 is similar to that of γ-FeOOH, lepidocrocite, EO_4-type. Chains [100] of strongly distorted octahedra (Cu—OH: 2×1.93 Å, 2×1.94 Å, 2×2.63 Å) linked through common short edges form cor-

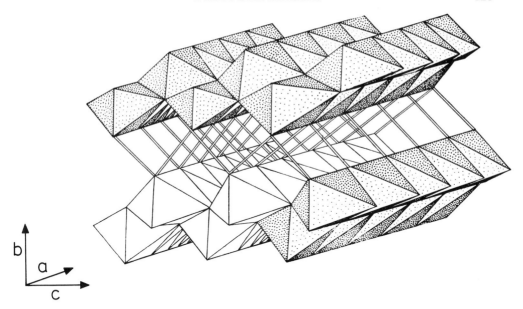

Fig. 17. Principle of the crystal structure of $Cu(OH)_2$. (Cu^{2+} in the centers, OH^- at the corners of the octahedra; Jaggi and Oswald [218].)

rugated layers (010) with an interlayer distance of $b/2 = 5.29$ Å. These layers are held together by hydrogen bonding.

In a later investigation, Schönenberger et al. [219] found evidence for a small but sharply defined reversible phase transition in $Cu(OH)_2$. As single crystals are still lacking, no better accurate structure determination of $Cu(OH)_2$ exists at present, although this would be highly desirable in order to reveal the proton positions.

8.3.2. Some Chemical Properties

The stability of $Cu(OH)_2$ strongly depends on its preparation. Fresh samples of small particle size and with disordered crystal structure have a marked tendency to transform under the mother liquor into brown or black CuO. This happens at room temperature and is accelerated by heating. Alkalies and H_2O_2 favour the reaction, and small amounts of built-in foreign metal ions like Mg^{2+} or Ni^{2+} have a stabilizing effect (see e.g. Feitknecht and Maget [216]). This can be important for the technical production of very finely dispersed but nevertheless stable $Cu(OH)_2$ preparations. On the other hand, suspensions of larger, perfectly ordered $Cu(OH)_2$ crystallites are much more stable and may resist dehydration up to about 100 °C.

The thermal dehydration of $Cu(OH)_2$ in a dry state begins with the release of adsorbed water, for good samples not more than about 1%. Slightly above 100 °C, depending on preparation, the samples begin to turn bluish-green, green, and later brown, as more water is lost. At 160 °C, the decomposition is practically complete, but traces of water are retained up to 200 °C and more. No other distinct phase between $Cu(OH)_2$ and CuO exists.

Günter and Oswald [220] studied the topotactic electron induced and thermal decomposition of $Cu(OH)_2$ in detail, and a quantitative investigation into the thermochemistry of $Cu(OH)_2$ besides other copper compounds was undertaken by Gedansky et al. [221].

Solubility data on $Cu(OH)_2$ were compiled by Feitknecht and Schindler [33]. They recommend for 25 °C and $J=0$:

	$\log Ks_0$	
$Cu(OH)_2$ (inactive)	−18.8	
CuO (active)	−19.7	(for comparison)
CuO (inactive)	−20.5	

The easy solubility of $Cu(OH)_2$ in diluted acids points to predominantly basic properties. As, however, $Cu(OH)_2$ is also dissolved by hot alkaline hydroxide solutions to at least some extent, it should rather be designated as weakly amphoteric. The structure of a hydroxo-cuprate(II), $Ba_2[Cu(OH)_6]$, was recently determined with high accuracy from single crystal X-ray data [222, 223].

Reliable values for thermodynamic data of $Cu(OH)_2$ are given by Latimer (cf. [105], p. 184): $\Delta H°_{298}=-106.1$ kcal mole^{-1}, $\Delta G°_{298}=-85.3$ kcal mole^{-1}, in good agreement with the -85.41 kcal mole^{-1} by Feitknecht [207]. The lattice energy is indicated in [97] with 764 kcal mole^{-1}.

9. Zinc Hydroxides, $Zn(OH)_2$

9.1. General information

Besides amorphous $Zn(OH)_2$, a number of polymorphic hydroxide phases called α-, β-, γ-, δ-, ε-$Zn(OH)_2$ are known which at present are not all sufficiently well characterized. Furthermore, the mutual relationships regarding stability and structure are difficult to establish.

It has to be stressed that these crystalline phases are clearly different from each other not only from the morphological but also from the structural point of view, as may be shown e.g. by X-ray crystallography, selected area electron diffraction and electron microscopy. Thus, the often encountered attempts to explain the facts with solid solutions between only two modifications, or with particle size and disorder effects, are definitely incorrect. One important reason for this variability is rooted in the fact that the Zn^{2+} ion with its completely filled 3d electron shell exhibits, contrary to Mn^{2+}, Fe^{2+}, Co^{2+}, etc., a marked tendency to adopt a tetrahedral coordination, but, on the other hand, is capable of occupying a regular octahedron as well.

The most stable compound in the whole range is doubtless ε-$Zn(OH)_2$, but from a purely thermodynamical point of view, even this is slightly less stable than ZnO besides H_2O at 25 °C.

TABLE VI
Different modifications of $Zn(OH)_2$

Phase	Composition	Unit cell	Lattice type	Ref.
amorphous $Zn(OH)_2$	$Zn(OH)_2$, aq	–	–	[226, 227, 228]
α-$Zn(OH)_2$	$Zn(OH)_2$ (contains foreign anions like Cl^-, NO_3^-)	hexagonal $a=3.11$ Å $c=7.8$ Å (for Cl^-)	layer lattice Zn^{2+} in octahedral and tetrahedral coordination	[126, 146, 227, 228, 229]
'β-$Zn(OH)_2$' (high pressure form)	$Zn(OH)_2$	hexagonal D_{3d}^3 3.194 Å 4.714 Å $Z=1$	CdI_2 layer lattice, with Zn^{2+} in octahedral coordination	[224]
β_1- and β_2-$Zn(OH)_2$	$Zn(OH)_2$	orthorhombic $a=13.17$ Å $b=6.42$ Å $c=24.1$ Å	unknown layer lattice	[234, 236, 237]
γ-$Zn(OH)_2$	$Zn(OH)_2$	orthorhombic Imm2 $a=23.07$ Å $b=8.08$ Å $c=3.30$ Å $Z=12$	chain structure with Zn^{2+} in tetrahedral coordination	[236, 237, 239]
δ-'$Zn(OH)_2$'	$Zn(OH)_2 \cdot \frac{1}{2} H_2O$	orthorhombic $a=13.24$ Å $b=6.42$ Å $c=34.0$ Å	unknown layer lattice, related to β-$Zn(OH)_2$	[236, 237, 238]
ε-$Zn(OH)_2$	$Zn(OH)_2$	orthorhombic $P2_12_12_1$-D_2^4 $a=5.17_0$ Å $b=8.54_7$ Å $c=4.93_0$ Å $Z=4$	space lattice with Zn^{2+} in tetrahedral coordination	[233, 242, 243]

Complete crystal structure determinations have only been carried out for ε- and γ-$Zn(OH)_2$, both of which contain tetrahedrally coordinated zinc ions and have no layer lattice. On the other hand, α-, β- and δ-$Zn(OH)_2$ and a quite new high pressure form also called 'β-$Zn(OH)_2$' [224] seem to have typical layer lattices, but have not yet been well explored structurally.

The older literature on $Zn(OH)_2$, which is particularly rich in respect to preparative details, is available up to about 1950 in *GMELINS Handbook* [225].

Some data on the different modifications of $Zn(OH)_2$ are assembled in Table VI. In the following text, only a few important facts are added for each phase, without claiming completeness.

9.2. Preparation and characterization of the $Zn(OH)_2$ modifications

9.2.1. Amorphous $Zn(OH)_2$

The amorphous $Zn(OH)_2$ phase is always formed when diluted zinc salt solutions are rapidly made weakly alkaline with strong bases. In addition, it can be obtained by topochemical transformation of α-$Zn(OH)_2$, or of zinc hydroxy-salts, or of hydrated zinc oxalate, upon suspension in aqueous alkaline media. Moreover, careful neutralization of solutions containing solved zinc hydroxo complexes will lead to the same product. For detailed information, the reader is referred to a number of reports by Feitknecht and his co-workers, for example [226, 227, 228].

The substance is described as a bluish-transparent, gel-like precipitate with a high degree of dispersion and extremely difficult to free from adsorbed anions. From the latter point of view, zinc nitrate is more favorable than e.g. zinc halide solutions. The designation 'amorphous' should not be understood only in a morphological sense, but characterizes, at least to a large extent, also the X-ray crystallographic behaviour of this phase. When amorphous $Zn(OH)_2$ is dried, or even by standing in aqueous suspension, it tends to transform into active ZnO and water at room temperature already.

9.2.2. α-$Zn(OH)_2$

α-$Zn(OH)_2$ never appears as large crystals, but always either in a gel-like state or pseudomorphous to basic zinc salts. It is formed by controlled, not fully complete precipitation from zinc salt solutions with strong alkalies, and by topochemical reaction of basic zinc salts, e.g. $Zn_5(OH)_8Cl_2$, with limited amounts of alkalies or water (otherwise, amorphous $Zn(OH)_2$ is obtained). For literature, see e.g. [126, 227, 228, 229].

To stabilize the crystal lattice of α-$Zn(OH)_2$, a certain small amount of foreign anions is indispensable, where the different anions do not have the same stabilizing capacity. Cl^- is for example much more readily replaced by OH^- under loss of the α-structure than CO_3^{2-} or silicate. A typical composition for α-$Zn(OH)_2$ is one mole of $ZnCl_2$ (or other salts) per 14 to 20 moles of $Zn(OH)_2$.

The structure of α-$Zn(OH)_2$ is a disordered double layer lattice (cf. also Section 1) with main layers containing Zn^{2+} ions in octahedral coordination and disordered intermediate layers where Zn^{2+} is tetrahedrally coordinated and some foreign anions replace OH^-. Structurally related phases, some of them with fully ordered double layer lattices, can be found in the field of hydroxy-salts; the reader is referred to e.g. the structures of 4 $Zn(OH)_2 \cdot ZnCl_2 \cdot H_2O$ resp. $Zn_5(OH)_8Cl_2 \cdot H_2O$ (Nowacki and Silverman [230]), or $Zn_5(OH)_8(NO_3)_2 \cdot 2 H_2O$ (Stählin and Oswald [11]). In α-$Zn(OH)_2$, Lotmar and Feitknecht [146] determined for the a-axis (i.e. the Zn-Zn distance within the main layers) a value of 3.11 Å. The main layer separation c depends on the nature of the stabilizing anions; for Cl^-, 7.8 Å are given, for NO_3^-, 9.5 Å, see Feitknecht [126]. Little can be said about the exact stacking mode.

α-$Zn(OH)_2$ forms mixed crystals with $Mg(OH)_2$, $Ni(OH)_2$, and $Co(OH)_2$, whereby a part of the octahedrally coordinated Zn^{2+} ions in the main layers are replaced by the other metal ion (see [131, 231]). A more recent description of a related Co-Zn double hydroxide $Co_3Zn_2(OH)_{10} \cdot 2 H_2O$ is given by Günter and Oswald [232].

9.2.3. $Zn(OH)_2$ of the C6 structure type

Several CdI_2- (or C6-) type hydroxides of metals other than zinc form mixed crystal series where a certain amount of Me^{2+} ions at octahedral sites are replaced by Zn^{2+} ions. This observation enabled Lotmar and Feitknecht [146] to determine by extrapolation the lattice dimensions of a hypothetical, pure $Zn(OH)_2$ phase in the C6-type: $a = 3.19$ Å, $c \approx 4.7$ Å. This hypothetical phase is – although related to α-$Zn(OH)_2$ in a structural chemical sense – not identical with α, as the latter is derived from the double layer lattice type, whereas C6 is a single layer lattice type with octahedrally coordinated metal ions only.

More recently, Baneeva and Popova [224] studied the behavior of $Zn(OH)_2$ at high temperatures and pressure, starting from an α-form precipitated from $ZnSO_4$ solution with NaOH. Between room temperature and 600 °C, the α-form was stable if the pressure did not exceed 110 kbar. But in the range above 120 kbar at 400 ± 40 °C, they obtained a new form of $Zn(OH)_2$ which was called 'β-$Zn(OH)_2$', although it has little in common with the β-phases of Feitknecht and his co-workers. Its X-ray diffraction powder pattern was resolved into a structural type of CdI_2, space group D_{3d}^3, with

$$a = 3.194 \text{ Å}, \quad c = 4.714 \text{ Å}.$$

These lattice dimensions come very near to the ones extrapolated for the hypothetical C6-type $Zn(OH)_2$ by Lotmar and Feitknecht. The density of the C6-phase reported by Baneeva and Popova [224] is found to be 4.0 g cm^{-3}, whereas ε-$Zn(OH)_2$ has a density of only 3.05 g cm^{-3}, according to Corey and Wyckoff [233]. It is easy to understand that the coordination number four in ε-$Zn(OH)_2$ leads to a much lower density than a sixfold coordination around the Zn^{2+} ions, thus additional evidence is gained that the high pressure form in question does really crystallize in the C6-type. Nevertheless, it should not be called 'β-$Zn(OH)_2$' in order to avoid confusion with the other β-phases described in literature, but rather p-$Zn(OH)_2$, where p stands for pressure.

9.2.4. β_1- and β_2-$Zn(OH)_2$

These phases were described by Feitknecht (cf. [234]) and older literature, where up till 1946 only β-$Zn(OH)_2$ is mentioned. The indices 1 and 2 refer to characteristic but rather small differences in preparation, habit of particles, and in X-ray powder patterns, but in principle, β_1 and β_2 can be considered as derived from the same type of crystal structure. Several methods, going back mainly to the work of Feitknecht on the preparation of β-$Zn(OH)_2$ free from other polymorphs and from ZnO, are presented in [235]. In principle, zinc nitrate solutions are precipitated under severe exclusion of carbon dioxide by addition of slightly sub-stoichiometric amounts of alkaline hydroxide, with subsequent aging of the product. β-Zn(OH)$_2$ has also been observed as a product of zinc corrosion in aqueous media [228, 229].

β-Zn(OH)$_2$ forms microcrystalline rhombic plates, the habit of which gives strong evidence for a layered type of structure (see Feitknecht [234]). Besides the γ- and δ-modifications, Giovanoli et al. [236] studied β-Zn(OH)$_2$ more recently with electron microscopy, electron diffraction and high resolution X-ray powder diffractometry, mainly as a base for their investigations of the thermal decomposition behaviour [237]. The very complex X-ray powder pattern was, supported by selected area electron diffraction, indexed with the orthorhombic unit cell listed in Table VI. The c-axis of 24.1 Å (or 4 × 6.02 Å) extends normally to the (001) plane which corresponds to the morphologically evident layers. Some relationship to the layered lattice basic salts is supposed, but the structure seems too complex to be solved without single crystal data.

9.2.5. γ-Zn(OH)$_2$

γ is not easy to obtain free from other forms of Zn(OH)$_2$, especially δ or ε, and from ZnO. The principle of its preparation is to dilute supersaturated alkali zincate solutions, prepared by reaction of zinc salt solutions with an excess of strong alkaline hydroxide, but the nature of the product is very dependent on details such as concentrations, velocity of stirring, aging etc. The preparative literature is reviewed in [238].

γ-Zn(OH)$_2$ forms needle- or rod-like crystals. Its orthorhombic unit cell was determined by Giovanoli et al. [236, 237] from X-ray powder data (cf. Table VI). Later, Christensen [239] succeeded in solving the structure from single crystal data. It consists of rings of three tetrahedral Zn(OH)$_4$ groups which are linked through their remaining vertices into infinite columns running parallel to the c-axis and linked to similar ones by hydrogen bonds. Therefore, the structure is not a typically layered one.

9.2.6. δ-'Zn(OH)$_2$'

In their older work, Feitknecht and his co-workers described the compound as a hydroxide (see literature cited in [238]). Giovanoli et al. [236, 237] obtained pure samples of δ by rapid deposition from solutions of the zinc tetrammine complex, whereas slow deposition leads to the more stable ε-Zn(OH)$_2$. The morphology of the δ-phase is quite similar to that of β-Zn(OH)$_2$, and the above mentioned authors proved by thermoanalytical methods that δ is in truth a hemihydrate, Zn(OH)$_2 \cdot \frac{1}{2}$ H$_2$O, the additional water being lost slightly below 100 °C and the OH$^-$ ions condensing near 120 °C. These facts agree well with the X-ray and electron diffraction results, which point to a layered structure very similar to that of β-Zn(OH)$_2$ with respect to the layers parallel to (001) (see unit cell in Table VI). The only difference is that the c-axis of δ is about 2.5 Å longer per layer than the c-axis of the β-type. Therefore, δ appears as a hemihydrate strongly related to β-Zn(OH)$_2$, with sheets of water molecules between the zinc hydroxide layers. The transformation $\delta \rightarrow \beta$ occurs topotactically, with perfect conservation of the morphology.

9.2.7. ε-Zn(OH)$_2$

This is the most stable Zn(OH)$_2$-modification. It is formed by transformation of other, less stable polymorphs of Zn(OH)$_2$ under their mother liquor, or directly as well, provided the solid phase is deposited at a very low rate. Therefore, ε is the only Zn(OH)$_2$ modification from which large single crystals can easily be produced. Standard procedures are either the slow deposition from alkali zincate solutions according to Scholder and Hendrich [240] or the slow removal of NH$_3$ from solutions of the zinc tetrammine complex (see Dietrich and Johnston [241], and further references in [242]).

The crystals are colourless, transparent double pyramids. They enabled the very early structure determination by Corey and Wyckoff [233] which was later made more precise by von Schnering [243]. The structure is built up from Zn^{2+} tetrahedrally coordinated by OH$^-$ ions, forming a three dimensional network resembling the cristobalite structure by linkage over all four corners of the tetrahedra.

9.3. PROPERTIES

9.3.1. *Stability Relations from Energetic and Solubility Data*

Schindler et al. [244] carried out detailed work on the solubility products and free enthalpies of formation of zinc oxide, amorphous Zn(OH)$_2$, β_1-, β_2-, γ-, δ-, and ε-Zn(OH)$_2$ in order to complete the knowledge on the relative stabilities of the solid phases. Their report contains a compilation of the most reliable older thermodynamic data on ZnO and some of the Zn(OH)$_2$-modifications. Information on their relative stabilities are listed in Table VII.

TABLE VII
Results of work on the relative stability of solid phases in the system Zn^{2+}-H$_2$O (refs. [225, 244])

Reaction	$\Delta G_{298.2}$[kcal]	$\Delta H_{298.2}$[kcal]
amorphous, fresh Zn(OH)$_2 \rightarrow \varepsilon$-Zn(OH)$_2$	−1.59	−2.95
amorphous, aged Zn(OH)$_2 \rightarrow \varepsilon$-Zn(OH)$_2$	−1.30	−2.72
β_1-Zn(OH)$_2 \rightarrow \varepsilon$-Zn(OH)$_2$	−0.35	−0.29
γ-Zn(OH)$_2 \rightarrow \varepsilon$-Zn(OH)$_2$	−0.29	−0.20
amorphous Zn(OH)$_2 \rightarrow$ ⎫	−1.2 to −1.6	−0.97
β_1-Zn(OH)$_2 \rightarrow$ ⎬ inactive ZnO +	−0.34	1.68
γ-Zn(OH)$_2 \rightarrow$ ⎨ H$_2$O$_{liq.}$	−0.27	1.77
ε-Zn(OH)$_2 \rightarrow$ ⎭	−0.27 (other: −0.11; 0.0)	1.97

The solubility of the chemical species was then studied by detecting the Zn^{2+} and H$^+$ concentrations of solutions in contact with the solid phases. [H$^+$] was measured by the e.m.f. method at constant ionic strength (0.2 M KNO_3 of $NaClO_4$) while [Zn^{2+}] was calculated from analytical data. Free enthalpies of formation were calculated from the solubility constants by means of a simple cycle. Table VIII contains the principal results of this work.

TABLE VIII

Solubility constants and free enthalpies of formation of zinc hydroxides, at 25 °C (according to Schindler et al. [244])

Solid phase	log K_{S0} (J=0.2)	$\Delta G°_{298.2}$ [kcal mole^{-1}]	log K_{S0} (J=0)
am. $Zn(OH)_2$	-14.70 ± 0.03	-131.54 ± 0.03	-15.52 ± 0.03
β_1-$Zn(OH)_2$	-15.42	-132.52	-16.24
β_2-$Zn(OH)_2$	-15.38	-132.46	-16.20
γ-$Zn(OH)_2$	-15.44	-132.55	-16.26
δ-'$Zn(OH)_2$'	-15.33	-132.40	-16.15
ε-$Zn(OH)_2$	-15.65	-132.83	-16.47
ZnO (active)	-15.84	-76.40	-16.66
ZnO (inactive)	-16.01	-76.63	-16.83

The order of increasing stability goes thus from amorphous $Zn(OH)_2$ via δ-'$Zn(OH)_2$', β_2-$Zn(OH)_2$, β_1-$Zn(OH)_2$, γ-$Zn(OH)_2$, ε-$Zn(OH)_2$, ZnO (active) to ZnO (inactive). This order is principally in agreement with the earlier observations of Feitknecht et al. and of other authors, but the absolute values of solubility constants and $\Delta G°$ show considerable differences compared with older measurements, where probably the influence of CO_2 was not sufficiently excluded. It is astonishing how small the differences in $\Delta G°$ of the solid phases in the system are. The authors have furthermore determined experimentally the dependence of the solubility of ZnO on its particle size, and from this the diameter D of ZnO-nuclei, which are in equilibrium with saturated solutions of the different hydroxides:

Phase:	am.$Zn(OH)_2$	δ-'$Zn(OH)_2$'	β_2-$Zn(OH)_2$
D[Å]:	41	79	86
Phase:	β_1-$Zn(OH)_2$	γ-$Zn(OH)_2$	ε-$Zn(OH)_2$
D[Å]:	92	95	150

If this 'critical' size of nuclei is a measure for the activation energy of the reaction $Zn(OH)_2 \rightarrow ZnO + H_2O$, it may be understood that the pure crystallized hydroxides can be relatively stable in aqueous suspensions.

The value of $\Delta H°_{298} = -153.5$ kcal mole^{-1} in Table I is by Rossini (ref. [202], p. 177).

9.3.2. Thermal Decomposition

Giovanoli et al. [236, 237] have studied in detail the thermal decomposition of β-, γ-, δ- and ε-$Zn(OH)_2$. This was done by normal heating in a furnace as well as, for comparison, by the impact of electrons within an electron microscope. The decomposition reaction $Zn(OH)_2 \rightarrow ZnO + H_2O$ takes place in the range from 70 to 140 °C, and the most stable ε-form decomposes at a higher temperature than e.g. γ-Zn(OH)$_2$. For the δ-modification, two steps can be distinguished on slow decomposition:

$$\delta\text{-Zn(OH)}_2 \cdot 0.5\,\text{H}_2\text{O} \xrightarrow[-0.5\,\text{H}_2\text{O}]{\approx 90\,°\text{C}} \beta\text{-Zn(OH)}_2 \xrightarrow[-\text{H}_2\text{O}]{\approx 115\,°\text{C}} \text{ZnO}$$

In high vacuum, and especially under the influence of an electron beam, the decomposition of the β-, γ- and δ-forms is topotactic in nature, yielding very fine but highly oriented crystallites of ZnO. For β and δ on the one hand and γ on the other hand the texture of the oxide pseudomorphs is distinctly different. The products contain significant traces of residual water. From 600 °C upwards the zinc oxides begin to sinter continuously. The influence of the original hydroxide's nature on the sintering rate is remarkable.

10. Cadmium Hydroxide, Cd(OH)$_2$

10.1. GENERAL INFORMATION

Three forms of cadmium hydroxide have been described: α-, β-, and γ-Cd(OH)$_2$. The normal form, called β-Cd(OH)$_2$, is a stable and well-defined modification which is easy to prepare, whereas α-Cd(OH)$_2$ – sometimes also named 'active' Cd(OH)$_2$ – consists of very small particles with a heavily disordered crystalline structure and is therefore much more difficult to characterize. Finally, γ-Cd(OH)$_2$ is a very rare polymorphic form probably less stable than β, but of unambiguously proven existence and known crystal structure. The following text mainly refers to β-Cd(OH)$_2$, the α- and γ-modifications being included in the section on special forms.

β-Cd(OH)$_2$ has – like several other hydroxides of bivalent metals – the CdI$_2$- (or C6-) structure type. Its crystals are thin, shiny white hexagonal plates. Apart from its CO$_2$ sensitivity, cadmium hydroxide is of good chemical stability. In particular, Cd(OH)$_2$ is thermodynamically stable compared to the system CdO-H$_2$O, and hence different from the zinc hydroxides. Although it shows quite clearly a basic character, Cd(OH)$_2$ is sometimes called amphoteric, due to its slightly acidic behaviour when in contact with concentrated alkali hydroxide solutions [245]. At least part of the interest Cd(OH)$_2$ has found, is related to its role in nickel-cadmium storage batteries with alkaline electrolytes.

10.2. COMPOUND PREPARATION

There is a relative large amount of literature available concerning the preparation of the hydroxide which in part is rather contradictory. This is due to the fact that on the one hand side the hydroxide has a high rate of crystallization, but on the other hand often precipitates first in the active α-form (see Section 10.3) and, especially in this form, strongly adsorbs foreign anions and water. Moreover, there is a marked tendency to basic salt formation. During all steps of preparation the hydroxide should be protected against CO$_2$ from the air as this will transform it into CdCO$_3$.

10.2.1. By Hydrolysis

No generally applicable procedure exists for the preparation of $Cd(OH)_2$ via hydrolysis of Cd compounds with water. The reaction of CdO with water is slow and barely complete [245, 246]. Another rather exclusive method utilizes dimethylcadmium as a source material [247] and yields preparations with a high degree of dispersion.

10.2.2. By Precipitation

The most common procedure is based on $Cd(OH)_2$ precipitation from a Cd^{2+} solution in basic environment. To avoid difficulties of the kind described below, it is of great importance to employ the right reaction partners. Problems arising are:
- Basic salt formation: basic nitrates [21, 139], basic chlorides [248], basic sulphates [249]; for general aspects cf. also [250].
- Adsorption of particles from the solution [251].
- Limited precipitation as a result of complex formation, for example $[CdX_4]^{2-}$ [252].
- Re-dissolving of the substance caused by complex formation such as in concentrated KOH [253] or in aqueous NH_3 [254].
- Formation of stabilized α-$Cd(OH)_2$; see Section 10.3 on special forms.

Pure $Cd(OH)_2$ is best obtained by precipitation from a $Cd(NO_3)_2$ solution [252]. If this proceeds slowly, formation of the stable β-modification will predominate but one may also observe enhanced aggregate formation, built up from numerous hexagonal plates [142]. Replacing the commonly used NaOH by triethylamine results in the formation of very pure $Cd(OH)_2$ [255].

The preparation of large β-$Cd(OH)_2$ crystals can proceed along the following two lines, which have been used successfully.

(a) To 10 g of CdI_2 in 150 ml water, 360 g KOH with 13% H_2O, is added at 135 °C. When dissolved it is kept for 12 h [246, 256].

(b) By means of a pipette, a 0.5 M solution of $Cd(NO_3)_2$ is carefully introduced at the bottom of a 200 ml beaker previously three-fourths filled with distilled water. The beaker is placed in a desiccator together with a second beaker containing 3 M ammonia. After some days crystals appear in a ring-shaped zone on the wall of the first beaker as a result of the diffusion process taking place. By this technique, de Haan [139] obtained hexagonal plates or prisms of β-$Cd(OH)_2$ ranging from a few tenths of a mm to a few mm.

10.2.3. By Electrolysis

The cathodic reduction of Cd^{2+} salt solutions not only leads to elemental cadmium but also to $Cd(OH)_2$ [138], which is formed as a gelatinous film [257]. The hydroxide can also be obtained electrochemically by the reverse reaction, i.e. oxidation of a cadmium anode in a NaCl or H_3PO_4 electrolyte [258, 259]. For this process – carried out in a NaOH solution – Huber [260] has investigated such parameters like concentration, current density and potential. On metallic cadmium, single crystal epitaxial growth of cadmium hydroxide was observed by Capdecomme and his co-workers [259, 261]. In this case the relative orientational relationship [0001] Cd(OH)$_2$// [0001] Cd was found. More recent detailed investigations on the kinetics of anodic oxidation of cadmium amalgam to cadmium hydroxide show clearly that hexagonal Cd(OH)$_2$ layers are formed parallel to the surface of the amalgam substrate right from Cd^{2+} salt solutions. Thereby NO$_3^-$ anions act in favour of α, but not Cl$^-$ and layer is the rate-determining step [262].

10.3. Special forms

10.3.1. α-Cd(OH)$_2$

This 'active' phase is the first product formed when Cd(OH)$_2$ is rapidly precipitated from Cd^{2+} salt solutions. Thereby NO$_3^-$ anions act in favour of α, but not Cl$^-$ and SO$_4^{2-}$. Normally, α-Cd(OH)$_2$ is transformed into the 'normal' β-Cd(OH)$_2$ by the aging process under the mother liquor. Addition of glucose or higher alcohols, however, can hamper this transformation, according to Feitknecht [130]. For more details about the α to β ratio as a function of time in the precipitation of cadmium nitrate solutions, the reader is referred to work by Feitknecht and Studer [142].

Structurally, the α-phase is related to the C6-type of β-Cd(OH)$_2$. From X-ray powder diagrams can be concluded that the lattice dimensions within the hexagonal (0001) planes are only a few % smaller for α than for β. The poor quality of the α powder patterns – they show hardly more than (hk.0) reflections – makes it difficult to decide whether α has something like a rather disordered 'two-dimensional' β-structure, or whether it is a unique form built up according to the double layer lattice principle similar to α-Co(OH)$_2$ and α-Ni(OH)$_2$, although with an extremely disordered intermediate layer as well as stacking. The appropriate investigations by Feitknecht and his co-workers [28, 126, 130, 131, 142] speak rather for the second possibility. Such laminar disperse forms are interesting as intermediates between the amorphous and the crystalline states. In the present case, their formation seems to be favoured by the possibility for the metal ion to adopt, besides its usual octahedral environment in the main layers, a tetrahedral coordination in the intermediate layers. This holds for the Cd^{2+} ion as well as for Co^{2+} and Zn^{2+}.

10.3.2. Colloidal Cd(OH)$_2$

Cd(OH)$_2$ has never been found in a fully X-ray amorphous state. This is due to the fact that in the α- as well as in the β-form there is a marked tendency to build up at least a planar hexagonal arrangement of edge-linked coordination octahedra. In the conversion process from α to β there is usually no significant increase in particle size [130]. There exists, however, literature on preparative procedures to produce colloidal Cd(OH)$_2$ in gel- or sol-like forms [263].

10.3.3. γ-Cd(OH)$_2$

Glemser [264] was the first to report on the γ-modification in 1957 and he also presented a special preparation technique, although for microcrystalline disordered material only. In the same year Feitknecht [265] recognized that a compound described much earlier, designated cadmium hydroxychloride VI, $Cd(OH)_{1.93}Cl_{0.07}$, gives the same X-ray powder pattern. Thus, part of the OH^- ions in γ-$Cd(OH)_2$ can be replaced by Cl^- (or also F^-) ions.

γ-$Cd(OH)_2$ is monoclinic. Its crystal structure as determined from X-ray powder data in 1966 by de Wolff [266] is unique among the bivalent metal hydroxides and consists of octahedral coordination groups linked into double chains by face sharing, these double chains sharing vertices to form a three dimensional framework of a new type. Whenever good single crystals become available, a refinement of this interesting result shall become possible.

By refluxing several hours with water, γ-$Cd(OH)_2$ is not changed, but yields β-$Cd(OH)_2$ on heating 4 days with H_2O in a bomb at 150 °C, and in vacuum at 115 °C and $p_{H_2O}=10$ torr decomposition to CdO and H_2O is observed [264].

10.4. Properties

10.4.1. *Physical*

The unit cell data are as follows:

α-$Cd(OH)_2$: hexagonal, $a=3.36$ Å, $c=$ not det. [126]

β-$Cd(OH)_2$: our own determination [95] agrees very well with data by de Haan [1] resp. Bertrand and Dusausoy [267]

$a=3.499\pm0.005$ Å $\quad c=4.701\pm0.008$ Å $\quad c/a=1.344$
$Z=1 \quad\quad\quad\quad\quad V=49.8_4$ Å^3 $\quad\quad\quad\quad D_x=4.87_7$ g cm^{-3}
$\quad\quad\quad\quad\quad\quad\quad\quad\quad\quad\quad\quad\quad\quad\quad\quad D_{exp.}=4.810$ g cm^{-3} [256]

free parameter $z_{OH}=0.241$ (according to unpublished results cited by de Haan [1]).

For the lattice energy of stable $Cd(OH)_2$ a value of 676 kcal mole^{-1} was calculated from thermochemical data which is in agreement with the results by the Born-Haber cycle, 673 kcal mole^{-1}, according to Yatsimirskij [96].

γ-$Cd(OH)_2$: monoclinic, space group Im

$a=5.67$ Å, $\quad\quad\quad b=10.25$ Å, $\quad\quad\quad c=3.41$ Å [266]
$\beta=91.4°$ $\quad\quad\quad\quad V=49.5$ Å^3 $\quad\quad\quad\quad D_x=4.908$ g cm^{-3}
$Z=4$ $\quad\quad\quad\quad\quad$ (per $Cd(OH)_2$) $\quad\quad D_{exp.}=4.81$ g cm^{-3} [264]

More physical data on β-$Cd(OH)_2$ can be found under [250].

10.4.2. *Solubility*

As $Cd(OH)_2$ can exist in various states of aging and purity, it is obvious that there is only moderate agreement among the different solubility data. Values for the solubility product lie between 1.6×10^{-13} – polarographically determined [268] – and 2.5×10^{-16} (electrochemically determined [269]). In their compilation Feitknecht and Schindler [33] recommend using the following values (for 25 °C, $J=0$):

log $Ks_0\beta$-Cd(OH)$_2$ (inactive) $= -14.4 \pm 0.1$
(calculated from the free enthalpy of formation, according to Schindler [270])

log $Ks_0\beta$-Cd(OH)$_2$ (active) $= -13.7$
(according to Feitknecht and Reinmann [271], see also [272, 273]).

It is stated that values for α- and γ-Cd(OH)$_2$ are not known to some accuracy.

10.4.3. Thermodynamic Data

Values for $\Delta H°_{298}$ and $\Delta G°_{298}$ are presented in Table IX. Whereas these data are in agreement with each other, there is some discrepancy among the values of the heat of reaction for the formation of Cd(OH)$_2$ from CdO and water:

$$CdO + H_2O \rightarrow Cd(OH)_2.$$

According to Maier [274] it is -1.288 at 25 °C and -4.0 at 0 °C, while Fricke et al. [277] report the value of -4.74 at 45 °C for CdO heated at 800 °C, and a value of -5.30 at 45 °C for CdO heated at 350 °C (all values in [kcal mole^{-1}]).

TABLE IX
Thermodynamic data of Cd(OH)$_2$

Method	$\Delta H°_{298}$ [kcal mole^{-1}]	$\Delta G°_{298}$ [kcal mole^{-1}]	Ref.
electrochemically	-131.85	-112.2 ± 0.5	[274]
thermochemically (20 °C)	-133.41		[275]
e.m.f. measurement		-112.469	[276]
activity product (active Cd(OH)$_2$)		-112.40	[272]
activity product (inactive β-Cd(OH)$_2$)		-113.19	[272]

11. Conclusions

The prospective statement made in the introduction that the entire group of bivalent metal hydroxides with layered lattices was highly problematic from the viewpoints of preparation, purity and general chemical stability has certainly been confirmed by this review. In fact these compounds behave in many respects quite contrary to the requirements for solid state physical investigations or technical applications. Instead of large, pure and stable single crystals we mainly get fine dispersed, even colloidal materials which are, as a consequence of their morphology and structure, hard to get free from built-in or adsorbed impurities and water. Moreover, the phases are of quite low chemical stability, reacting readily with carbon dioxide from the air, some of them even with molecular oxygen, and decomposing to oxides and water upon even moderate heating.

In what respects are these hydroxides of interest then? First they are worth consideration from the structural chemical viewpoint, in the framework of both comparative studies among hydroxides of different metals themselves, and of work extending towards the wide field of hydroxy-salts. In the latter case, the C6 structure type of the hydroxides plays the role of a general building principle which may be varied and modified in a multiple, sometimes very complex manner. Second, the above-mentioned apparent 'disadvantages' should be focused on. It should be considered an enrichment to have systems at our disposal which are somewhere in between the amorphous and the crystalline state, as for example the α-hydroxides with ordered main layers and amorphous intermediate layers, or the extremely thin $Ni(OH)_2$ plates which are practically two-dimensional crystals. It is quite easy nowadays to collect a broad knowledge on physical and chemical details of a large single crystal, but this is much more difficult for these fine dispersed phases. In this context, the morphology on a very small scale should certainly not be neglected, and the hydroxides or hydroxy-salts are indeed excellent systems to start investigations into the field of the relations between structure and morphology on one hand and chemical reactivity on the other hand.

Due to the dominant layer or chain structural character, almost all of the reactions these hydroxides undergo – thermally or with solutions – are topotactic in nature, this means that the product crystallites are formed in well-defined orientations relative to the crystal lattice of the decomposing initial crystals. Whereas the geometrical aspects of topotaxy are quite well known – for an attempt at a systematic classification see e.g. Günter and Oswald [278] –, this does not yet hold for the quantitative elucidation of the related solid state reaction kinetics. Such work is not only interesting from the viewpoint of pure science, but may also be of technical importance, for example when hydroxides are considered as particularly reactive initial substances for mixed metal oxide powders. Another interesting point is how small the free energy differences between the whole number of $Zn(OH)_2$ modifications in fact are, if we consider on the other hand that they are quite different in their synthesis and crystal structure.

This article may not be concluded without stressing our deep indebtedness – of course among others – to the late Professor W. Feitknecht and his school. During several decades he dedicated a good part of his forces to the hydroxides and hydroxy-salts, and his work is considered in many respects as very important for following developments.

Acknowledgements

The authors are indebted to Dr R. M. A. Lieth (Eindhoven) for his helpful advice, encouragement and patience. Furthermore they wish to thank Miss L. Guyer, Mrs M. Wirz, Miss E. Gehrer and Mr A. Portmann (University of Zürich) very much for their excellent secretarial work and technical assistance.

References

1. Y. M. de Haan: 'Kristalbouw en roosterenergie van laagstructuren van het type cadmiumhydroxyde', PhD-thesis Technische Hogeschool Delft, Netherlands, 1965.

2. W. Feitknecht and H. R. Oswald: *Helv. Chim. Acta* **49** (1966), 344.
3. W. Feitknecht, P. Brunner, and H. R. Oswald: *Z. Anorg. Allg. Chem.* **316** (1962), 154.
4. H. R. Oswald, W. Feitknecht, and P. Brunner: in *From Molecule to Cell: Symposium on Electron Microscopy, Modena, 1963*, (ed. by P. Buffa), CNR Roma, 1964, p. 141.
5. F. Freund: *Fortschr. chem. Forsch.* **10** (1968), 347.
6. W. Feitknecht: *Fortschr. chem. Forsch.* **2** (1953), 670.
7. H. R. Oswald and W. Feitknecht: *Helv. Chim. Acta* **44** (1961), 847.
8. H. R. Oswald and W. Feitknecht: *Helv. Chim. Acta* **47** (1964), 272.
9. Y. Iitaka, S. Locchi, and H. R. Oswald: *Helv. Chim. Acta* **44** (1961), 2095.
10. H. R. Oswald, Y. Iitaka, S. Locchi, and A. Ludi: *Helv. Chim. Acta* **44** (1961), 2103.
11. W. Stählin and H. R. Oswald: *Acta Cryst.* **B 26** (1970), 860.
12. H. R. Oswald, *Helv. Chim. Acta* **52** (1969), 2369.
13. T. Shirasaki, *Denki Kagaku* **30** (1962), 43; cf. *CA* **62** (1965), 11204f.
14. *GMELINS Handbuch der anorganischen Chemie*, 8th ed., Verlag Chemie, Weinheim, Syst. nr. 27, Magnesium, Vol. B1, 1939, p. 55.
15. G. O. Rees: *Phil. Mag.* **10** (1837), 454.
16. M. Le Blanc and K. Richter: *Z. Phys. Chem.* **107** (1923), 400.
17. W. Feitknecht and H. Braun: *Helv. Chim. Acta* **50** (1967), 2040.
18. R. Giovanoli, W. Feitknecht, and W. Fahrer: *J. Microsc.* **7** (1968), 177.
19. W. Feitknecht: *Helv. Chim. Acta* **9** (1926), 1036.
20. G. Friedel: *Bull. Soc. Min.* **14** (1891), 74.
21. A. Berton: *Bull. Soc. Chim.* (1947), 289.
22. H. Muroya, T. Shirosaki, and H. Kodaira: Japan. 10, 784 ('58) Dec. 26; cf. *CA* **53** (1959), 22785f.
23. N. I. Soboleva, A. G. Bol'shakov, and A. V. Kortnev: *Kolloid. Zh.* **20** (1958), 742.
24. G. Brauer: *Handbuch der präparativen anorganischen Chemie*, vol. I, 2nd ed., Enke, Stuttgart, 1960, p. 805.
25. T. Hideo and S. Katsuichi: *Yogyo Kyokai Shi* **76** (871) (1968), 81; cf. *CA* **71** (1969), 7269u.
26. D. P. Zosimovich and O. I. Kirichenko: *Mem. Inst. Chem. Ukrain. Acad. Sci.* **3** (1936), 257.
27. R. Fricke, R. Schnabel, and K. Beck: *Z. Elektrochem.* **42** (1936), 881.
28. W. Feitknecht: *Kolloid Z.* **136** (1954), 52.
29. J. Mering and A. Oberlin: *Bull. Soc. Min.* **80** (1957), 158.
30. H. E. Swanson, N. T. Gilfrich, and M. I. Cook: 'Standard X-Ray Patterns', *Circ. Nat. Bur. Stand.* **539** [6] (1956), 30, 36.
31. A. Finch and P. J. Gardner: *J. Phys. Chem.* **69** (1965), 384.
32. A. Travers: *Compt. Rend. Acad. Sci.* **188** (1929), 501.
33. W. Feitknecht and P. Schindler: *Pure Appl. Chem.* **6** (1963), 130.
34. J. K. Gjaldbaek: *Z. Anorg. Allg. Chem.* **144** (1925), 269.
35. R. Näsänen: *Z. Phys. Chem.* **A 190** (1942), 183.
36. R. Fricke and J. Lüke: *Z. Elektrochem.* **41** (1935), 181.
37. A. F. Gill: *Canad. J. Res.* **10** (1934), 709.
38. D. F. Richardson: *Ind. Eng. Chem.* **19** (1927), 626.
39. N. S. Kurnakow and V. V. Cernyh: *N. Jb. Min.* **1** (1927), 315.
40. S. J. Gregg and R. K. Packer: *J. Chem. Soc.* (1955), 51.
41. L. Hackspill and A. P. Kiefer: *Ann. Chim.* **14** (1930), 249.
42. L. G. Berg, M. G. Altykies, and A. P. Pirmatov: *Zh. Neorg. Khim.* **12** (1967), 1751; cf. *CA* **67** (1967), 87363a.
43. J. F. Goodman: *Proc. Roy. Soc.* **A 247** (1958), 346.
44. F. Freund and H. Naegerl: *Thermal Analysis, Proc. 2nd Int. Conf., Worcester, Mass. U.S.A., 1968* (ed. by R. F. Schwenker and P. D. Garn) Academic Press, New York, vol. 2, 1969, p. 1207.
45. H. Naegerl and F. Freund: *J. Therm. Anal.* **2** (1970), 387.
46. J. Thomsen: *J. pr. Chem.* **11** (1875), 250.
47. W. A. Roth and P. Chall: *Z. Elektrochem.* **34** (1928), 192.
48. *GMELINS Handbuch der anorganischen Chemie*, 8th ed., Verlag Chemie, Weinheim, Syst. nr. 28, Calcium, vol. B2, 1971, p. 306.
49. E. Weinschenk: *C. Min.* (1905), 581.
50. F. W. Ashton and R. Wilson: *Am. J. Sci.* **13** (1927), 209.
51. L. M. Clark and W. F. Gerrard: *J. Soc. Chem. Ind.* **57** (1938), 301.

52. J. L. Gay-Lussac: *Ann. Chim. Phys.* **1** (1816), 334.
53. G. Rose: *Ber. Berl. Akad.* (1860), 575.
54. T. Selivanov: *Anorg. Allg. Chem.* **85** (1914), 329.
55. V. J. Burdakov: *J. Russ. Chem. Ges.* **44** (1912), 1325.
56. V. Kohlschütter and W. Feitknecht: *Helv. Chim. Acta* **6** (1923), 337.
57. J. Johnston: *J. Am. Chem. Soc.* **36** (1914), 16.
58. Riffault and Chompré: *Gilb. Ann.* **28** (1808), 115.
59. Ceskoslovenska Akademie Ved (by Cestmir Barta and Jan Zemlicka) Fr. 1,456,872 (C1.B 01b, k), Jan. 13, 1967; Czech. Appl. Jan. 29 and Dec. 8, 1965; cf. *CA* **67** (1967), P112137q.
60. W. R. Busing and H. A. Levy: *J. Chem. Phys.* **26** (1957), 563.
61. H. E. Petch: *Acta Cryst.* **14** (1961), 950.
62. C. W. Bunn, L. M. Clark, and I. L. Clifford: *Proc. Roy. Soc.* **A151** (1935), 141.
63. H. Remy: *Lehrbuch der anorganischen Chemie*, vol. I, 12/13th ed., Akad. Verl. Ges., Leipzig, 1970, p. 305.
64. H. Bassett: *J. Chem. Soc.* (1934), 1270.
65. R. Nacken and R. Mosebach: *Z. Anorg. Allg. Chem.* **223** (1935), 161.
66. J. W. Shipley and I. R. McHaffie: *Trans. J. Soc. Chem. Ind.* **42** (1923), 321.
67. R. S. Mikhail, St. Brunauer, and L. E. Copeland: *J. Colloid Interf. Sci.* **21** (1966), 394.
68. N. G. Dave and S. K. Chopra: *J. Amer. Ceram. Soc.* **49** (1966), 575.
69. N. H. Brett: *Min. Mag.* **37** (1969), 244.
70. O. V. Krylov, M. Ya. Kushnerev, and V. V. Kiryushkin: *Izv. Akad. Nauk S.S.S.R., Ser. Khim* **10** (1971), 2155.
71. F. D. Rossini et al.: *Circ. Nat. Bur. Stand.* **500** (1952), 386.
72. W. E. Hatton et al.: *J. Am. Chem. Soc.* **81** (1959), 5028.
73. K. K. Kelley: *Bull. Bur. Mines* **350** (1932), 18; *ibid.* **477** (1950), 32, 102.
74. A. F. Kapustinkij and K. B. Yatsimirskij: *Zh. Fiz. Khim.* **22** (1948), 1271.
75. K. Taylor and L. S. Wells: *J. Res. Nat. Bur. Stand.* **21** (1938), 133.
76. *GMELINS Handbuch der anorganischen Chemie*, 8th ed., Verlag Chemie, Weinheim, Syst. nr. 56, Manganese, vol. C1, 1973, p. 366.
77. R. G. Lepilina et al.: *Zh. Prikl. Khim.* **40** (1967), 2462; *J. Appl. Chem. U.S.S.R.* **40** (1967), 2359.
78. Esso Research and Engineering Co., Fr. 1,524,505 (C1.C 01b), 10 May 1968, Appl. 10 May 1967; cf. *CA* **71** (1969), 62823.
79. R. Scholder and A. Kolb: *Z. Anorg. Allg. Chem.* **264** (1951), 209.
80. C. Klingsberg and R. Roy: *Am. Min.* **44** (1959), 819.
81. I. G. Berikashvili: *Tr. Inst. Prikl. Khim. i Elektrokhim. Akad. Nauk. Gruz. S.S.R.* **1** (1960), 63; cf. *CA* **56** (1962), 1288e.
82. T. V. Kashcheeva and A. L. Tseft: *Izv. Akad. Nauk Kaz. S.S.R. Ser. Met. Obogasch. i. Ognenporov* **2** (1961), 33; cf. *CA* **56** (1962), 8364c.
83. H. T. S. Britton: *J. Chem. Soc.* **127** (1925), 2110.
84. I. M. Korenman: *Zh. Obshch. Khim.* **21** (1951), 10.
85. Ref. [76], 1973, p. 372.
86. J. Heubel: *Ann. Chim.* **4** (1949), 699.
87. G. Brauer: *Handbuch der präparativen anorganischen Chemie*, vol. II, 2nd ed., Enke, Stuttgart, 1962, p. 1272.
88. L. G. Berg and V. P. Kovyrzina: *Zh. Neorgan. Khim.* **8** (1963), 2041; *Russ. J. Inorg. Chem.* **8** (1963), 1065.
89. A. N. Christensen and G. Ollivier: *Solid State Comm.* **10** (1972), 609.
90. A. N. Christensen: *Acta Chem. Scand.* **19** (1965), 1765.
91. Ref. [76], 1973, p. 373.
92. J. Ambrose and G. W. D. Briggs: *Electrochim. Acta* **16** (1971), 111.
93. Ref. [76], 1973, p. 374.
94. G. V. Elmore and H. A. Tanner: *J. Phys. Chem.* **60** (1956), 1328.
95. H. R. Oswald: PhD-thesis, Univ. of Bern, Switzerland, 1960.
96. K. B. Yatsimirskij: *Izv. Akad. Nauk S.S.S.R. Otd. Khim.* (1948), 590; cf. *CA* **43** (1949), 2829a.
97. K. B. Yatsimirskij: *Zh. Neorgan. Khim.* **3** (1958), 2244; *Russ. J. Inorg. Chem.* **3** (1958), 26.
98. R. K. Fox, D. F. Swinehart, and A. B. Garrett: *J. Am. Chem. Soc.* **63** (1941), 1779.
99. S. S. Zavodnov and N. G. Fesenko: *Girodkhim. Materialy* **36** (1964), 148; cf. *CA* **63** (1965), 2437f.

100. H. R. Oswald and W. Feitknecht: *Proc. 5th Int. Conf. on Electron Microscopy*, Academic Press, New York, 1962, H-9.
101. O. Bricker: *Am. Min.* **50** (1965), 1296.
102. Ref. [76], 1973, p. 382.
103. D. D. Wagman, W. H. Evans, V. B. Parker et al.: *Nat. Bur. Stand. Tech. Note* **270-4** (1969), 107.
104. F. D. Rossini, D. D. Wagmann, W. H. Evans et al.: *Circ. Nat. Bur. Stand.* **500** (1952), 274.
105. W. M. Latimer: *The Oxidation States of the Elements and their Potentials in Aqueous Solutions*, 2nd ed., Prentice-Hall, New York, 1952, p. 235.
106. C. V. Schwarz: *Arch. Eisenhüttenw.* **24** (1953), 285.
107. L. J. Thénard: *Ann. Chim.* **56** (1805), 67.
108. J. J. Berzelius: *Lehrbuch der Chemie*, vol. 2, Dresden, 1826, p. 356.
109. *GMELINS Handbuch der anorganischen Chemie*, 8th ed., Verlag Chemie, Weinheim, Syst. nr. 59, Iron, vol. B1, 1932, p. 114.
110. E. Deiss and G. Schikorr: *Z. Anorg. Allg. Chem.* **172** (1928), 35.
111. S. Rihl and R. Fricke: *Z. Anorg. Allg. Chem.* **251** (1943), 405.
112. G. Natta and E. Casazza: *Atti Accad. Nazl. Lincei* **5** (1927), 803.
113. L. M. Clark, A. G. M. Hedley, and J. G. Robinson: *J. Soc. Chem. Ind. (London)* **63** (1944), 208.
114. J. D. Bernal, D. R. Dasgupta, and A. L. Mackay: *Clay Min. Bull.* **4** (1959), 15.
115. W. Feitknecht and G. Keller: *Z. Anorg. Allg. Chem.* **262** (1950), 61.
116. M. H. Francombe and H. P. Rooksby: *Clay Min. Bull.* **4** (1959), 1.
117. Y. Oka: *J. Chem. Soc. Japan* **59** (1938), 971.
118. D. L. Leussing and I. M. Kolthoff: *J. Am. Chem. Soc.* **54** (1948), 971.
119. B. O. Hedström: *Arkiv Kemi* **5** (1953), 457.
120. K. H. Gayer and L. Woontner: *J. Phys. Chem.* **60** (1956), 1569.
121. R. Scholder, H. Weber, and H. Dolge: *Angew. Chem.* **49** (1936), 255.
122. *JANAF Thermochemical Tables*, 2nd ed. (ed. by D. R. Stull and H. Prophet), *Nat. Stand. Ref. Data Ser., Nat. Bur. Stand. (U.S.)* **37** (1971).
123. R. Fricke and S. Rihl: *Z. Anorg. Allg. Chem.* **251** (1943), 414.
124. *GMELINS Handbuch der anorganischen Chemie*, 8th ed., Verlag Chemie, Weinheim, Syst. nr. 58, Cobalt, vol. A, 1932, p. 237.
125. *GMELINS Handbuch der anorganischen Chemie*, 8th ed., Verlag Chemie, Weinheim, Syst. nr. 58, Cobalt, vol. A, Suppl., 1961, p. 498.
126. W. Feitknecht: *Helv. Chim. Acta* **21** (1938), 766.
127. W. Feitknecht and W. Bédert: *Helv. Chim. Acta* **24** (1941), 676.
128. H. B. Weiser and W. O. Milligan: *J. Phys. Chem.* **36** (1932), 722.
129. W. Feitknecht: *Kolloid Z.* **92** (1940), 257.
130. W. Feitknecht: *Kolloid Z.* **93** (1940), 66.
131. W. Feitknecht: *Angew. Chem.* **52** (1939), 202.
132. R. Scholder and H. Weber: *Z. Anorg. Allg. Chem.* **216** (1933), 159.
133. C. A. MacKenzie and K. C. Edson: *J. Chem. Educ.* **18** (1941), 332.
134. Y. Oka: *Bl. Tokyo Univ. Engg. [Jap.]* **8** (1939), 411; Engl. abstr., p. 21.
135. R. Näsänen: *Ann. Acad. Fenn.* **A59** (Nr. 2) (1943), 3; cf. *CA* **41** (1947), 21g.
136. W. Feitknecht and H. Fischer: *Helv. Chim. Acta* **19** (1936), 1242.
137. G. Brauer: *Handbuch der präparativen anorganischen Chemie*, vol. II, 2nd. ed., Enke, Stuttgart, 1962, p. 1326.
138. A. Nicol: *Ann. Chim.* **2** (1947), 670.
139. Y. M. de Haan: *Nature* **200** (1963), 876.
140. Y. M. de Haan: private communication to the author H.R.O.
141. T. Moeller: *J. Chem. Educ.* **17** (1940), 5.
142. W. Feitknecht and H. Studer: *Kolloid Z.* **115** (1949), 13.
143. S. Prakash and A. D. Pandey: *Kolloid Z.* **181** (1962), 46.
144. B. Kandelaki and I. Setašvili: *Kolloid Zh.* **2** (1936), 807; cf. *CA* **30** (1036), 13[4].
145. I. I. Žukov and Z. D. Pigarera: *Kolloid Zh.* **6** (1940), 491; cf. *CA* **35** (1941), 6854[9].
146. W. Lotmar and W. Feitknecht: *Z. Krist.* **93** (1936), 368.
147. P. W. Selwood, M. Ellis, and C. F. Davis: *J. Am. Chem. Soc.* **72** (1950), 3549.
148. T. Katsurai: *Bull. Chem. Soc. Japan* **18** (1943), 277.
149. W. Feitknecht and W. Bédert: *Helv. Chim. Acta* **24** (1941), 694.

150. Ref. [125], 1961, p. 504.
151. J. Besson: *Ann. Chim.* **2** (1947), 527.
152. R. H. Griffith: *Trans. Faraday Soc.* **33** (1937), 405.
153. M. Figlarz and F. Vincent: *Comt. Rend. Acad. Sci., Ser. C* **266** (1968), 376.
154. C. W. F. T. Pistorius: *Z. Phys. Chem.* **34** (1962), 287.
155. T. M. Ovčinuikova, É. Š. Ioffe, and A. L. Rotinjan: *Dokl. Akad. Nauk S.S.S.R.* **100** (1955), 469; cf. *CA* **49** (1955), 12170f.
156. W. Feitknecht and L. Hartmann: *Chimia* **8** (1954), 95.
157. K. H. Gayer and A. B. Garrett: *J. Am. Chem. Soc.* **72** (1959), 3921.
158. L. M. Gedansky, G. L. Bertrand, and L. G. Hepler: *J. Chem. Eng. Data* **12** (1967), 135.
159. Ref. [125], 1961, p. 507.
160. Ju. D. Kondrašev and N. N. Fedorova: *Dokl. Akad. Nauk S.S.S.R.* **94** (1954), 229; cf. *CA* **49** (1955), 2146b.
161. *Dana's System of Mineralogy*, vol. 1, Wiley, London, 1958, p. 650; and unpublished results by the author H.R.O.
162. P. C. Milner and U. B. Thomas: 'The Nickel-Cadmium Cell', in: P. Delahay and C. W. Tobias (eds.), *Advances in Electrochemistry and Electrochemical Engineering*, vol. 5, New York, 1967, p. 1.
163. J. Labat: *J. Chim. Phys.* **60** (1963), 1253; *Ann. Chim.* **9** (1964), 399.
164. R. Konaka, S. Terabe, and K. Kurama: *J. Org. Chem.* **34** (1969), 1334; *J. Am. Chem. Soc.* **91** (1969), 5655.
165. B. T. Golding and D. R. Hall: *Chem. Commun.* (1970), 1574.
166. E. Gehrer, A. Portmann, and H. R. Oswald: unpublished, 1976.
167. O. Glemser, in G. Brauer (ed.): *Handbuch der präparativen anorganischen Chemie*, vol. II, 2nd ed., Enke, Stuttgart, 1962, p. 1347.
168. *GMELINS Handbuch der anorganischen Chemie*, 8th ed., Verlag Chemie, Weinheim, Syst. nr. 57, Nickel, vol. B2, 1966, p. 434.
169. W. J. Singley and J. T. Carriel: *J. Am. Chem. Soc.* **75** (1953), 778.
170. W. Feitknecht and A. Collet: *Helv. Chim. Acta* **22** (1939), 1428; *Helv. Chim. Acta* **23** (1940), 180.
171. C. Cabannes-Ott: *Ann. Chim.* **5** (1960), 905.
172. R. Scholder and E. Giesler: *Z. Anorg. Allg. Chem.* **316** (1962), 237.
173. L. M. Volchkova, L. G. Antonova, and A. I. Krasil'shchikov: *Zh. Fiz. Khim.* **23** (1948), 714.
174. O. Glemser and J. Einerhand: *Z. Anorg. Allg. Chem.* **261** (1950), 26.
175. A. Merlin and S. Teichner: *Comt. Rend. Acad. Sci.* **236** (1953), 1892.
176. S. Teichner and J. A. Morrison: *Trans. Faraday Soc.* **51** (1955), 961.
177. Nickel, Fr. 1,441,749, June 10, 1966; Appl. April 30, 1965; cf. *CA* **66** (1967), 25489s.
178. International Nickel Co. Inc., W. J. Kirkpatrik, U.S.P. 2602070 [1950/2].
179. Mond Nickel Comp., D.P. 903692 [1951/4].
180. Ref. [168], 1966, p. 444.
181. H. R. Oswald, unpublished results.
182. P. Franzen, J. J. B. van Eyk van Voorthuijsen: *Trans. 4th Int. Congr. Soil Sci. Amsterdam, 1950*, vol. 3, p. 34; cf. *CA* **46** (1952), 5475d.
183. L. A. Romo: *J. Phys. Chem.* **60** (1956), 1021.
184. M. A. Aia: *J. Electrochem. Soc.* **113** (1966), 1045.
185. W. Feitknecht, R. Signer, and A. Berger: *Kolloid Z.* **101** (1942), 12.
186. W. Feitknecht, H. Studer, and H. Meyer: *Kolloid Z.* **139** (1954), 131.
187. Ref. [168], 1966, p. 445.
188. S. Le Bihan, J. Guenot, and M. Figlarz: *Compt. Rend. Acad. Sci., Ser. C* **270** (1970), 2131.
189. S. Le Bihan and M. Figlarz: *J. Cryst. Growth* **13** (1972), 458.
190. G. Natta and L. Passerini: *Gazz. Chim. Ital.* **58** (1928), 597.
191. A. Wyttenbach: *Helv. Chim. Acta* **44** (1961), 418.
192. W. Feitknecht, A. Wyttenbach, and W. Buser: *Proc. Int. Symp. Reactivity Solids, Amsterdam, 1960*, Elsevier, Amsterdam, 1961, p. 234.
193. W. Feitknecht, W. Buser, and A. Wyttenbach: *Angew. Chem.* **72** (1960), 594.
194. J. T. Richardson and W. O. Milligan: *Phys. Rev.* **102** (1956), 1289.
195. K. H. Gayer and A. B. Garrett: *J. Am. Chem. Soc.* **71** (1949), 2973.
196. R. Paris: *Compt. Rend. Acad. Sci.* **232** (1951), 840; *Ann. Chim.* **10** (1955), 353.
197. W. Feitknecht, H. R. Christen, and H. Studer: *Z. Anorg. Allg. Chem.* **283** (1956), 88.

198. *GMELINS Handbuch der anorganischen Chemie*, 8th ed., Verlag Chemie, Weinheim, Syst. nr. 57, Nickel, vol. A II-2, 1968, p. 475.
199. G. W. D. Briggs, G. W. Scott, and W. F. K. Wynne-Jones: *Electrochim. Acta* **7** (1962), 249.
200. W. L. Latimer: *The Oxidation State of the Elements and their Potentials in Aqueous Solutions*, Prentice-Hall, New York, 1952, p. 200.
201. L. G. Sillén and A. E. Martell: *Stability Constants of Metal-Ion Complexes*, London, 1964, p. 56.
202. F. D. Rossini, D. D. Wagmann, W. H. Evans, S. Levine, and I. Jaffe: *Circ. Nat. Bur. Stand.* **500** (1952), 245.
203. K. Murata: *Bull. Chem. Soc. Japan* **3** (1928), 57, 267.
204. F. R. Bichowsky and F. D. Rossini: *The Thermochemistry of the Chemical Substances*, New York, 1936, p. 84, 302.
205. Ref. [168], 1966, p. 465.
206. H. Bode: *Angew. Chem.* **73** (1961), 553.
207. W. Feitknecht: *Helv. Chim. Acta* **27** (1944), 771.
208. *GMELINS Handbuch der anorganischen Chemie*, 8th ed., Verlag Chemie, Weinheim, Syst. nr. 60, Copper, vol. B1, 1958, p. 98.
209. R. Scholder: *Z. Anorg. Allg. Chem.* **220** (1934), 209.
210. W. Feitknecht and H. W. Lenel: *Helv. Chim. Acta* **27** (1944), 775.
211. H. R. Oswald and H. Jaggi: *Chimia* **14** (1960), 22.
212. R. Fricke and J. Kubach: *Z. Elektrochem.* **53** (1949), 76.
213. T. Labanukrom: *Kolloidchem. Beih.* **29** (1929), 80.
214. H. R. Oswald and P. Brunner: *Proc. 5th Int. Symp. Reactivity Solids, Munich 1964* (ed. by G. M. Schwab), Elsevier, Amsterdam, 1965, p. 382.
215. Kennecott Copper Corp. Fr. 1,441,900, June 10 (1966); W. H. Furness, U.S. 3,194,749, July 13 (1965); J. Baker (Cities Service Co.), U.S. 3,635,668, Jan. 18 (1972).
216. W. Feitknecht and K. Maget: *Z. Anorg. Allg. Chem.* **258** (1949), 150.
217. H. R. Oswald: unpublished results, 1965.
218. H. Jaggi and H. R. Oswald: *Acta Cryst.* **14** (1961), 1041.
219. U. W. Schönenberger, J. R. Günter, and H. R. Oswald: *J. Solid State Chem.* **3** (1971), 190.
220. J. R. Günter and H. R. Oswald: *J. Appl. Cryst.* **3** (1970), 21.
221. L. M. Gedansky, P. J. Pearce, and L. G. Hepler: *Can. J. Chem.* **48** (1970), 1770.
222. P. Korber: Ph.D.-thesis, Univ. of Zürich, Switzerland, 1974.
223. E. Dubler, P. Korber, and H. R. Oswald: *Acta Cryst.* **B29** (1973), 1929.
224. M. I. Baneeva and S. V. Popova: *Geokhimiya* **8** (1969), 1014; cf. *CA* **71** (1969), 85569f.
225. *GMELINS Handbuch der anorganischen Chemie*, 8th ed., Verlag Chemie, Weinheim, Syst. nr. 32, Zinc, suppl. vol., 1956, p. 819.
226. W. Feitknecht and E. Häberli: *Helv. Chim. Acta* **33** (1950), 922.
227. W. Feitknecht and R. Weidmann: *Helv. Chim. Acta* **26** (1943), 1911.
228. W. Feitknecht and R. Petermann: *Korros. Metallschutz* **19** (1943), 181.
229. W. Feitknecht: *Métaux Corros.* **22** (1947), 192.
230. W. Nowacki and J. N. Silverman: *Z. Krist.* **115** (1961), 21.
231. W. Feitknecht and W. Lotmar: *Helv. Chim. Acta* **18** (1935), 1369.
232. J. R. Günter and H. R. Oswald: *Helv. Chim. Acta* **51** (1968), 1775.
233. R. B. Corey and R. W. G. Wyckoff: *Z. Krist.* **85** (1933), 8.
234. W. Feitknecht: *Helv. Chim. Acta* **32** (1949), 2294.
235. Ref. [225], 1956, p. 821.
236. R. Giovanoli, H. R. Oswald, and W. Feitknecht: *J. Micr.* **4** (1965), 711.
237. R. Giovanoli, H. R. Oswald, and W. Feitknecht: *Helv. Chim. Acta* **49** (1966), 1971.
238. Ref. [225], 1956, p. 822.
239. N. A. Christensen: *Acta Chem. Scand.* **23** (1969), 2016.
240. R. Scholder and G. Hendrich: *Z. Anorg. Allg. Chem.* **241** (1939), 76.
241. H. G. Dietrich and J. Johnston: *J. Am. Chem. Soc.* **49** (1927), 1419.
242. Ref. [225], 1956, p. 823.
243. H. G. Schnering: *Z. Anorg. Allg. Chem.* **330** (1964), 170.
244. P. Schindler, H. Althaus, and W. Feitknecht: *Helv. Chim. Acta* **47** (1964), 982.
245. R. Scholder and E. Staufenbiel: *Z. Anorg. Allg. Chem.* **247** (1941), 259.
246. J. Piater: *Z. Anorg. Allg. Chem.* **174** (1928), 321.

247. A. S. Carson, H. Hartley, and H. A. Skinner: *Nature* **161** (1948), 725.
248. W. Feitknecht: *Experentia* **1** (1945), 230.
249. J. Bye: *Bull. Soc. Chim.* (1947), 198.
250. *GMELINS Handbuch der anorganischen Chemie*, 8th ed., Verlag Chemie, Weinheim, Syst. nr. 33, Cadmium, suppl. vol., 1959, p. 439.
251. D. Balarev et al.: *Z. Anal. Chem.* **120** (1940), 393.
252. T. Moeller and P. W. Rhymer: *J. Phys. Chem.* **46** (1942), 477.
253. A. de Schulten: *Compt. Rend. Acad. Sci.* **101** (1885), 73.
254. H. Remy: *Lehrbuch der anorganischen Chemie*, vol. II, 12/13th ed., Akad. Verl. Ges., Leipzig, 1973, p. 612.
255. F. V. Denisov, R. G. Lepilina, and Yu. P. Ballo: *Issled. v Obl. Khim. i Tekhnol. Mineral'n. Solei i Okislov, Akad. Nauk S.S.S.R. Sb. Statei* (1965), 272; cf. *CA* **65** (1966), 4975e.
256. G. F. Hüttig and R. Mytyzek: *Z. Anorg. Allg. Chem.* **190** (1930), 353.
257. E. S. J. Hedges: *J. Chem. Soc.* (1927), 1077.
258. B. A. Kosobrynkhov and N. P. Zhil'tsov: Russ. Pat. 44971, Nov. 30, 1935; cf. *CA* **32** (1938), 2851[2].
259. L. Capdecomme and Y. Schwob: *Métaux Corros. Usure* **18** (1943), 173.
260. K. Huber: *Compt. rend. réunion comité thermodynam. et cinét. électrochim.* (1951), 117 (publ. 1952).
261. L. Capdecomme: *Comm. Techn. Etats Propr. Surf. Métaux Paris* (1945), 247.
262. M. Fleischmann, K. S. Rajagopalan, and H. R. Thirsk: *Trans. Faraday Soc.* **59** (1963), 741.
263. W. Dans and O. F. Tower: *J. Phys. Chem.* **33** (1929), 605.
264. O. Glemser, U. Hauschild, and H. Richert: *Z. Anorg. Allg. Chem.* **290** (1957), 58.
265. W. Feitknecht: *Chimia* **11** (1957), 166.
266. P. M. de Wolff: *Acta Cryst.* **2** (1966), 432.
267. G. Bertrand and Y. Dusausoy: *Compt. Rend. Acad. Sci.* **270** (1970), 612.
268. Z. Zabransky: *Coll. Czech. Chem. Comm.* **24** (1959), 2426.
269. Y. Marcus: *Acta Chem. Scand.* **11** (1957), 690.
270. P. Schindler: *Helv. Chim. Acta* **42** (1959), 2736.
271. W. Feitknecht and R. Reinmann: *Helv. Chim. Acta* **34** (1951), 2255.
272. R. Reinmann: Ph.D. thesis, Univ. of Bern, Switzerland, 1948.
273. W. Feitknecht and R. Reinmann: *Compt. Rend. CITCE* **93** (1951).
274. C. G. Maier: *J. Am. Chem. Soc.* **51** (1929), 194.
275. G. Becker and W. A. Roth: *Z. Phys. Chem.* **A 167** (1933), 1.
276. F. Ishikawa and E. Shibata: *Sci. Rep. Tohoku* **I 21** (1972), 499.
277. R. Fricke and F. Blaschke: *Z. Elektrochem.* **46** (1940), 46.
278. J. R. Günter and H. R. Oswald: *Bull. Inst. Chem. Res. Kyoto Univ.* **53** (1975), 249.

TRANSITION METAL DICHALCOGENIDES

R. M. A. LIETH and J. C. J. M. TERHELL

Chemical Physics Division, Solid State Group, Dept. of Physics,
Technische Hogeschool Eindhoven, The Netherlands

Table of Contents
- 1.1. Introduction
- 1.2. Description of TCh_2 Structures
- 1.3. Polymorphism and Polytypism
- 1.4. Bond Character
- 2. Group IV Compounds
  - 2.1. TiS_2
    - 2.1.1. Compound Preparation
    - 2.1.2. Crystal Growth
  - 2.2. $TiSe_2$
    - 2.2.1. Compound Preparation
    - 2.2.2. Crystal Growth
  - 2.3. $TiTe_2$
    - 2.3.1. Compound Preparation
    - 2.3.2. Crystal Growth
  - 2.4. ZrS_2
    - 2.4.1. Compound Preparation
    - 2.4.2. Crystal Growth
  - 2.5. $ZrSe_2$
    - 2.5.1. Compound Preparation
    - 2.5.2. Crystal Growth
  - 2.6. $ZrTe_2$
    - 2.6.1. Compound Preparation
    - 2.6.2. Crystal Growth
  - 2.7. HfS_2
    - 2.7.1. Compound Preparation
    - 2.7.2. Crystal Growth
  - 2.8. $HfSe_2$
    - 2.8.1. Compound Preparation
    - 2.8.2. Crystal Growth
  - 2.9. $HfTe_2$
    - 2.9.1. Compound Preparation
    - 2.9.2. Crystal Growth
  - 2.10. Solid Solutions
    - 2.10.1. Crystal Growth
- 3. Group V Compounds
  - 3.1. VCh_2
    - 3.1.1. Compound Preparation
    - 3.2.2. Crystal Growth
  - 3.2. NbS_2
    - 3.2.1. Compound Preparation
    - 3.2.2. Crystal Growth
  - 3.3. $NbSe_2$
    - 3.3.1. Compound Preparation
    - 3.3.2. Crystal Growth
  - 3.4. $NbTe_2$
    - 3.4.1. Compound Preparation

   3.4.2. Crystal Growth
 3.5. TaS$_2$
   3.5.1. Compound Preparation
   3.5.2. Crystal Growth
 3.6. TaSe$_2$
   3.6.1. Compound Preparation
   3.6.2. Crystal Growth
 3.7. TaTe$_2$
   3.7.1. Compound Preparation
   3.7.2. Crystal Growth
 3.8. Solid Solutions
   3.8.1. Compound Preparation and Crystal Growth
4. Group VI Compounds
 4.1. MoS$_2$
   4.1.1. Compound Preparation
   4.1.2. Crystal Growth
 4.2. MoSe$_2$
   4.2.1. Compound Preparation
   4.2.2. Crystal Growth
 4.3. MoTe$_2$
   4.3.1. Compound Preparation
   4.3.2. Crystal Growth
 4.4. WS$_2$
   4.4.1. Compound Preparation
   4.4.2. Crystal Growth
 4.5. WSe$_2$
   4.5.1. Compound Preparation
   4.5.2. Crystal Growth
 4.6. WTe$_2$
   4.6.1. Compound Preparation
   4.6.2. Crystal Growth
 4.7. Solid Solutions
   4.7.1. Compound Preparation
5. Group VII Compounds
 5.1. TcCh$_2$
   5.1.1. Compound Preparation and Crystal Growth
 5.2. ReCh$_2$
   5.2.1. Compound Preparation
   5.2.2. Crystal Growth
6. Group VIII Compounds
 6.1. CoTe$_2$, RhTe$_2$, IrTe$_2$, NiTe$_2$, PdTe$_2$, and PtCh$_2$
   6.1.1. Preparation Procedures
References

1.1. INTRODUCTION

Since the early 60's a group of materials called the transition metal dichalcogenides have received a rapidly growing interest. Of the sixty compounds of this family, about two-thirds assume layer structures. These TCh$_2$ compounds (Ch=S, Se or Te) form a structurally and chemically well-defined family. The basic structure of loosely coupled Ch-T-Ch sheets makes such materials extremely interesting. Within a layer, the bonds are strong, while between adjacent layers they are remarkably weak. As a consequence the crystals have facile basic cleavage, lubricity and marked anisotropy in many physical properties which accounts for the great interest in this family of materials. Electrically, however, they cover a wide spectrum of properties. There are insulators like HfS_2, semiconductors like MoS_2, semi-metals like WTe_2 and TcS_2 and true metals like NbS_2 and VSe_2. All the $NbCh_2$ and $TaCh_2$ are superconductors and the true metallic ones also show band antiferromagnetism below about 150 K [1].

The diversity of properties arises through the existence of non-bonding d-bands – which are quite wide (~ 1 eV) in these compounds – and the degree to which they are filled. Some of the compounds like $TaTe_2$, TcS_2 and $ReSe_2$ show structural distortion which seems to make their properties different from the other members. The anisotropy in chemical behaviour has also attracted attention; particularly important have been the discoveries that many atoms and molecules can be inserted between the adjacent layers, thus forming intercalation compounds.

Stoichiometry and crystalline order are of importance; the occurrence of superconductivity in the Nb and Ta compounds for example was found to be strongly dependent on stoichiometry.

The preparation of stoichiometric compounds is usually possible, although there are significant difficulties in certain cases.

In some systems the stoichiometric dichalcogenide represents the boundary of a continuum of materials best written as $T_{1+x} Ch_2$ where x varies continuously from zero upward. This is especially severe in the group IV dichalcogenides and in the VCh_2. In the Nb and Ta disulfides and diselenides metal-rich regimes have been found and care should be taken especially with NbS_2 [2]. In general the group V compounds under consideration seem to be difficult to prepare as pure single phases.

The group VI compounds seem to possess only limited ranges of homogeneity; they can usually be prepared stoichiometrically and as single phase compounds. The difficulty, however, lies in the preparation of large single crystals, a problem that also occurs in connection with the group VII compounds. Of the group VIII compounds with layered structures, little synthetic work is known and so far no crystal growth experiments have been reported.

In certain cases no compounds higher than the dichalcogenide can be synthesized from the elements. For these systems an excess of the chalcogen may be used. Examples are $TiSe_2$, $TiTe_2$, VSe_2, VTe_2. In the other systems trichalcogenides can be found during preparation of the substances and in such cases special procedures have to be followed to obtain the TCh_2 when working with an excess of chalcogen.

For the preparation of single crystals, the discovery of the chemical vapour transport technique has been of great influence. It enabled investigators to grow single crystals of substances hitherto known as 'difficult materials'. This technique is discussed in detail by Schäfer [3].

This development in synthetic work was also largely responsible for the aforementioned exponential growth in interest and one can say that it opened the door to careful examination of the extremely fascinating properties.

Future progress in this large field will now largely be determined by the effort put into the materials problem.

In the following sections the various TCh_2 will be discussed according to the group

– in the periodic table – of the transition metal. Compound preparation and crystal growth will be approached mainly from the practical side.

In Section 2 only an introductory description of the crystal structure will be presented, since this topic will be treated in full detail in volume V where attention will be focussed on the crystallography and crystal chemistry of these materials.

In subsequent sections, first the group IV compounds will be treated. Data on solid solutions formed among the various compounds will be presented at the end of that section.

Thereafter attention will be focussed on the group V materials and again existing data on solid solutions are given at the end of that section. Then the group VI compounds and the group VII will be discussed. Finally the layered members of group VIII will briefly be mentioned.

Where possible thermochemical data will be presented in the sections. As much data concerning one group as possible will be tabulated to enable the reader to compare the various group members.

1.2. Description of TCh_2 structures

The TCh_2 structures fall into two classes: layered and non-layered.

The non-layered members are found in groups VII and VIII; they are $MnCh_2$, $FeCh_2$, $RuCh_2$, $OsCh_2$, $CoCh_2$, $RhCh_2$, IrS_2, $IrSe_2$, NiS_2 and $NiSe_2$. The ditellurides of Co and Rh can also adopt a CdI_2-type of structure, the others occur in one or more of the following structure types: the pyrite, the marcasite, the $IrSe_2$ and the PdS_2-type. A discussion of the differences between these four types and the manner in which the marcasite type is related to the CdI_2-type is presented in [1].

In the layered compounds under consideration, two types of structures are dominant although a mixture of the two is sometimes found. In the CdI_2-type (and the $CdCl_2$-type), a 6:3 coordination exists in which the metal atom is surrounded octahedrally by the chalcogens. Another 6:3 coordination, in which the metal atom is surrounded in a trigonal prismatic fashion is found in the MoS_2-type of structure. See Figure 1. The 'mixtures' contain alternatingly prismatic and octahedral coordinated layers.

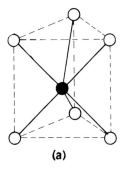

Fig. 1a. Trigonal prism, stacking sequence AbA.

According to Wijckhoff [14] there is a distinction between the CdI_2 and the $CdCl_2$ structure; the iodine structure has an almost perfect hexagonal close packing of the anions, with the metal in octahedral interstices while in the chloride structure – where the octahedral surrounding is retained – the close packing of the anions is cubic rather than hexagonal. For a detailed treatment of crystal characteristics by Balchin, the reader is referred to volume II.

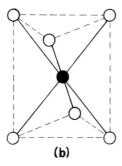

Fig. 1b. Octahedron, stacking sequence AbC.

Whereas in the layered compounds the bonding in one slab is strong, the interlayer forces are weaker and this leads to the marked cleavage perpendicular to the hexagonal/trigonal symmetry axis, making such materials very anisotropic both mechanically and electrically and enabling one to introduce metal atoms or molecules of organic complexes into the van der Waals gap between the slabs.

In several compounds the layers are distorted; the metal atoms are actually displaced from the centre of the coordination units. As a result the chalcogenide sheets buckle somewhat to accommodate these displacements and chains of metal atoms with short metal-metal distances are found in $NbTe_2$, $TaTe_2$, WTe_2, β-$MoTe_2$, TcS_2, $TcSe_2$ and in the $ReCh_2$ [1].

Table I contains the lattice parameters and the relevant structure data.

TABLE I
Structure type and Lattice parameters

Compound, character, and colour	Group	Thermodynamic data ‡ [137]	Coordination of the metal atom and Space group	Parameter (Å) a	Parameter (Å) c	Refs.	Density (g/cm³)	z-value formula units/cell
TiS_2 Metallic character. golden metallic lustre. dark bronze. greenish-gold. brassy. [14] [51] [28] [33] [34]	IV, d^2	$\Delta H°_{298} = -97.3$ $S°_{298} = 18.7$	Octahedral $P\bar{3}m1$	3.397 3.4073 3.39 3.404 3.4095 3.405	5.691 5.6953 5.70 5.699 5.6898 5.687	[19] [21,30] [28] [14] [33] [34]	3.31 [19] 3.242 [21] 3.24 [28]	1
$TiSe_2$ Metallic character. purplish-brown. metallic-bronze. metallic. [14] [28] [33] [34]	IV, d^2	$\Delta H°_{298} = -92$ $S°_{298} = 24$	Octahedral $P\bar{3}m1$	3.533 3.535 3.5345 3.537 3.542	5.995 6.004 6.0039 6.00 6.015	[19] [28] [33] [34] [14]	5.26 [28]	1
$TiTe_2$ Metallic charater. grey-metallic. black. metallic. [14] [33] [28] [34]	IV, d^2	$\Delta H°_{298} = -50.9$ $S°_{298} = 28$	Octahedral $P\bar{3}m1$	3.774 3.76 3.691 3.773 3.768 [a]	6.539 6.48 6.529 6.516 6.500 [a]	[19] [28] [33] [34] [14]	6.24 [19] 6.28 [28]	1
ZrS_2 Ionic character anticipated. violet-brown. deep red, tending to metallic black. violet metallic. [14] [28] [33] [34]	IV, d^2	$\Delta H°_{298} = -138$ $S°_{298} = 18.7$	Octahedral $P\bar{3}m1$	3.662 3.661 5.6673 3.66 3.667	5.813 5.829 5.8178 5.85 5.817	[28] [14] [33] [34] [50]	3.78 [28] 3.72 3.81 (m) } [50]	1

TRANSITION METAL DICHALCOGENIDES 147

Table I (Continued)

Compound, character and colour	Group	Thermodynamic data ‡ [137]	Coordination of the metal atom and Space group	Parameters (Å) a	b	Refs.	Density (g/cm^3)	z-value formula units/cell
$ZrSe_2$ Ionic character anticipated. [14] dark purplish-brown. [28] metallic bronze, with green lustre. [33] metallic. [34] dark bronze-green. [41] black. [42] dark-violet. [43]	IV, d^2	$\Delta H°_{298} = -102$ $S°_{298} = 22.6$	Octahedral P$\bar{3}$m1	3.773 ^{b} 3.770 3.7742 3.76 3.771 3.774	6.133 ^{b} 6.137 6.1314 6.15 6.148 6.131	[14] [28] [33] [34] [43] [50]	5.36 5.47 (m) $\}$ [50]	1
$ZrTe_2$ Metallic character. [14] purplish brown. [28]	IV, d^2	$\Delta H°_{298} = -72$ $S°_{298} = 26.0$	Octahedral P$\bar{3}$m1	3.951 ^{c} 3.952 3.950 3.9507 ^{d}	6.627 ^{c} 6.660 6.630 6.6238 ^{d}	[14] [28] [44] [45]	6.07 [24] 6.38 [28] 6.21 [45]	1
HfS_2 Ionic character. [14] red. [48] light-purplish brown. [28] dark-red metallic. [34] transparant reddish-orange. [33]	IV, d^2	$\Delta H°_{298} = -140$ $S°_{298} = 20$	Octahedral P$\bar{3}$m1	3.635 3.6233 3.62 3.630	5.837 5.8418 5.88 5.854	[28] [33] [34] [14]	6.03 [28]	1
$HfSe_2$ Ionic character. [14] dard-red. [48] very dark-brown. [28] metallic bronze. [33] dark-red metallic. [34]	IV, d^2	$\Delta H°_{298} = ?$ $S°_{298} = ?$	Octahedral P$\bar{3}$m1	3.748 3.7413 3.733 3.747	6.159 6.1439 6.146 6.158	[28] [33] [34] [14]		1
$HfTe_2$ Metallic character. [14] golden. [47]	IV, d^2	$\Delta H°_{298} = ?$ $S°_{298} = ?$	Octahedral P$\bar{3}$m1	3.951 3.949 ^{d} 3.95 ^{e} 3.9492 ^{f}	6.659 6.651 ^{d} 6.65 ^{e} 6.6514 ^{f}	[14] [18] [46] [47]		

Table I (Continued)

Compound, character and colour	Group	Thermodynamic data ‡ [137]	Coordination of the metal atom and Space group	Parameters (Å) a	Parameters (Å) c	Refs.	Density (g/cm³)	z-value formula units/cell
VS_2 character unknown dark [48]	V, d^3	$\Delta H°_{298} = ?$ $S°_{298} = ?$	?	3.29 ^{d}	5.66 ^{d}	[18]		
VSe_2 character unknown dark [48]	V, d^3	$\Delta H°_{298} = ?$ $S°_{298} = 23$	Octahedral	3.36 3.352 ^{d}	6.05 6.104 ^{d}	[48] [18]		1
VTe_2 charakter unknown dark grey [55]	V, d^3	$\Delta H°_{298} = ?$ $S°_{298} = ?$	Distort. Octah.	3.6	6.45	[18]		
NbS_2 Metallic character grey-black } [2,29]	V, d^3	$\Delta H°_{298} = -84,8$ $S°_{298} = 17$	Trigon. Prism. $P6_3/mmc - D_{6h}^4$ Trigon. Prism. $R\bar{3}m - C_{3v}^5$	2H: 3.31 3R: 3.33	11.89 17.81	[60,63] [60,63]		2 3
$Nb_{1+x}S_2$ metallic lustre for 2H black for 3R	g [2] [2]		Trigon. Prism. $P6_3/mmc$ Trigon. Prism. $R\bar{3}m - C_{3v}^5$	2H: 3.309 3R: 3.329	12.71 17.89	[2,63] [2,63]		
$NbSe_2$ Metallic character metallic-black [65, 74] [2]	V, d^3	$\Delta H°_{298} = ?$ $S°_{298} = 23$	Trigon. Prism. $P\bar{6} - D_{6h}^4$ Trigon. Prism. $P6_3/mmc - D_{6h}^4$ Trigon. Prism. $P6_3/mmc - D_{6h}^4$ Trigon. Prism. $R\bar{3}m$ Trigon. Prism. $P\bar{6}m2$	2H: 3.449 4H: 3.439 2H: 3.45 3R: 3.45 4H: 3.44	12.998 25.188 12.54 18.88 25.24	[65] [65] [11] [11] [11]	6.22 [65] 6.46 [65]	2 4 2 3 4

Table I (Continued)

Compound, character and colour	Group	Thermodynamic data ‡ [137]	Coordination of the metal atom and Space group	Parameters (Å) a	c	Refs.	Density (g/cm³)	z-value formula units/cell
NbSe₂ (Continued)			Trigon. Prism.	2H: 3.442	12.54	[72]		2
			Trigon. Prism.	3R: 3.459	18.77	[72]		3
			Trigon. Prism. $P\bar{6}m2 - D_{3h}^1$	4H: 3.444	25.23	[72]		4
			Trigon. Prism. $P6_3/mmc - D_{6h}^4$	2H: 3.446	12.55	[74]		2
			Trigon. Prism.	2H: 3.442	12.54	[12]		2
			Trigon. Prism.	4Ha: 3.444	25.23	[12]		4
			Tr. Pris. + Octah.	4Hd: 3.48	25.45	[12]		4
			Octahedral	1T: 3.53	6.29	[12]		1
Nb₁₊ₓSe₂ $0.0 \leq x < 0.29$			Trigon. Prism. $P6_3/mmc$	2H: 3.446 to 3.450	12.55 to 13.02	[66]	6.40–6.75 [66]	
				2Hb: 3.451 to 3.457	12.57 to 12.53	[70]		
				2Ha: 3.450 to 3.448	12.93 to 12.99	[70]		
				3R: 3.457 to 3.450	18.81 to 18.75	[70]		
NbTe₂ Metallic character grey black	V, d³ [65] [65] [64]	$\Delta H°_{298} = ?$ $S°_{298} = 26.5$	Distort. Octahdr. C2/m $D_{3d}^5 - R\bar{3}m$	a = 19.39 b = 3.64 10.904	c = 9.375 $\beta = 134° 35'$ 19.888 ᵏ	} ᵏ [67, 75] [65]	6.8 [75] 7.37 ʰ 7.62 [65]	27

Table I (Continued)

Compound, character and colour	Group	Thermodynamic data ‡ [137]	Coordination of the metal atom and Space group	Parameters (Å)		Refs.	Density (g/cm³)	z-value formula units/cell
				a	c			
TaS$_2$	V, d^3							
Metallic character for 2H yellow-1T form	[84]	$\Delta H°_{298} = -84.6$	Octahedral P$\bar{3}$m	1T: 3.36	5.90	[77]		1
metallic grey-2H form		$S°_{298} = 18$	Trigon. Prism. P6$_3$/mmc	2H: 3.315	12.10	[77]		2
			Trigon. Prism. R3m	3R: 3.32	17.9	[77]		3
			Trigon. Prism. +Octahedral	6R: 3.335	35.85	[77]		6
			Octahedral P$\bar{3}$m	1T: 3.365	5.853	[84]^k		1
			Trigon. Prism. P6$_3$/mmc	2H: 3.316	12.070	[84]		2
			Trigon. Prism. +Octahedral	4Hb: 3.332	23.62	[78]^k		4
			Superstructure P$\bar{3}$—C$_3^1$ or P$\bar{3}$—C$_{3i}$	1T: 3.365	5.897	[79]^k		1
Ta$_{1+x}$S$_2$			Trigon. Prism.	2H: 3.295 ←sulfur-rich→ 12.45 to 3.29 ←metal-rich→ 12.65		[77]		
			Trigon. Prism.	3R: 3.31	18.2	[77]		
			Trigon. Prism. +Octahedral.	6R: 3.315	36.2	[77]		

Table 1 (Continued)

Compound, character and colour	Group	Thermodynamic data ‡ [137]	Coordination of the metal atom and Space group	Parameters (Å) a	c	Refs.	Density (g/cm³)	z-value formula units/cell
TaSe₂ Metallic character [65]	V, d^3	$\Delta H°_{298}=?$ $S°_{298}=24$	Octahedral. $P\bar{3}m1$	1T: 3.4769	6.2722	[86]	8.569ᵐ [86]	1
			Trigon. Prism. $P6_3/mmc$	2H: 3.436	12.696ᵏ	[86]	8.669ᵐ [86]	2
			Trigon. Prism. $R\bar{3}m$	3R: 3.4348	19.177	[86]	8.164ᵐ [86]	3
			Trigon. Prism.	4Ha: 3.4362	25.399	[86]	8.665ᵐ [86]	4
			Trigon. Prism. +Octahedral. $P6_3/mmc$	4Hb: 3.4575	25.143	[86]		4
			Trigon. Prism. +Octahedral.	6R: 3.4558	37.826	[86]		6
			Trigon. Prism. $P6_3/mmc-D_{6h}^4$	2H: 3.431	12.737ᵏ	[65]	8.66ᵐ [65]	2
			Trigon. Prism. $R\bar{3}m-D_{3d}^5$	3R: 3.428	19.100	[65]	8.6ᵐ [65]	3
			Trigon. Prism. $P6_3/mmc$	2H: 3.437	12.72	[72]		2
			Trigon. Prism.	3R: 3.437	19.21	[72]		3
			Trigon. Prism.	4Ha: 3.43	25.5	[9, 72]		4
			Trigon. Prism. +Octahedral.	6R: 3.455	37.77	[72]		6
			Octahedral. $P\bar{3}m1$	1T: 3.44	6.27	[88]		1

Tabel I (Continued)

Compound, character and colour	Group	Thermodynamic data ‡ [137]	Coordination of the metal atom and Space group	Parameters (Å) a	c	Refs.	Density (g/cm³)	z-value formula units/cell
TaSe₂ (Continued)	V, d³		Trigon. Prism. P6₃/mmc	2H: 3.43	12.71 [k]	[11, 67]		2
			Trigon. Prism. +Octahedral.					
			P6₃/mmc	4Hb: 3.46	25.18 [k]	[11, 67]		4
				4Hb: 3.455	25.148	[9]		4
			Trigon. Prism.	4Hc: 3.436	25.532	[9]		4
Ta₁₊ₓSe₂			Trigon. Prism.	2Ha: 3.434 to 3.432	12.87 to 12.93			
			Trigon. Prism.	2Hb: 3.440 to 3.443	12.64 to 12.59	[70]		
			Trigon. Prism.	3R: 3.437 to 3.442	19.09 to 18.91 see also [86]			
TaTe₂	V, d³	ΔH°₂₉₈ = ?	Distort. Octahed.					
Metallic character [65]		S°₂₉₈ = 27.5	C2/m	a = 19.31 b = 3.651	c = 9.377 β = 134° 13'	[67, 75] [k]	8.4; 9.16 [h] [75]	
			D₃d⁵ — R3̄m	10.904	20.075 [k]	[65]	9.4 [65]	27

Table I (Continued)

Compound, character and colour	Group	Thermodynamic data ‡ [137]	Coordination of the metal atom and Space group	Lattice parameters (Å) a	Lattice parameters (Å) c	Refs.	Density (g/cm³)	z-value formula units/cell
MoS₂ covalent character grey-black [14] [48]	VI, d^4	$\Delta H°_{298} = -65.8$ $S°_{298} = 14.96$	Trigonal Prism. P6₃/mmc R3m—C_{3v}^5	2H: 3.160 3R: 3.163 2H: 3.16 3R: 3.17 3R: 3.16 2H: 3.162 2H: 3.14 3R: 3.165	12.294 18.37 12.29 18.38 18.37 12.29 12.327 k 18.371 k	[91, 92] [91, 92] [60] [60] [93] [14] [103] [103]	4.96 [93] 4.92 [94] 5.024 [93]	2 3 2 3 3 2 2 3
MoSe₂ covalent character grey-black [14]	VI, d^4	$\Delta H°_{298} = -47$ $S°_{298} = 21$	Trigonal Prism. P6₃/mmc R3m P6₃/mmc—D_{6h}^4	2H: 3.288 3R: 3.296 2H: 3.288 2H: 3.288 3R: 3.292	12.903 19.392 12.900 12.92 19.392	[91, 136] [91, 105] [65] [103] [103]	6.98 ᵐ [65]	2 3 2 2 3
MoTe₂ semi-conducting character at low temp. and metallic character at high temp. grey-black [110]	VI, d^4	$\Delta H°_{298} = -32$ $S°_{298} = 24.5$	Trigonal Prism. P6₃/mmc—D_{6h}^4 P2₁/m monoclinic	α-type =2H } 3.517 β-type a = 6.33 b = 3.469	13.949 c = 13.86 β = 93° 55'	[65] [108]	7.80 ᵐ [65] 7.5 ᵐ 7.67 ₕ [108]	2 4

Table I (Continued)

Compound, character and colour	Group	Thermodynamic data ‡ [137]	Coordination of the metal atom and Space group	Parameters (Å) a	c	Refs.	Density (g/cm³)	z-value formula units/cell
MoTe₂ (Continued)	VI, d^4		P6₃/mmc	2H or α-type } 3.519	13.964	[107]		2
				α-type: 3.517 β-type:	13.949	[109]		4
				a = 6.33 b = 3.469	c = 13.86 β = 93° 55′			
				α-type: 3.522	13.967	[103]		2
WS₂ grey-black	VI, d^4	$\Delta H°_{298} = -62$ $S°_{298} = 15.5$	Trigonal Prism. P6₃/mmc R3̄m	2H: 3.155 3R: 3.162	12.35 18.35	[91, 92] [91, 92]		2 3
WSe₂ grey-black	VI, d^4	$\Delta H°_{298} = -45$ $S°_{298} = 21.5$	Trigonal Prism. P6₃/mmc	2H: 3.286 2H: 3.280 2H: 3.282 3.2860	12.976 12.950 12.937 12.979	[117, 119] [65] [103] [115]	9.35 ᵐ [117, 119] 9.40 ᵐ [65]	2 2 2
WTe₂	VI, d^4	$\Delta H°_{298} = -32$ $S°_{298} = 25$	Distorted Octahed. Pnm2₁ Pnmn–D_{2h}^{13}	a = 6.282 b = 3.496 a = 14.028 b = 3.495	c = 14.07 c = 6.270	[108, 109] [65]	9.51 ᵐ [108] 9.44 ʰ 9.49 ᵐ [65]	4

Table I (Continued)

Compound, character and colour	Group	Thermodynamic data ‡ [137]	Coordination of the metal atom and Space group	Parameters (Å) a	Parameters (Å) c	Refs.	Density (g/cm^3)	z-value formula units/cell
TcS$_2$	VII, d^5	$\Delta H°_{298} = -53.5$ $S°_{298} = 17$	Distorted Octahedr. triclinic sym.	a = 6.465 b = 6.375	c = 6.659 $\alpha = 103.61°$ $\beta = 62.97°$ $\gamma = 118.96°$	[91, 121]	5.006* [121]	
TcSe$_2$	VII, d^5	$\Delta H°_{298} = ?$ $S°_{298} = ?$	Also triclinic sym. no other data available.			[91, 121]		
TcTe$_2$	VII, d^5	$\Delta H°_{298} = ?$ $S°_{298} = ?$	Distorted. monoclinic sym. Cc or C2/c	a = 12.522 b = 7.023	c = 13.828 $\alpha = 90°$ $\beta = 101.26°$ $\gamma = 90°$	[91, 121]	7.890 * [121]	
ReS$_2$	VII, d^5	$\Delta H°_{298} = -42.7$ $S°_{298} = 14.5$	Distorted CdCl$_2$ type. triclinic sym. P$\bar{1}$	a = 6.455 b = 6.362	c = 6.401 $\alpha = 105.04°$ $\beta = 91.60°$ $\gamma = 118.97°$	[91, 121]	7.506 [121, 124] 7.42 [121, 127]	
ReSe$_2$	VII, d^5	$\Delta H°_{298} = -34$ $S°_{298} = 20.5$	Distorted CdCl$_2$ type.	a = 6.716 b = 6.602	c = 6.728 $\alpha = 104.90°$ $\beta = 91.82°$ $\gamma = 118.94°$	[91, 121]		
			Distorted Cd(OH)$_2$ type. P$\bar{1}$	a = 6.7272 b = 6.6065	c = 6.7196 $\alpha = 118.93°$ $\beta = 91.82°$ $\gamma = 104.93°$	[122]	9.237 [122]	

Table 1 (Continued)

Compound, character and colour	Group	Thermodynamic data ‡ [137]	Coordination of the metal atom and Space group	Parameters (Å) a	Parameters (Å) c	Refs.	Density (g/cm³)	z-value formula units/cell
CoTe$_{1.8}$ silver white [128]	VIII, d^7	$\Delta H°_{298} = ?$ $S°_{298} = ?$	Octahedral	3.8017	5.4094	[128]	7.739 [128]	1
RhTe$_2$	VIII, d^7	$\Delta H°_{298} = -19$ $S°_{298} = 29.5$	Octahedral $P\bar{3}m - D_{3d}^3$	3.92	5.41	[129]	8.26 ᵐ [129]	1
IrTe$_2$	VIII, d^7	$\Delta H°_{298} = -17$ $S°_{298} = 29.5$	Octahedral $P\bar{3}m1 - D_{3d}^3$	3.930	5.393	[132]	10.30 ᵐ [129]	1
NiTe$_2$	VIII, d^8	$\Delta H°_{298} = -21$ $S°_{298} = 28.76$	Octahedral $P\bar{3}m1$	3.8542	5.2604			1
PdTe$_2$	VIII, d^8	$\Delta H°_{298} = -13$ $S°_{298} = 30.25$	,,	4.0365	5.1262			1
PtS$_2$	VIII, d^8	$\Delta H°_{298} = -26.4$ $S°_{298} = 17.85$	,,	3.5432	5.0388	[133, 134]	7.83 [134]	
PtSe$_2$	VIII, d^8	$\Delta H°_{298} = -19$ $S°_{298} = 24.5$	,,	3.7278	5.0813		9.53 [134]	1
PtTe$_2$	VIII, d^8	$\Delta H°_{298} = -14$ $S°_{298} = 28.92$	,,	4.0259	5.2209		10.16 [134]	1

a = data for TiTe$_{1.17}$; b = data for ZrSe$_{1.98}$; c = data for ZrTe$_{1.76}$; d = metal-rich; e = data for Hf$_{1+x}$Te$_2$ with $0.0 < x < 0.25$; f = data for HfTe$_{1.939}$; g = x is variable as function of preparation temperature; h = theoretical value; k = data are specified for single crystals, all other data are powder data or not specified; m = X-ray density value; ‡ = $\Delta H°_{298}$ in kcal. mol⁻¹; $S°_{298}$ in cal. mol⁻¹. K⁻¹; * = based on ⁹⁹Tc.

1.3. POLYMORPHISM AND POLYTYPISM

Where so many variations in stacking order of the sandwiches exist, it is not surprising that polytypism is found in many of the compounds. It is not our intention to treat the phenomenon of polymorphism or polytypism in detail in this volume or in this section, as it will receive full attention in Volume II. At this stage the various modifications of one compound – found for some of the TCh_2 – will be presented for the sake of information and we will use the word polytype. In literature polymorphism and polytypism are used freely, it is therefore perhaps useful to draw attention to the definition as presented by Verma et al. which shows the distinction between the two.

Polymorphism is defined by Verma and Krishna [5] as the ability of the same chemical compound to exist in more than one crystalline form and polytypism is a special kind of one-dimensional polymorphism.

In his discussion of this topic, Balchin [6] emphasizes the fact that polymorphs usually exhibit a stability governed by the phase rule and the criterion of minimum Gibbs free energy, having at normal pressure a distinct temperature range over which they exist. It must be remarked that Balchin also uses the name polymorph for one-dimensional modifications.

Polytypism is mainly observed in the groups V and VI and the information regarding their structure type and lattice parameters is presented in Table I. In Table II the various modifications are assembled in three groups, i.e. those having a prismatic surrounding of the metal by the chalcogen atoms, those having an octahedral surrounding, and those having alternating layers with prismatic and octahedral surrounding.

In all cases the thickness of the repeat unit in the c-direction is indicated by the number of slabs 1, 2, 3, 4, 6, and if necessary higher. The symmetry is denoted by T (trigonal), H (hexagonal), or R (rhombohedral). If – like for instance in the 2H-modification (see Table II) – these two are not sufficient to identify the polytype uniquely, it is common in literature to add a third lower case letter a, b, c,... etc. to a newly discovered stacking sequence. This can be seen in the cases of 2H-types and 4H-types.

The use of a, b, c and so on seems to be entirely dependent on the order of discovery of that modification. The stacking sequence denoted 4H(a) observed for a certain compound means therefore that this one has been detected before the one denoted 4H(b) etc. The third lower case letter does not indicate a certain compound but a certain stacking order (see Table II)*. In group IV it seems that polytypism has been observed in the Ti—S system. From the study of Tronc and Hubert [7] at 800° and 900 °C it appears that there are four structures that have been worked out, i.e. 8H, 10H, 12H and 48R, and that four others have been identified, i.e. 12 Ha, 24 Hb, 40 H and 696 R and lately a 24 R [8]. According to their report, however, the modi-

---

* In the opinion of the authors the use of further different notations – I, II like in $4H(d_I)$, $4H(d_{II})$ – to identify a new discovery, will lead to more confusion when in the future new stackings are found.

fications mentioned are not of TiS_2 but of a composition near $TiS_{1.70}$ and part of the excess metal lies between the layers. So far no other polytypes of group IV compounds have been reported, neither is this the case for groups VII and VIII. A survey of the polytypes found in groups V and VI is presented in Table II, where Figures 2.1 to 2.15 are $(11\bar{2}0)$ sections of the various polytypes. The following division can be seen.

Layers with a prismatic surrounding of the metal atom

A one slab structure is not known and not likely to exist, due to the very unstable lattice that will be formed. For a two layer structure there are three stacking sequences possible. These are presented in Figures 2.1, 2.2 and 2.3. The structure of Figure 2.1 is sometimes called the TaS_2 structure and is generally indicated as 2H(a), while Figure 2.2 represents the structure of molybdenite, MoS_2, and is generally indicated with 2H(b). The structure of Figure 2.3 is observed in $TaSe_2$, it is a 2H-modification but it has not received a lower case letter. It would be reasonable to call it 2H(c)*.

For a three layer structure only two possible ways of stacking the slabs exist. Only one is found and that is the structure of Figure 2.4. Generally this one is indicated as 3R. Theoretically there are ten possible ways to form a four layered unit cell. Up to the present only two of these 4H-modifications have been observed. These are given in Figure 2.5 and 2.6; in literature they are denoted as 4H(a) and 4H(c). They correspond respectively to the stackings 4 and 9 reported by Brown and Beerntsen [11].

A stacking of six layers leading to a 6 R structure is only possible in two ways; only one is found, it is presented in Figure 2.7 and observed in $Ta_{1+x}S_2$ where intercalated Ta lies between the slabs.

No higher polytypes, built up from layers which have an exclusively prismatic surrounding of the metal atoms, are known.

Layers with an octahedral surrounding of the metal atom

Here a one slab structure is known to give a stable lattice and several compounds are known to crystallize in this so-called 1T-modification presented in Figure 2.8.

Only one 2H-modification is possible, Figure 2.9, and only one 3R-type is possible. This is presented in Figure 2.10.

The compounds TaS_2 and $TaSe_2$, respectively, adopt these structures. Theoretically several 4H-stackings can be derived, but none is found. From the higher polytypes, only one 6R-modification is found, presented in Figure 2.11. The compounds TaS_2 is found to crystallize in this structure.

---

* Whereas the notation 2H(a) seems to be used by all authors for the stacking sequence presented in Figure 2.1, it seems that Wilson et al. [10] use the notation 2H[b] for the stacking presented in Figure 2.2, while the notation 2H(b) used by Huisman and Jellinek [9] does not correspond with Figure 2.2 but with the stacking of Figure 2.3, the one proposed to be called 2H(c).

TRANSITION METAL DICHALCOGENIDES   159

TABLE II

Structures with layers having a prismatic surrounding of the metal atom in the layers

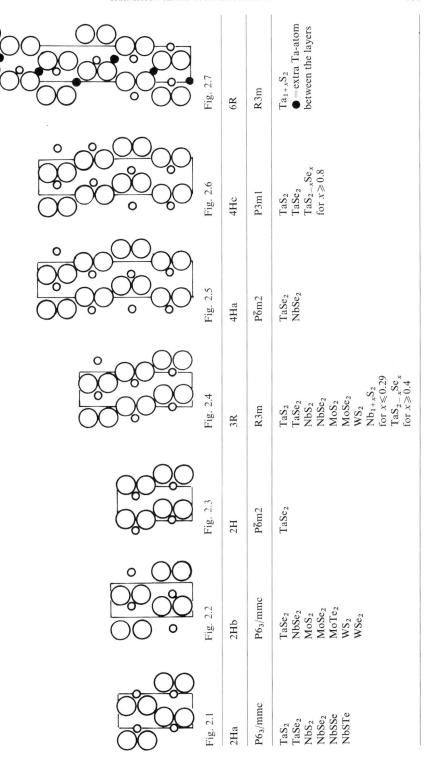

	Fig. 2.1	Fig. 2.2	Fig. 2.3	Fig. 2.4	Fig. 2.5	Fig. 2.6	Fig. 2.7
	2Ha	2Hb	2H	3R	4Ha	4Hc	6R
	$P6_3/mmc$	$P6_3/mmc$	$P\bar{6}m2$	$R3m$	$P\bar{6}m2$	$P3m1$	$R3m$
	TaS_2	$TaSe_2$	$TaSe_2$	TaS_2	$TaSe_2$	TaS_2	$Ta_{1+x}S_2$
	$TaSe_2$	$NbSe_2$		$TaSe_2$	$NbSe_2$	$TaSe_2$	● = extra Ta-atom between the layers
	NbS_2	MoS_2		NbS_2		$TaS_{2-x}Se_x$	
	$NbSe_2$	$MoSe_2$		$NbSe_2$		for $x \geq 0.8$	
	$NbSSe$	$MoTe_2$		MoS_2			
	$NbSTe$	WS_2		$MoSe_2$			
		WSe_2		WS_2			
				$Nb_{1+x}S_2$			
				for $x \leq 0.29$			
				$TaS_{2-x}Se_x$			
				for $x \geq 0.4$			

Table II (Continued)

Structures with layers having an octahedral surrounding of the metal atom in the layer.

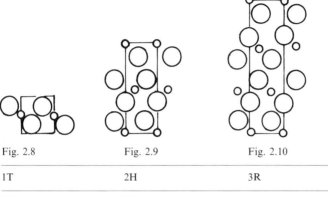

Fig. 2.8	Fig. 2.9	Fig. 2.10	Fig. 2.11
1T	2H	3R	6R
$P\bar{3}m1$	$P6_3/mmc$	$R3m$	$R3m$
TaS_2	TaS_2	TaS_2	TaS_2
$TaSe_2$	$TaSe_2$	$TaSe_2$	
$NbSe_2$		$Ta_{1+x}S_2$	
$TaS_{2-x}Se_x$			

Structures with alternating prismatically and octahedrally surrounded metal atoms in the layers

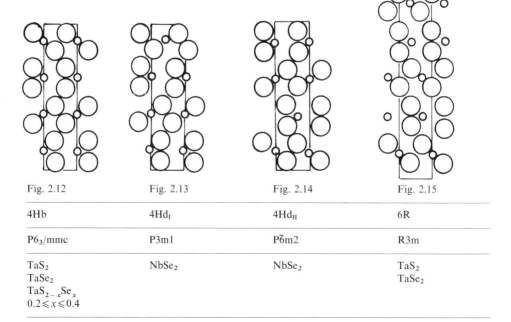

Fig. 2.12	Fig. 2.13	Fig. 2.14	Fig. 2.15
4Hb	4Hd$_I$	4Hd$_{II}$	6R
$P6_3/mmc$	$P3m1$	$P\bar{6}m2$	$R3m$
TaS_2	$NbSe_2$	$NbSe_2$	TaS_2
$TaSe_2$			$TaSe_2$
$TaS_{2-x}Se_x$			
$0.2 \leqslant x \leqslant 0.4$			

Structures having alternating prismatic and octahedral layers

Theoretically the packing of four layers to 4H-polytypes can take place in 18 different ways [12]. To date, only three have been found. These are presented in Figures 2.12, 2.13 and 2.14 respectively and denoted 4H(b), 4H(d_I) and 4H(d_{II}). Of the 6R-type, one form is known to exist (Figure 2.15). This structure is adopted by TaS_2 and $TaSe_2$.

1.4. BOND CHARACTER

A number of investigators have called attention to differences between the compounds that imply substantial differences in ionicity. White and Lucovsky [13] and later Lucovsky and co-workers [14] have pointed out that the differences in the compounds of the IV and VI group can easily be understood by recognizing that although their structures may be similar, the nature of their bonding may be very different. Such a difference in bonding precludes a rigid band scheme, as was attempted by Wilson and Yoffe in their comprehensive study of the physical properties of the TCh_2 [1]. White *et al.* [13] and Lucovsky *et al.* [14] start with the assumption that the chalcogenides of the group IV and VI members are ionic since the electronegativity of the chalcogenides exceeds that of the metals. Investigating the consequence of this assumption they arrive at the conclusion that TiS_2, $ZrTe_2$ and $HfTe_2$ are metals due to an overlap of the chalcogenide s- and p-states with the d-states of the metal atoms. A similar metallic band scheme is assumed in $TiSe_2$ and $TiTe_2$.

ZrS_2, HfS_2 and $HfSe_2$ are thought to be ionic – and an ionic type of bonding is anticipated in $ZrSe_2$ – which ionic bonding picture should be completely in accord with a tight binding calculation of Murray, Bromley and Yoffe [15]. For MoS_2, $MoSe_2$, WS_2 and WSe_2 they find predominantly covalent bonding, which is also in agreement with calculations of Bromley and co-workers [16]. Huisman and co-workers [17] have suggested that in d^0, d^1 and d^2 members, d-covalency provides a stabilizing factor for trigonal prismatic coordination observed in group VI and V members vs the more symmetric, electrostatically favored octahedral coordination observed in the IV and some V compounds and in one VI compound. Gamble [18] shows that the variation in crystal parameters, observed in the different compounds, can be viewed as a natural and remarkably smooth function of the elemental electronegativity differences of the bonding atoms. Furthermore it is shown that the distribution of charge in these compounds – the ionicity – to first order is determined by the elemental electronegativities and not by the structure. Also, the shorter intralayer and interlayer Ch—Ch distances in the group VI members do not imply Ch—Ch bonds but are the consequence of the smaller size of the almost neutral chalcogens. The concept of metallic bonding in the metallic members of this class is unnecessary in a consideration of the structure or lattice parameters of any of these materials.

Gamble also points out that the radii of the atoms in this series of substances cannot be assigned from any set of ionic or covalent radii, but will vary as a function of the electronegativity difference of the bonding atoms. Only after taking this variation into account can one determine which crystal structures are geometrically allowed, and it becomes clear that while all the compounds can adopt octahedral coordination, only certain members can adopt trigonal prismatic coordination, and these do so.

It appears that the trigonal prismatic structure is maintained until the chalcogens above and below the metal atom come into contact, and then in the case of Nb- and Ta-compounds apparently deform to some extent. Finally the octahedral structure is adopted, permitting a closer approach to the cation by large anions and hence the attainment of the desired bond lengths [18].

2. Group IV compounds

2.1. TiS_2

Titanium disulfide is a golden coloured material with a metallic lustre, not known to occur naturally but readily synthesized. It is the only known layered disulfide in the first row of transition elements and has a metal-like character [14].

In dry air TiS_2 is stable but it reacts slowly in moist air at room temperature; the reaction with oxygen proceeds slowly above 375 °C and complete oxidation to TiO_2 takes place after heating at 1000 °C for several hours (see also [35]). The crystal structure [19] is of the CdI_2 type (see 1.2), the Ti being octahedrally surrounded by sulfur.

After the experiments of Benard and Jeannin [20] at 800° and 1000 °C, the stoichiometry of TiS_2 has been a point of controversy for some time. In their report these authors expressed their doubt about the ideal $z=\frac{1}{4}$ position of the sulphur in TiS_2; according to their report this z-position was less than $\frac{1}{4}$.

Newer data [21], however, have established the fact that careful preparation at specified temperatures will yield stoichiometric TiS_2 (see 2.1.1). Four polytypes have been worked out [7] and some other have been identified near the composition $TiS_{1.70}$. See also [8].

2.1.1. *Compound Preparation*

The group IV transition metals are known to react with the chalcogens S, Se and Te in various stoichiometric ratios, and in addition non-stoichiometric phases of considerable width occur. Numerous papers exist about this field and the most extensive are presented in the reference list [22–26].

In the earliest report on the preparation of TiS_2, the substance was obtained by reduction of TiO_2 with CS_2 at elevated temperatures. More important was the open flow system employed by Biltz and his co-workers [22] to prepare TiS_2 from $TiCl_4$ and H_2S.

This mixture was heated at temperatures between 400° and 600 °C and the resulting TiS_2 was strongly contaminated with incorporated chlorine and had to be purified by a heat treatment with excess sulfur. This procedure, however, has been widely in use to produce commercially large quantities of material (see also [35]). In the modification briefly indicated by Biltz [22] and extensively described by Brauer [27], the chlorine-rich product is heat treated for 3 days at 600 °C with excess sulfur and the resulting TiS_2—TiS_3 mixture is subsequently heat treated at ~500 °C and yields pure TiS_2. The method successfully employed by McTaggert and Wadsley [28] is based on the decomposition of the trisulfide. The product is dark bronze coloured and according to their analyses the value n in TiS_n is 1.95 and not 2. The dark colour should be an indication for non-stoichiometry. The primary disadvantage of this procedure therefore seems to be the lack of adequate control over the stoichiometry of the resulting product.

Presently the most successful preparation procedure is the direct reaction of the elements under very specified conditions of temperature and reaction time. In their report Thompson and co-workers [21] heated the high purity elemental constituents – Ti in wire form and S powder – in sealed evacuated quartz ampoules to 500 °C for one week, thus avoiding rupture of the reaction tube by the high sulfur vapour pressure. Thereafter the temperature was raised to 700 °C and kept there for several days to ensure complete reaction. At this temperature the Ti-rich TiS_2 is in equilibrium with S_2. The ampoule was then annealed for another week at 600 °C and subsequently slowly cooled in one or two days to room temperature. At this stage no excess sulfur should be present. The duration of the heat-treatment at 700 °C seems to affect the crystallite size; an increase in size was observed when the material was kept for a longer period of time at this temperature. Excess sulfur in this preparation procedure has also yielded good material [29]; in such a case, however, cooling from 600 °C to room temperature should proceed more quickly to prevent TiS_3 formation – this black fibrous compound is stable at low temperatures – and the resulting TiS_2 should be stoichiometric to within 0.10%.

The lattice parameters of stoichiometric TiS_2 should be $a=3.4073\pm0.0002$ Å and $c=5.6953\pm0.0002$ Å according to [21, 30].

Larger values would indicate the presence of an excess Ti in the material which accordingly is darker and more brittle. Non-stoichiometric compounds could be formed on the Ti-rich side of TiS_2 only, on the S-rich side no deviations of stoichiometry occur. This confirms the suggestion of Takeuchi [31] that interlayer Ti is present – even when TiS_2 is stoichiometric – together with Ti vacancies in the metal layers.

2.1.2. Crystal Growth

Crystal growth with chemical vapour transport using either iodine or sulfur [32] as an agent will give very large crystals, sometimes of more than one centimeter in basal extent. Sulfur is preferable [29] due to the fact that iodine as a transporting agent always contaminates the material. Incorporation of the halogen up to about 0.1% by weight has been observed. This affects the electrical conductivity of the produced substance as it possesses a higher concentration of electronic carriers than those grown in sulfur vapour.

Iodine transport. In reported work, iodine assisted transport takes place from $T_H=800° \to T_L=720$ °C in vitreous silica tubes of 200 mm length and 16 mm bore

[33]; reaction times are 500 h. In their work Greenaway and Nitsche [34] use essentially the same amount of iodine as Rimmington and his co-workers [33] – 5 mg cm^{-3} tube volume – but their reaction time of 60 h is much shorter. Likewise their ampoules differ in length (length = 100 mm) and their temperature gradient in the two-zone tubular furnace – with $T_H = 900\,°C$ and $T_L = 800\,°C$ respectively – was also different.

The experiments of [34] showed the influence of the tube diameter on the crystal size: in 16 mm diameter ampoules crystals up to 0.5 cm^2 were obtained but in 30 mm ampoules larger samples were grown. Reported dimensions for TiS$_2$ are 6×6 mm^2. Single crystal rotation photographs showed the c-axis to be perpendicular to the plane of the plates. Iodine transport has also been used by Lucovsky and his group [14].

Sulfur transport. Transport in sulfur vapour is an exotherm process and takes places from $T_L = 750° \rightarrow T_H = 900\,°C$ [29], but according to [32] from $T_H \rightarrow T_L$. An excess sulfur of 4–5 mg cm^{-3} tube volume is required in this process. An important facet is that growth conditions seem to be quite critical and deviations of more than 50 °C from those specified hamper transport.

In Table I structure data and lattice parameters are presented, in Table III the information on compound preparation and crystal growth are compiled and also the available data on heats of formation.

2.2. TiSe$_2$

Titanium diselenide is a purplish-brown substance with a metallic lustre which does not occur naturally. Large crystals can easily be prepared. At room temperature it is easily oxidized and, according to Bear *et al.* [35], it is in general a less stable compound than TiS$_2$ with properties almost similar to those of TiS$_2$. See also [36] and [37].

The crystal structure is the same as described for TiS$_2$ [19]. Superlattices are not found in TiSe$_2$; however, it is probable that Schottky defects similar to those assumed in TiS$_2$ [21, 31] exist. No polytypism has been found yet, and stoichiometric TiSe$_2$ seems to be easier to prepare than stoichiometric TiS$_2$, according to the results obtained by McTaggart *et al.* [28]. This may reflect the fact that TiSe$_2$ is the selenium-rich phase limit as TiSe$_3$ is not found. According to Lucovsky *et al.* [14], TiSe$_2$ is probably metallic.

2.2.1. Compound Preparation

As described for the disulfide, the reaction between TiCl$_4$ and H$_2$Se can be used to prepare TiSe$_2$ [35]. Most widely used for laboratory work, however, is the direct synthesis from the elements. The procedure is essentially the same as discussed in 2.1.1 for TiS$_2$ preparation. Following [28] the temperature is not raised above 580 °C, according to [29] the substance is kept at 900 °C. Owing to the lower Se-vapour pressure at elevated temperatures – as compared to that of sulfur – precautions to prevent bursting of the reaction ampoule are not stringent. This then allows faster heating of the entire ampoule.

Following [29] the reaction mixture, enclosed in sealed, evacuated ampoules, is placed in a furnace which is held at 500 °C. The temperature is then raised at 100 °C per hour to 900 °C and maintained for a week. Thereafter the system is cooled at 100° or 200 °C per week to 400 °C and then cooled rapidly to room temperature. The product contains no excess selenium.

2.2.2. Crystal Growth

Either the halogen assisted transport or the selenium assisted transport may be used.

Using iodine, the transport is–like that for TiS_2–from $T_H \rightarrow T_L$. Rimmington, Balchin and Tanner [33] used 800° and 720 °C for T_H and T_L respectively and produced good crystals in about 340 hours with 200 mm × 16 mm ampoules.

Greenaway and Nitsche [34] got very good results with 900° and 800 °C. Their tube bores were between 16 mm and 30 mm with a lenght of 100 mm. Effects of the tube diameter have been discussed in 2.1.2; the iodine concentration was 5 mg cm^{-3} and reaction times 60 h. According to [29] an increased transport rate was observed when excess chalcogen ($\sim$ 5 mg cm^{-3}) was added to the tube carrying the $TiSe_2$ and iodine. It also seemed to improve the material purity with respect to iodine. Lucovsky's group also used iodine [14]. Optimum conditions for $TiSe_2$ transport in selenium vapour have not yet been established (See also [32]).

Table I gives structure data and in Table III the information on compound preparation and crystal growth is presented.

2.3. $TiTe_2$

Titanium ditelluride is a grey metallic, highly crystalline material with properties similar to $TiSe_2$. The chemical and physical properties have also been studied by several investigators like Ehrlich [36], McTaggart et al. [28], Bear et al. [35] and Raaum and co-workers [38]. According to Raaum et al., in the Ti/Te system the $Ti_{2-x}Te_2$ phase has a homogeneity range within the limits 55.4–66.67 atomic % Te. Up to 60.0 at % Te, the structure is monoclinic and related to the NiAs-type, while between 60.0 and 66,67 at % Te the $Ti_{2-x}Te_2$ phase has a hexagonal NiAs–CdI_2 type of structure.

According to [34], $TiTe_2$ is probably a semi-metal, according to [14] a metal. From the results of their work, Bear et al. [35] assumed that the stability of the ditelluride was less then that of the diselenide and the disulfide.

Structurally $TiTe_2$ is similar to TiS_2 and $TiSe_2$ [19]. A defect structure reported by [36] has not been substantiated in $TiTe_2$, which is easy to prepare stoichiometrically, but it is likely that interlayer Ti constitutes a defect in material prepared at high temperatures [29].

2.3.1. Compound Preparation

The direct synthesis from the elements in sealed evacuated quartz ampoules–as discussed for TiS_2 and $TiSe_2$–is a successful procedure for $TiTe_2$ too. Following McTaggart et al. [28] reaction temperatures are in the range 580°–620 °C. The resulting material is black and stoichiometric. According to [29], the reaction proceeds best at 800° to 900 °C and after one week the temperature is decreased at a rate of 100° or 200 °C per week and kept at 400 °C for 30 days to ensure sample uniformity and stoichiometry. The tellurium vapour pressure is low and does not lead to tube rupture. No reports exist about a tritelluride in literature and consequently the degradation procedure for obtaining $TiTe_2$ is not valid.

2.3.2. Crystal Growth

Chemical vapour transport with the aid of iodine along the same lines as discussed for TiS_2 and $TiSe_2$ produces large single crystals. Greenaway and Nitsche [34] use the same conditions as mentioned for $TiSe_2$ and TiS_2 (Sections 2.1.2 and 2.2.2) while the conditions in Rimmington's work [33] are different for the three compounds. With tubes of 200 mm × 16 mm and 5 mg cm^{-3} iodine, they produce crystals of $TiTe_2$ in a run of 300 h. Their temperatures are 750° and 690 °C respectively.

In the experiments reported by Gamble and Thompson [29], the temperatures employed are 800° and 750 °C respectively. The iodine concentration is 5 mg cm^{-3} tube volume but their tubes are larger – 300 mm × 15 mm – and their growth times on the order of 1 or 2 weeks. Centimeter large crystals are the result. $TiTe_{1.17}$ has also been grown by Lucovsky and his co-workers via the iodine vapour transport [14]. Optimum conditions for transport with the aid of tellurium vapour have not yet been fully established, but according to [32] there are indications for success. For a comparison of the preparation and growth procedures of the three titanium compounds the reader is referred to Table III. Structure data and lattice parameters are presented in Table I.

2.4. ZrS_2

Zirconium disulfide is a violet metallic substance, which according to most investigators can be prepared as a stoichiometric compound. According to Lucovsky and his co-workers [14], ZrS_2 is an ionic compound.

In addition to the disulfide, other phases of considerable width occur, like in all group IV members; investigations in the Zr-S system have been reported by Strotzer and co-workers [39], Hahn and co-workers [40] and McTaggart et al. [28].

The experiments of Bear et al. [35] and McTaggart et al. [28] on the oxidation rate of TCh_2 members have shown that ZrS_2 absorbs oxygen rapidly, thereafter a protective layer is formed which slows the rate of absorption to almost zero. In [35] the oxidation rate of several group IV compounds is compared.

As far as known today, polytypism seems not to occur in ZrS_2.

2.4.1. Compound Preparation

The older preparation techniques mentioned for the $TiCh_2$ members are in general also valid for the $ZrCh_2$ members [25, 39]. Newer procedures are the reaction of zirconium hydride with hydrogen sulfide, the degradation of ZrS_3 and direct synthesis from the elements.

According to Bear et al. [35], for the preparation of chalcogenides of high purity for research purpose, direct synthesis from the elements is to be preferred. Stringent precautions, however, to avoid reaction between the metal and the tube material seem to be necessary [41]. McTaggert et al. [28] prepared ZrS_2 by degradation of ZrS_3 at 900 °C in evacuated and sealed ampoules. Prior to the experiments the quartz ampoules were carefully outgassed. The decomposition was complete when no further sublimation occured. This was checked by pulling the tube an inch or more out of the furnace, followed – after a suitable interval – by an inspection for newly formed sublimate. The direct synthesis technique is also employed by McTaggart et al. [28]. The elements are heated slowly and homogeneously – from 6 h to 2 days – in carefully outgassed silica tubes which were evacuated and sealed.

It is noteworthy that Gleizes et al. [41] in their report on $ZrSe_2$ preparation emphasize the use of small sintered alumina boats to contain the metal, in order to avoid reaction between Zr and the ampoule wall. In McTaggart's experiments the ampoules were held at the desired temperature for 5 to 7 days and then cooled. The contents of the reaction ampoule were subsequently ground and reheated for a similar period of time. Additional grinding and heating cycles were made if necessary, until equilibrium was reached. These authors preferred prolonged heating at relatively low temperatures to prevent oxygen diffusion through the silica walls of the tube.

2.4.2. Crystal Growth

Greenaway et al. [34] and Rimmington and co-workers [33] have reported on growth conditions during crystal growth with the aid of iodine. Their growth experiments are essentially the same as discussed for the $TiCh_2$ compounds and a brief discussion is therefore sufficient.

[34] uses 100 mm × 16 mm transport tubes and employs 5 mg cm^{-3} iodine. In 60 h crystals up to 0.5 cm^2 are grown, while in 100 mm × 30 mm ampoules the crystals reach dimensions up to 5 cm^2.

The temperatures are $T_H = 900$ °C and $T_L = 800$ °C respectively and all crystals showed good cleavage perpendicular to the c-axis.

No analysis was carried out, but the deviation from stoichiometry – if existing – must be small as in many cases complete transport of the whole material was observed without a visible residue.

The conditions reported by [33] were almost similar; 5 mg iodine cm^{-3}, tubes of 200 mm × 16 mm and $T_H = 900°$ and $T_L = 820$ °C. Only their growth times, 500 h, differed from those reported by [34].

Rimmington and co-workers were careful not to exceed scource temperatures of 950 °C to avoid reactions between zirconium and the tube material. Iodine was also used to transport and grow ZrS_2 by Lucovsky and co-workers [14].

For the $ZrCh_2$ no sufficiently accurate thermodynamic data seem to exist. According to Bear et al. [35] the difficulties encountered in determining heats of combustion of this class of compounds in a bomb calorimeter are considerable. It is often impossible to burn all the sample even in the presence of combustible material.

The higher compounds, especially tellurides, tend to lose chalcogen by degradation and volatilization during burning. In the case of sulfur and sulfides, the formation of SO_2 as well as SO_3 and of sulphuric acid – if any moisture is present – can lead to large errors and similar problems are encountered with Se- and Te-containing compounds.

In Table I structure data are presented, in Table III the preparation and growth data.

2.5. $ZrSe_2$

Zirconium diselenide, a dark bronze-green material, was first synthesized in 1924 by Van Arkel [42] who also determined its structure (CdI_2-type). In the system Zr-Se many other compounds exist and other non-stoichiometric phases have been reported [43]. The question of $ZrSe_2$ stoichiometry seems to be still very actual; the work of McTaggart et al. [28] and of Hahn and Ness [43] mention sample densities which correspond to $ZrSe_2$ but the elemental analysis of McTaggert and Wadsley correspond to $ZrSe_{1.95}$, which is in agreement with the work of Gleizes and Jeannin [41]. On the other hand, Hahn and Ness found a homogeneity region for $ZrSe_2$ ranging from $ZrSe_{1.7}$ to $ZrSe_2$. The colour is also a point of disagreement; according to [28] it is dark purplish brown, according to [43] it is dark violet, Van Arkel [42] calls it black and in the reports [33] and [34] it is called metallic bronze with green lustre and metallic respectively. Gleizes et al. [41] finally report a dark bronze-green microcrystalline powder.

The calculations of Lucovsky anticipate ionic bonding in $ZrSe_2$ too [14]. Polytypism seems not to occur in this diselenide.

2.5.1. *Compound Preparation*

The more dense materials prepared by Hahn et al. and McTaggart et al. were prepared by direct synthesis from the elements at 800 °C [43] and by degradation of the triselenide [28] at 900 °C respectively. Gleizes and Jeannin [41] also followed the procedure of direct reaction between elemental constituents, taking the necessary precautions to avoid reactions between Zr. and the silica tube wall. In their experiments the metal was placed in alumina boats and their thermal treatment was made in two steps. First the ampoule and its contents were heated for two days at 500 °C followed by a heat treatment at 800 °C for a whole week. The substance was then powdered and subsequently reheated at 800 °C for another period of three weeks and finally water quenched. In their investigations devoted to the non-stoichiometry of $ZrSe_2$ and the defects responsible for it, Gleizes et al. found the limiting Se/Zr ratio to be 1.85 and 1.95 for the non-stoichiometric diselenide phase. If a Se/Zr ratio greater than 1.95 was observed, their X-ray photographs indicated the presence of $ZrSe_3$. They therefore conclude that stoichiometric $ZrSe_2$ cannot be prepared at 800 °C and that $ZrSe_{1.94}$ is the compound with the highest selenium content. This is in agreement with the results of McTaggart's experiments although the preparation technique differed [28].

The work of Gleizes et al. [41] emphasizes the complicated defect nature since two types of defects occur simultaneously; the Se-rich compound contains only Se-vacancies but when the Se/Zr ratio decreases, vacancies continuously disappear since additional Zr atoms fill them up. The phase limit is reached when all Se-vacancies are occupied by Zr.

2.5.2. Crystal Growth

Chemical vapour transport of $ZrSe_2$ has been applied with success by Greenaway et al. [34] and Rimmington and co-workers [33] and by Lucovsky and co-workers [14]. Essentially the same conditions as for ZrS_2 are reported and they are tabulated in Table III, while Table I presents the structure type and lattice parameters.

2.6. $ZrTe_2$

Zirconium ditelluride is a purplish-brown substance which was first synthesized in 1957 by Hahn and Ness [44]. They assigned it the CdI_2 structure and reported a continuous transition from ZrTe (NiAs-structure) to $ZrTe_2$ (CdI_2-type).

In 1959 these authors reported on an extensive study of the Zr-Te system and mentioned the preparation procedure for $ZrTe_2$. Also in the groups of McTaggart [28] and Jeannin [45] efforts were made to investigate this zirconium compound. From their reports one may conclude that they did not observe a stoichiometric compound $ZrTe_2$.

It seems that there is a discrepancy in the results of the various workers in this field. Furthermore, Bear and Mc Taggart [35] reported on the oxidation rate of the various tellurides and from the reported calculations of Lucovsky and co-workers [14] it is known that $ZrTe_2$ should be a metal.

Polytypism has not been reported as far as known today.

2.6.1. Compound Preparation

In the preparation procedure of Hahn et al. [44], stoichiometric amounts of the elements are heated to 850 °C for a period of 2 weeks. The material is then cooled but the cooling rate is not specified. The densities found by these workers was 10% below X-ray densities. McTaggart and Wadsley [28] degradated $ZrTe_3$ at 900 °C in the manner as described for ZrS_2 and $ZrSe_2$ and obtained $ZrTe_{1.7}$, but according to their density measurements they suggest the formula $Zr_{1.12}Te_{1.90}$ for the obtained substance. This then requires an excess metal and a deficit of tellurium which would be in good agreement with the experiences of Gleizes and Jeannin in their report (1972) on the Zr-Te system near $ZrTe_2$ [45]. Their samples were prepared by direct reaction between Zr and Te at 800 °C in a similar procedure as reported for $ZrSe_2$. The tubes were carbon coated to prevent reactions between the contents of the tube and the tube wall, but they were not able to protect the tube competely; some ZrSiTe was found by X-ray detection techniques. The ampoules containing the elemental constituents were heated to 500 °C and kept at this temperature for two days. Thereafter the temperature was raised to 800 °C and kept at it for a week. The non-homogeneous product was ground in an atmosphere of pure nitrogen and reheated for 3 weeks at 800 °C and then cooled in air. Apart from the non-stoichiometric material, their X-ray pictures revealed the silico compound ZrSiTe as a product of the aforementioned reaction with the tube material.

The limits of the homogeneous phase determined from lattice constant measurements – by X-ray powder pattern photographs – are $ZrTe_{1.735}$ and $ZrTe_{1.45}$. If $1.735 > Te/Zr > 1.60$ the crystal contains Te-vacancies and Zr-atoms substituting some Te atoms. If $1.60 > Te/Zr > 1.45$, all Te-vacancies seem to have been filled up and additional Zr-atoms are inserted between Te-planes, untill the lowest Te-limit is reached. The French investigators thus concluded that it is impossible to prepare stoichiometric $ZrTe_2$ at 800 °C, and give the ditelluride the general formula $Zr_{1+x}Te_{1.85}$. It seems therefore necessary that further studies be directed on the effect of cooling rates on the non-stoichiometry of the end product.

2.6.2. *Crystal Growth*

Chemical vapour transport with iodine as an agent seems to be of no success according to Greenaway and Nitsche [34]. Neither was $ZrTe_2$ mentioned in the growth studies of Rimmington and his group [33]. Relevant data are presented in Tables I and III.

2.7. HfS_2

Hafniumdisulfide according to McTaggart *et al.* [28] who were the first to prepare it in 1958, is a light purplish-brown material. The metal is octahedrally surrounded by sulfur and the calculations of Lucovsky and his group [14] indicated an ionic character for this substance.

A variety of colours is reported in literature: light purplish-brown as mentioned in McTaggart's report [28], transparent reddish-orange as reported by Rimmington's group [33] and dark red metallic in Greenaway and Nitsches's work [34]. As far as is known no polytypism has been reported.

2.7.1. *Compound Preparation*

HfS_2 was prepared by McTaggart and Wadsley [28] via the degradation of HfS_3 at 900 °C and their analyses indicated a stoichiometric compound. This is in agreement with the investigation of Lucovsky and co-workers [14]. Their compounds – including the mixed-anion member $HfS_{1.0}Se_{1.0}$ – were single phases and the chemical analysis indicated compounds very close to the ideal stoichiometry TCh_2 for all Hf-compounds. No reports have been found of synthesis by direct reaction of elemental constituents or via the reaction between $HfCl_4$ and H_2S.

2.7.2. *Crystal Growth*

Chemical transport as a method to grow single crystalline HfS_2 is reported by Rimmington and co-workers [33], Greenaway *et al.* [34] and Lucovsky and his group [14].

Rimmington's group report transport from $T_H = 1010° \rightarrow T_L = 1000$ °C and growth times of 1000 h with 5 mgI$_2$ cm^{-3} tube volume in 200 mm × 16 mm ampoules.

In the experimental work of Greenaway and Nitsche, transport takes place from $T_H = 900° \rightarrow T_L = 800$ °C in about 60 h with 5 mg cm^{-3} iodine in 100 mm × 16 mm tubes.

In none of the reports cited has there been an indication about non-stoichiometry and the available data on chemical analysis all mention stoichiometry. Although crystal thickness may be responsible for the colour differences referred to in the various reports, these differences could indicate slight variations in the composition or perhaps be the result of incorporated impurities like iodine. Structure data and lattice parameters are collected in Table I, growth and preparation techniques in Table III.

2.8. HfSe$_2$

Like hafniumdisulfide, the very dark hafniumdiselenide was first prepared by the Australian group of McTaggart and Wadsley [28] in 1958 by degradation of HfSe$_3$.

Their analyses suggested that the material was non-stoichiometric, having a composition HfSe$_{1.85}$. The substance is very dark brown, but other investigators found different colours like metallic bronze [33] and dark red metallic [34].

Lucovsky assigned an ionic character to it [14]. As far as is known today, no polytypism has been observed for HfSe$_2$.

2.8.1. Compound Preparation

The only procedure reported at the present time seems to be the degradation of HfSe$_3$ at 900 °C, a method successfully employed by McTaggart et al. [28] The commonly used direct reaction between the elemental constituents is not mentioned for the production of HfSe$_2$.

2.8.2. Crystal Growth

Chemical vapour transport is used by Rimmington's group [33], by Lucovsky's group [14] and by Greenaway et al. [34] to grow single crystals. In Rimmington's group the conditions are similar to those mentioned for HfS$_2$: transport is from $T_H = 900° \rightarrow T_L = 860$ °C in 675 h, while Greenaway et al. report a slightly different growth temperature, $T_H = 900° \rightarrow T_L = 800$ °C and shorter growth times : 60 h. The conditions are also presented in Table III; structure data are given in Table I.

2.9. HfTe$_2$

Hafnium ditelluride cannot be prepared according to the investigations of McTaggart et al. [28] in 1958. Their experiments in the hafnium telluride field led to the identification of only one single phase Hf$_2$Te$_3$, a surprising fact in view of what has been found for closely related systems.

A systematic study of the Hf-Te system was reported in 1972 by Brattås and

Kjekshus [46], who prepared a metal-rich $Hf_{1+x}Te_2$ from the elements and in 1973 Smeggil and Bartram [47] reported their study of the $HfTe_{2-x}$ phase (with $x=0.061$), in which they present a structure based on non-stoichiometry in the tellurium sublattice and random but unequal distribution of hafnium atoms on two possible sites. This seems to be a rather unusual situation among these compounds. In the reports the colour is given as brassy or gold-like. According to the results of the work of Lucovsky and his group [14], the character is that of a metal. So far no reports of polytypism exist.

2.9.1. Compound Preparation

Preparation of $HfTe_2$ by heating a mixture of the elements in the atomic ratio 1 : 2, gave no satisfactory results [28]; diffraction patterns indicated additional tellurium and after removal of excess non-metal the upper composition limit was found to be $HfTe_{1.6}$. In their analyses of the Hf-Te system, Brattås and Kjekshus [46] found it impossible to make use of one single technique for synthesis throughout the whole system. Their study revealed the existence of a hexagonal phase in samples containing ~ 58 to ~ 75 at .% Te. A more accurate estimate of the Te-poor phase limit is impossible due to the reaction between Hf, Te and the tube wall. At $< \sim 58$ at .% Te, reactions lead to the formation of HfSiTe. A convenient procedure for preparing pure $Hf_{1+x}Te_2$ is to heat the components to 600 °C for a period of 5 days. The substance is cooled, ground and reannealed at 600 °C for another 5 day period followed by a second grinding procedure. A final heat treatment is performed at 800 °C for 7 days. After the heat treatments a considerable amount of single crystals was found in the ampoules. Thermal decomposition at 600 °C of the $HfTe_3$ phase – also prepared by Brattås et al. – led to the formation of the $Hf_{1+x}Te_2$ phase.

2.9.2. Crystal Growth

Smeggil and Bartram examined the Hf-Te system by utilizing the vapour phase technique [47]. The elements were weighed out in 1:1 stoichiometry, sealed in evacuated 8 mm tubes and heated for 5 days at 500 °C. Thereafter, 200 mm × 25 mm quartz tubes were filled with this prereacted compound together with 10 mg thoroughly dried NH_4I and sealed after evacuation to 10^{-2} torr. Material transport took several days and occurred from $T_H=900° \rightarrow T_L=680$ °C. The resulting crystals were golden coloured, plate-like, and several mm in diameter. Smeggil et al. were apparently unaware that their lattice parameters for the phase $HfTe_{1.93}$ ($a=3.9492$ Å and $c=6.6514$ Å) coincided with the values reported by the Oslo group of Brattås [46], which are $a \sim 3.95$ Å and $c \sim 6.65$ Å for the phase $Hf_{1+x}Te_2$ with a having values between 0.25 and 0.0 or Te contents ranging from ~ 61.5 to ~ 66.7 at. % Te.

Using iodine as the transporter, Greenaway and Nitsche [34] reported being unable to grow $HfTe_2$ single crystals and nothing is reported about this substance by Rimmington either [33]. Lucovsky's group [14] mentions the growth of $HfTe_2$ single crystals by utilizing the halogen vapour transport with iodine as a transporter in the gradient $T_H=900° \rightarrow T_L=800$ °C. Data on $HfTe_2$ can be found in Tables I and III.

TABLE III

Compound preparation and crystal growth techniques

	Compound preparation procedures		Crystal growth techniques	
TiS_2	a. $TiCl_4 + H_2S$ in open flow system; 400°–600 °C	[22, 27, 35]	I_2 transport, 5 mg cm^{-3} $T_H = 800° \rightarrow T_L = 720 °C$; tubes of 200 mm × 16 mm; 500 h $T_H = 900° \rightarrow T_L = 800 °C$; tubes of 100 mm × 16 mm, or 100 mm × 30 mm; 60 h	[33]
	b. Degradation of TiS_3 at 900 °C.	[28]	See also [14].	[34]
	c. Direct synthesis from the elements, at <800 °C	[21, 30]	Sulfur transport, 4–5 mg cm^{-3} $T_L = 750° \rightarrow T_H = 900 °C$. See also [32].	[29]
$TiSe_2$	a. Same as for TiS_2.	[35]	I_2 transport, 5 mg cm^{-3} $T_H = 800° \rightarrow T_L = 720 °C$; tubes of 200 mm × 16 mm; 340 h $T_H = 900° \rightarrow T_L = 800 °C$; tubes of 100 mm × 16 mm or 100 mm × 30 mm; 60 h	[33]
	b. Not valid, $TiSe_3$ not existing.			[34]
	c. Direct synthesis at temperatures of 550°–580 °C. or at 900 °C.	[28] [29]	See also for I_2 transp. [14] and for Se-$transp$. [32].	
$TiTe_2$	a. Not valid.		I_2 transport, 5 mg cm^{-3} $T_H = 750° \rightarrow T_L = 690 °C$; tubes of 200 mm × 16 mm; 300 h $T_H = 900° \rightarrow T_L = 800 °C$; tubes of 100 mm × 16 mm or 100 mm × 30 mm; 60 h	[33]
	b. Not valid.			[34]
	c. Direct synthesis from elements, at 580°–620 °C. or at 800°–900 °C, annealing at 400 °C.	[28] [29]	[29] reports $T_H = 800° \rightarrow T_L = 750 °C$; 300 mm × 15 mm tubes; 1–2 weeks. See also [14], and for Te-$transport$ [32].	
ZrS_2	a. $ZrCl_4 + H_2S$ in open flow system, similar to TiS_2.	[27, 39]	I_2 transport, 5 mg cm^{-3} $T_H = 900° \rightarrow T_L = 820 °C$; tubes 200 mm × 16 mm; 500 h $T_H = 900° \rightarrow T_L = 800 °C$; tubes of 100 mm × 16 mm or 100 mm × 30 mm; 60 h	[33]
	b. Degradation of ZrS_3 at 900 °C.	[28]		[34]
	c. Direct synthesis from the elements.	[28]	See also [14].	

Table III (Continued)

Compound preparation procedures		Crystal growth techniques	
$ZrSe_2$ a. Not valid.		I_2 transport, 5 mg cm^{-3}	
b. Degradation of $ZrSe_3$ at 900 °C.	[28]	$T_H = 850° \to T_L = 800$ °C; tubes 200 mm × 16 mm; 500 h	[33]
	[43]	$T_H = 900° \to T_L = 800$ °C; tubes 100 mm × 16 mm or	
c. Direct synthesis at 800 °C. various a- and c-values for the Se/Zr ratios.	[41]	100 mm × 30 mm; 60 h	[34]
		See also [14].	
$ZrTe_2$ a. Not valid.		Crystal growth with iodine has not been reported, according to [34], they were not able to obtain single crystals with this technique.	
b. Degradation of $ZrTe_3$ at 900 °C.	[28]		
	[44]		
c. Direct synthesis at 850 °C for periods of two weeks at 800 °C for one week, followed by annealing for 3-weeks. various a- and c-values.	[45]		
HfS_2 a. No reports.		I_2 transport, 5 mg cm^{-3}	
b. Degradation of HfS_3 at 900 °C.	[28]	$T_H = 1010° \to T_L = 1000$ °C; tubes 200 mm × 16 mm; 1000 h	[33]
c. No reports.		$T_H = 900° \to T_L = 800$ °C; tubes 100 mm × 16 mm or	
		100 mm × 30 mm; 60 h	[34]
		See also [14].	
$HfSe_2$ a. No reports.		I_2 transport, 5 mg cm^{-3}	
b. Degradation of $HfSe_3$ at 900 °C.	[28]	$T_H = 900° \to T_L = 860$ °C; tubes 200 mm × 16 mm; 675 h	[33]
c. No reports.		$T_H = 900° \to T_L = 800$ °C; tubes 100 mm × 16 mm or	
		100 mm × 30 mm; 60 h	[34]
		See also [14].	
$HfTe_2$ a. No reports.		I_2 transport, 5 mg cm^{-3}	
b. Degradation of $HfTe_3$ at 600 °C gives $Hf_{1+x}Te_2$		$T_H = 900° \to T_L = 800$ °C.	[14]
c. Direct synthesis at 600 °C, followed by annealing at 800 °C. Data are for $Hf_{1+x}Te_2$ with $0.0 < x < 0.25$.	[46]	[33] does not report crystal growth, [34] was not able to prepare single crystals of $HfTe_2$.	

2.10. Solid Solutions

Reports of solid solution formation have been presented by Nitsche [48], Thompson and co-workers [49], Whitehouse and co-workers [50], Rimmington et al. [51] and Lucovsky [14]. In all cases the investigators made use of the versatile vapour transport growth to produce single crystals of the various materials.

Nitsche was able to grow mixed crystals of the following composition: $TiS_2.TiSe_2$; $TiS_2.TiTe_2$; $ZrS_2.ZrSe_2$; $TiS_2.VS_2$; $TiS_2.NbS_2$ and $TiS_2.TaS_2.$; while the work of Thompson, directed at investigating the interesting compound TiS_2, showed that $Ti_xTa_{1-x}S_2$ could be prepared by vapour phase growth.

From the work reported in [50] and [51] it is known that crystals of ZrS_xSe_{2-x} and TiS_xSe_{2-x}, TiS_xTe_{2-x} and $TiSe_xTe_{2-x}$ can be prepared with the aid of iodine as a transporter in the vapour phase. Lucovsky prepared the mixed anion compound $HfS_{1.0}Se_{1.0}$ [14].

2.10.1. Crystal Growth

In his report Nitsche [48] gives the following specific data for the growth of monocrystalline material of at least $5 \times 5 \times 0.1$ mm^3. Tube dimensions are 120 mm in length and 10 mm in diameter, the iodine concentrations are in the range 0.05–0.1 millimoles cm^{-3} and growth times lie between 48 and 72 h.

His experiments were considered successful if the obtained crystals were of the order of several mm^3 and could be used for physical measurements. These data are given in Table IV.

Thompson and his co-workers [49] found that the substitutional replacement of Ti for Ta in TaS_2 stabilized the octahedral 1-T phase in $Ti_xTa_{1-x}S_2$ for x between 0.1 and 0.9. The properties at room temperature vary systematically from a diamagnetic, semiconducting behaviour for $1T-TaS_2$ to a paramagnetic metallic behaviour for TiS_2. At the composition $Ti_{0.1}Ta_{0.9}S_2$ a phase transition was observed.

In Thompson's experiments prereacted TiS_2 and TaS_2 powders are mixed in the proper stoichiometric ratio's to get the various mixed cation compounds. The pressed pellets of these mixtures are heated to about 1000 °C in evacuated quartz ampoules for 2 weeks, followed by annealing for 1 week at 500 °C, and are then slowly cooled to room temperature. This annealing and slow cooling should ensure that only the phase stable at room temperature is produced. Cooling rate, halogen type, growth temperatures, and duration of the run are not specified in [49]. Their powder X-ray diffraction studies revealed that the lattice parameters vary linearly in compositions. For $1T-TaS_2$, $a=3.36$ Å and $c=5.90$ Å while the values for TiS_2 are $a=3.410$ Å and $c=5.702$ Å respectively. No evidence of disorder or superlattice ordering was detected.

Whitehouse and co-workers [50] prepared mixed crystals of at least 5 mm in width of the composition ZrS_xSe_{2-x} with $0<x<2.$ by way of vapour transport with iodine as the transporter.

These investigators did not use prereacted compounds but filled their quartz ampoules (length = 200 mm, diameter = 17 mm) with the high purity elemental constituents Zr (wire 5N purity or powder 2N7), S(6N), Se (5N) and iodine (4N8 purity), with iodine in concentrations between 2.5–5 mg cm^{-3} tube volume. Total ampoule charges were in general 2 g; only with sulfur-rich end compounds was the total charge reducted to 1.2 g to avoid rupture of the reaction tube by the high vapour pressure of the chalcogenide constituents. After evacuation to 10^{-5} torr the ampoules were sealed off and heated for a period between 8–25 days. Size and quality of the crystals depended on the dimensions of the growth ampoules, on T_H and T_L, on the total pressure and on the I_2 concentration. Furthermore the presence of water vapour is reported to have a marked effect on the behaviour of iodine as a transporter.

The relevant growth parameters reported by [50] are given in Table IV. For the growth of ZrS_2, T_L is found to be critical; changes of T_L of only 10 °C from the values quoted lead to ZrS_3 growth. The red ZrS_3 needles are readily distinguishable from the ZrS_2 platelets. According to [50] two conditions should be obeyed if ZrS_2 is to be grown while avoiding ZrS_3 formation:

Firstly T_H (the reaction temperature) must lie within the range 900°–950 °C. Secondly T_L (the growth temperature) must be greater than 820 °C. At T_H values above 950 °C, Zr reacts with the silica tube wall and mixed sulpho-silicide crystals are the result. The use of Zr-powder instead of wire reduces growth times by about 25% below those required when Zr-wire is employed, because the powder offers a large surface area for reaction.

The mixed solid solutions form isomorphous hexagonal compounds over the whole range of composition and are isostructural with both ZrS_2 and $ZrSe_2$. Values of a and c are listed in Table IV. The length of the a-axis varies linearly with composition according to the equation $a(x) = 3.733 - 0.061 \, x$ Å, where x defines the solid solution ZrS_xSe_{2-x}. Variation of the c-axis length is appreciably non-linear and is given by $c(x) = 6.135 - 0.095x - 0.036x^2$ Å. The c/a axial ratio values are related by the equation: $(c/a)_x = 1.626 + 0.001x - 0.010x^2$.

The compositions of a number of grown materials have been confirmed using an oxide reduction method and the results are in good agreement with those obtained from X-ray fluorescence analysis.

Rimmington et al. [51] prepared single crystals of the series TiS_xSe_{2-x}, TiS_xTe_{2-x} and $TiSe_xTe_{2-x}$ with $0 < x < 2$ along the same lines as discussed above in Whitehouse's work. They did not use prereacted compounds as starting material for mixed powders, but filled their reaction tubes with the high purity elemental constituents under consideration together with the appropriate amount of transporter.

Apart from iodine, other transporting agents like HCl, bromine, redistilled water, redistilled iodine, HIO_3 and others were tried, but iodine was found to be very effective with a growth rate of the order of 5 mg h^{-1}. To obtain this rate, it was essential to have water vapour present in the transporter. In the absence of water, growth rates were strongly reduced. Pure water has some transporting properties comparable to those of water-free iodine. Their experiments showed the existence of an upper limit to the water fraction in the transporter, beyond which the water begins to have an adverse effect upon the growth. TiS$_2$ growth rates as a function of iodine concentration show that 4 mg cm^{-3} is a favourable concentration. At concentrations below 2 mg cm^{-3}, the rate was too slow to be of practical use. For a detailed discussion of this topic and of the distinct processes involved in iodine vapour transport of the aforementioned materials, the reader is referred to [51]. The small linear variations of lattice parameters with composition in the TiCh$_2$ compounds can – according to [51] – be traced to non-stoichiometry in these substances, which may be up to 7% deficient in chalcogen. The lattice parameters show slight deviations from linearity, especially in the c-direction, which is smallest in the series TiS$_x$Te$_{2-x}$ and largest in TiS$_x$Se$_{2-x}$.

In work devoted to an investigation of the type of bonding, the group of Lucovsky [14] also prepared the mixed anion member HfS$_{1.0}$Se$_{1.0}$ by way of iodine transport. In Table IV the relevant data are presented.

TABLE IV
Solid solutions of various TiCh$_2$ and ZrCh$_2$, their growth conditions and lattice constants

Compound	Transp. agent	Lattice	a (Å)	c (Å)	T_H (°C)	T_L (°C)	React. time (h)	Col.	Ref.
TiS$_2$.TiSe$_2$	I$_2$	Hex.	a	a	850	800	48–72	metal.	[48]
TiS$_2$.TiTe$_2$	,,	,,	a	a	780	740	,,	,,	,,
ZrS$_2$.ZrSe$_2$	,,	,,	a	a	850	800	,,	,,	,,
TiS$_2$.VS$_2$	,,	a	a	a	850	800	,,	brassy	,,
TiS$_2$.NbS$_2$	,,	Hex.	3.37	5.78	850	800	,,	,,	,,
TiS$_2$.TaS$_2$	,,	,,	3.34	5.80	850	800	,,	,,	,,
Ti$_x$.Ta$_{1-x}$S$_2$	c	,,	b	b	a	a	a	a	[49]
ZrSe$_2$	,,	,,	3.774	6.131	800	800	350	bronze-green	[50]
ZrS$_{0.1}$Se$_{1.9}$	,,	,,	3.762 ^{d}	6.120 ^{d}	852	802	200	a	,,
ZrS$_{0.2}$Se$_{1.8}$	,,	,,	3.756 ^{d}	6.104 ^{d}	855	803	700	a	,,
ZrS$_{0.3}$Se$_{1.7}$	,,	,,	3.753 ^{d}	6.100 ^{d}	858	803	200	a	,,
ZrS$_{0.4}$Se$_{1.6}$	,,	,,	3.743 ^{d}	6.082 ^{d}	860	804	250	a	,,
ZrS$_{0.5}$Se$_{1.5}$	,,	,,	3.743 ^{d}	6.071 ^{d}	863	805	210	a	,,
ZrS$_{0.6}$Se$_{1.4}$	,,	,,	3.733 ^{d}	6.057 ^{d}	865	806	200	a	,,
ZrS$_{0.7}$Se$_{1.3}$	,,	,,	3.729 ^{d}	6.048 ^{d}	868	807	500	a	,,
ZrS$_{0.8}$Se$_{1.2}$	,,	,,	3.719 ^{d}	6.036 ^{d}	870	808	310	a	,,
ZrS$_{0.9}$Se$_{1.1}$	,,	,,	3.718 ^{d}	6.002 ^{d}	872	809	250	a	,,
ZrS$_{1.0}$Se$_{1.0}$	,,	,,	3.715 ^{d}	6.013 ^{d}	875	810	400	a	,,
ZrS$_{1.1}$Se$_{0.9}$	,,	,,	3.704 ^{d}	5.985 ^{d}	878	811	240	a	,,
ZrS$_{1.2}$Se$_{0.8}$	,,	,,	3.702 ^{d}	5.971 ^{d}	880	812	240	a	,,
ZrS$_{1.3}$Se$_{0.7}$	,,	,,	3.697 ^{d}	5.952 ^{d}	883	813	260	a	,,
ZrS$_{1.4}$Se$_{0.6}$	,,	,,	3.690 ^{d}	5.947 ^{d}	885	814	350	a	,,
ZrS$_{1.5}$Se$_{0.5}$	,,	,,	3.685 ^{d}	5.925 ^{d}	888	815	200	a	,,
ZrS$_{1.6}$Se$_{0.4}$	,,	,,	3.677 ^{d}	5.894 ^{d}	890	816	200	a	,,
ZrS$_{1.7}$Se$_{0.3}$	,,	,,	3.673 ^{d}	5.887 ^{d}	893	817	290	a	,,
ZrS$_{1.8}$Se$_{0.2}$	,,	,,	3.665 ^{d}	5.856 ^{d}	895	818	290	a	,,
ZrS$_{1.9}$Se$_{0.1}$	,,	,,	3.662 ^{d}	5.841 ^{d}	898	819	220	a	,,
ZrS$_2$	,,	,,	3.667 ^{d}	5.817 ^{d}	900	820	250	violet-metallic.	,,

Table IV (Continued)

Compound	Transp. agent	Lattice	a (Å)	c (Å)	T_H (°C)	T_L (°C)	React. time (h)	Col.	Ref.
TiS_2	I_2	Hex.	3.407	5.689	800	720	400	golden-metallic	[51]
$TiS_{1.7}Se_{0.3}$	,,	,,	3.419	5.747	800	740	400	a	,,
$TiS_{1.5}Se_{0.5}$	,,	,,	3.432	5.780	800	740	350	a	,,
$TiS_{1.1}Se_{0.9}$	,,	,,	3.454	5.852	800	730	400	a	,,
$TiS_{1.0}Se_{1.0}$	,,	,,	3.457	5.868	790	730	400	a	,,
$TiS_{0.9}Se_{1.1}$	,,	,,	3.464	5.882	790	740	400	a	,,
$TiS_{0.7}Se_{1.3}$	,,	,,	a	a	790	730	400	a	,,
$TiS_{0.5}Se_{1.5}$	,,	,,	3.494	5.944	780	730	300	a	,,
$TiS_{0.3}Se_{1.7}$	,,	,,	3.508	5.967	780	750	350	a	,,
$TiSe_2$	,,	,,	3.528	6.003	780	740	300	metallic-bronze	,,
$TiSe_{1.5}Te_{0.5}$	,,	,,	3.587	6.159	780	720	300	a	,,
$TiSe_{1.2}Te_{0.8}$	,,	,,	3.623	6.236	760	720	300	a	,,
$TiSe_{1.0}Te_{1.0}$	,,	,,	3.643	6.291	760	720	300	a	,,
$TiSe_{0.8}Te_{1.2}$	,,	,,	3.672	6.351	760	700	300	a	,,
$TiSe_{0.5}Te_{1.5}$	,,	,,	3.702	6.425	760	700	300	a	,,
$TiTe_2$	,,	,,	3.763	6.529	650	690	300	grey-metallic	,,
$TiS_{1.8}Te_{0.2}$	,,	,,	3.442	5.808	780	720	350	a	,,
$TiS_{1.5}Te_{0.5}$	,,	,,	3.501	5.961	790	720	350	a	,,
$TiS_{1.0}Te_{1.0}$	,,	,,	3.586	6.192	780	700	400	a	,,
$TiS_{0.8}Te_{1.2}$	,,	,,	3.623	6.285	770	700	300	a	,,
$TiS_{0.5}Te_{1.5}$	,,	,,	3.674	6.387	770	700	350	a	,,
$HfS_{1.0}Se_{1.0}$	,,	,,	3.694	6.061	~900	~800	a	a	[14]

T_H = reaction temp.; T_L = growth temp.; [a] = not specified; [b] = only specified for TiS_2 and TaS_2; [c] = not specified, probably iodine; [d] = for a-and c-variation with x, see reference [50].

3. Group V compounds

3.1. VCh_2

The vanadium chalcogenides are substances characterized as being highly non-stoichiometric, displaying wide ranges of homogeneity within each crystal structure. According to Hoschek and Klemm – who presented the earliest comprehensive report on the vanadium-selenium system – it is senseless to speak of VSe or VSe_2 'compounds' and the meaning of the word stoichiometry is of no use for these materials [52]. Although Nitsche [48] reports the growth of VS_2 crystals by way of iodine transport, it seems commonly accepted that VS_2 does not exist or that it is not preparable with the commonly employed preparation techniques, and that V_5S_8 probably presents the sulfur-rich limit.

The systems V-Se and V-Te have been the object of extensive investigations by Hoschek et al. [52], Röst et al. [53], [54] and Grönvold et al. [55]. In the V-Se system three phases – denoted α-phase, β-phase and γ-phase – are reported by Hoschek et al. The so-called α-phase is a hexagonal one, with the NiAs type of lattice, and was observed in materials with compositions between $VSe_{0.98}$ and $VSe_{1.13}$. Between

VSe$_{1.25}$ and VSe$_{1.60}$ the so-called β-phase exists, which is not hexagonal but clearly of less symmetry and finally between VSe$_{1.62}$ and VSe$_{1.97}$ again a hexagonal phase is observed – CdI$_2$-like type of lattice – called the γ-phase (cf. the γ4-phase in [55]). This picture of the different regions in the V-Se system shows the gradual transformation from the NiAS (α-phase) to the CdI$_2$ (γ-phase) type which is due to the subtraction of metal atoms from the NiAS-type of structure. The study of the Oslo group [53], primarily directed at investigating the low symmetry phase found by Hoschek (β-phase), confirmed the existence of low symmetry material in samples quenched from 750 °C, and indexed it as a monoclinic homogeneity range, ranging from VSe$_{1.12}$ to VSe$_{1.70}$. At the ratio Se/V = 1.70 it reverts to the hexagonal CdI$_2$-like lattice. A continuous change in the hexagonal lattice constants is observed between VSe$_{1.70}$ and VSe$_2$.

This substance VSe$_2$ contains free selenium in addition to the hexagonal phase and precise measurements showed the phase boundary to correspond to VSe$_{1.96}$. Röst et al. [53] performed their experiments with substances which were water quenched from 750 °C, from 1000 °C and with samples slowly cooled (one week period) to room temperature. VSe$_2$ samples quenched from 1000 °C, had lattice constants corresponding to VSe$_{1.90}$ material quenched from 750 °C, namely $a = 3.374$ Å and $c = 6.088$ Å. Finally the constants of the VSe$_{2.0}$ lattice (slowly cooled) are equal, within experimental error limits, to those obtained after quenching from 750 °C.

In their study on the V—Te system which appeared in 1958, Grönvold and his co-workers [55] refer to the investigations of Ehrlich [56] dating back to 1949, who indicated the existence of a Cd(OH)$_2$-type structure in the V—Te system without giving any further evidence. In the previously discussed work of Hoschek et al. [52], the preparation of tellurides is also mentioned, but these authors were not able to establish the composition or structure of any of the tellurides formed in their experiments.

In Grönvold's experiments, material quenched from 750 °C showed a hexagonal VTe$_{1.75}$ phase at 63.64 at % Te. The V—Te system shows the same characteristic feature as has been found in the V—Se system and some other chalcogen systems of the 3d-transition elements. Depending on the temperature, between 50 and 66.7 at % chalcogen one or more phases exist which, at concentrations low in chalcogen, possess a hexagonal NiAS-like lattice. At intermediate concentrations a monoclinic structure and at higher concentrations a hexagonal CdI$_2$-like structure is derived from the NiAS-type of lattice by subtraction of metal atoms. In Grönvold's samples – quenched from 750 °C – with 64.29 to 64.91 at .% Te, a so-called γ4 phase was found. The corresponding X-ray powder photograph was tentatively indexed as orthorhombic [55]. This γ4 phase should not contain more than 64.91 at .% Te, corresponding with a composition VTe$_{1.85}$. In slowly cooled material this γ4-phase was found – free from other phases – at 64.29 at % Te (VTe$_{1.8}$) and mixed with free Te at 66.7 at % Te corresponding with VTe$_{2.0}$. These investigations conclude that stoichiometric VTe$_{2.0}$ does not exist. The most tellurium-rich phase obtained (γ4) differs markedly from VTe_2; changes in heat treatment apparently make no difference as to the non-existence of the stoichiometric ditelluride.

Röst and co-workers [54] reexamined the phase relationships in the V—Te system in 1964. At 64.3 at % Te ($VTe_{1.80}$) the structure could be indexed as a CdI_2-type in disagreement with the orthorhombic type found by Grönvold. At 66.10 at % Te ($VTe_{1.95}$), excess Te was found, in agreement with [55]. Röst, however, concluded that at still higher concentrations of tellurium a monoclinic deformation occurs, in disagreement with the aforementioned work of Grönvold [55].

This rather extensive discussion devoted to the VCh_2 shows that in these systems a wide variety of non-stoichiometric materials can be prepared and that the resulting product to a large extent depends on heat treatment and cooling rate. Finally it must be mentioned that nothing is found about the character of the substances denoted as VCh_2. No reports about polytypism are found either.

3.1.1. Compound Preparation

Hoschek and Klemm [52] made use of the reaction between V_2O_3 and H_2Ch to prepare their samples in the systems V—S, V—Se and V—Te. For the sulfides a temperature of 850 °C was convenient, for the selenides they worked at 600°–900 °C and for the tellurides at 900°–1000 °C. In the telluride system, however, they were not able to obtain reproducible results at reaction temperatures of 1000 °C. In the experiments of the Oslo group [53–55] the various samples were obtained by direct synthesis of calculated amounts of the elements. To avoid reactions between the metal and tube material, metal-rich samples were prepared in small alumina crucibles which were enclosed in evacuated silica ampoules.

Samples with more than 50 at % V were heated shortly at 1550 °C, those richer in Se at 1000 °C for one week. The most Se-rich materials were prepared by adding calculated amounts of Se to finely crushed samples $VSe_{1.3}$ followed by annealing the mixture at 750 °C [53]. Te-poor samples with 23–50 at % Te were heated to 1500–1550 °C for a short period to complete the reaction, materials with 53–63 at % Te were heated to 1050 °C for 2 days [54].

Reported colours are grey with metallic lustre [52] for VSe_x with $x > 1.30$ and dark grey [55] for all tellurides.

3.1.2. Crystal Growth

Nitsche [48] was able to grow VS_2 and VSe_2 single crystals with iodine as the transporter. Transport took place from $T_H = 900° \rightarrow T_L = 850$ °C for VS_2 and from $T_H = 850° \rightarrow T_L = 800$ °C in the case of VSe_2. Black plates of $8 \times 8 \times 0.1$ mm^3 in size were the result of a 48–72 h run. Iodine concentrations used varied between 0.005–0.1 millimoles cm^{-3}. The VSe_2 lattice is hexagonal with $a = 3.36$ Å and $c = 6.05$ Å; the VS_2 lattice is not reported.

Van Maaren et al. [57] have grown $V_{1.01}Te_2$ crystals by using iodine transport from $T_H = 900° \rightarrow T_L = 700$ °C and they claim a far better stoichiometry than in the isomorphous $VTe_{1.85}$ obtained by Grönvold's group [55].

Structure data and preparation conditions are presented in Tables I and V respectively.

3.2. NbS_2

Niobium disulfide is a grey-black metallic material which does not occur naturally. In 1938 Biltz and Köcher [58] reported the first systematic study of many T—Ch systems by means of X-ray diffraction and vapour pressure analyses. They found two phases, each with a broad range of homogeneity; e.g. $NbS_{0.5}$–$NbS_{1.0}$ and $NbS_{1.5}$–$NbS_{4.0}$. Hägg and Schönberg [59] proposed a rhombohedral $CdCl_2$-type of structure for the substance with composition NbS_2 in 1954, and in 1960 Jellinek et al. [60], reinvestigating the Nb—S system, distinguished seven different phases. Among them was the so-called 3s-NbS_2 phase, stable below 800 °C and probably identical with the $CdCl_2$-type disulfide of Hägg et al., and also a 2s-NbS_2 phase stable above 850 °C to ~ 1050 °C. In both polytypes, stacking disorder was found in the c-axis direction which in the 3s-structure could be reduced by annealing at 700 °C. Furthermore they found the corresponding Nb-rich phases like 3s- and 2s-$Nb_{1+x}S_2$ with temperature sensitive homogeneity ranges. The 3s form is stable at 1100 °C for $x \approx 0.12$ to ≈ 0.25 and stable at 800° for $x \approx 0.50$ while the 2s-form is stable above ~ 1000 °C in the range $x \approx 0.30$ to ≈ 0.43. The reader is further referred to [60] and [61]. The above-mentioned studies reveal the difficulty of preparing well-defined, stoichiometric materials.

In 1966 van Maaren et al. [62] observed superconductivity in NbS_2, below 6 K, with a composition-sensitive transition temperature. 3s-NbS_2 became superconducting between 5.0 K and 5.5 K and 2s-NbS_2 had a transition from about 5.8 K to 6.2 K. They suggested that the presence of $Nb_{1+x}S_2$ caused such a broad transition.

It was experimentally verified – as in the case of $NbSe_2$ – that addition of Nb lowered the T_c rather sharply. Stoichiometry problems may thus be expected in NbS_2. For example, one defect common to layered crystals is the presence of metal atoms between the layers. Such a defect has been identified by Thompson et al. [21], in their study of intercalation in TiS_2, as an inhibitor of intercalation due to the fact that the interlayer metal pins the layers together, thus slowing down the rate of intercalation. Analogous defects could play a role in NbS_2. As mentioned above, two polytypes are known at present: the so-called 2s and 3s-forms or alternatively the 2H-(hexagonal lattice type) and the 3R-(rhombohedral) type. Also the metal-rich materials show a 2H and 3R-type of stacking.

3.2.1. Compound Preparation

Direct synthesis from the elements seems to be the best procedure for obtaining stoichiometric or nearly stoichiometric NbS_2. The phase transition between 2H and 3R polytypes at 800°–850 °C observed by Jellinek et al. [60] can serve as an indication for the right preparation procedure. Kadijk et al. [63] prepared samples by direct reaction of the elements in stoichiometric quantities at 1000 °C, followed by either water quenching or slow cooling. The 3R-$Nb_{1+x}S_2$ material was obtained by heating

NbS$_2$ at 1000 °C followed by quenching. Gamble et al. [29] suggest a reaction at 750 °C for 2 weeks' time in ampoules just large enough to accommodate the mixture. At the end of this period the material was ground and reheated to assure homogeneity and subsequently annealed for another 2 weeks' period at 600 °C. This product is pure 3R-NbS$_2$.

The 2-H form can be produced by a reaction of the element at 900 °C for a period of 1 week, followed by water quenching the ampoule and contents. The stoichiometry of this end product, however, seems to need further study.

3.2.2. *Crystal Growth*

Growth experiments with iodine as well as with sulfur have been described by Schäfer and co-workers in 1965 and 1968 [64, 32].

The 1965 report [64] on the iodine transport process, carried out on various Nb compounds, had the interesting conclusion that NbCh$_n$ compounds poor in chalcogen will be transported from $T_L \rightarrow T_H$, those rich in chalcogen from $T_H \rightarrow T_L$. For the 3R-form specific conditions are: transport takes place from $T_H = 900° \rightarrow T_L = 700$ °C in 20 ml. ampoules in a period of 10 days. The compound Nb$_{1.78}$S$_2$ is transported in the gradient $T_L = 840° \rightarrow T_H = 1050$ °C in a 3 week period. Transport of the 2H-form seems to be more difficult; this should be done at $T > 850$ °C due to the transition temperature.

According to Schäfer and Fuhr [64], however, a competing reaction seems to exist in this temperature range. Above 850 °C the monosulfide is transported with the aid of iodine from $T_L = 850° \rightarrow T_H = 1050$ °C. This is an exothermic reaction and it is not surprising that attempts to transport and grow 2H-NbS$_2$ often yield a multiphase product. Furthermore, above 900 °C niobium also reacts with the SiO$_2$ of the tube and forms silicides and oxides.

In [32] a short description mentions possible transport with sulfur and the convenient temperatures $T_H = 800° \rightarrow T_L = 700$ °C although for TaS$_2$ also $T_L = 800° \rightarrow T_H = 1000$ °C is given. Experiments transporting with selenium and tellurium seem to have been promising. Data on structure and lattice parameters are presented in Table I, data on preparation techniques and crystal growth experiments are summarized in Table V.

3.3. NbSe$_2$

Niobium diselenide is a metallic black substance, sometimes with a somewhat greenish appearance [2]. In 1962, Brixner was the first to publish unit-cell parameters of niobium selenides and to propose structure models [65]. Thereafter, in quick succession, several other groups reported their investigations [11, 12, 66–71]. Brown and Beerntsen [11] discussed polytypism among Nb- and Ta-selenides and the growth of single crystals of some two-, three- and four layer selenides, giving cell dimensions and space group data. Revolinsky et al. [67] reported on the identification of selenide and telluride phases with X-ray techniques in the vicinity of NbCh$_2$ and TaCh$_2$. Their study was limited to the composition range TCh-TCh$_2$. Selte et al. [66, 68]

reviewed, among others, data available on intermediate phases in the selenide and telluride systems of Nb and Ta and Kadijk summarized in 1971 the reported data on polytypism of $NbSe_2$ [12]. Huisman et al. [70] discussed non-stoichiometric $Nb_{1+x}Se_2$ and $Ta_{1+x}Se_2$ phases. In their work on the Nb-Se compounds, Revolinsky and co-workers [71] observed the dependence of the superconducting transition temperature on the change of composition and structure in the two-layered hexagonal $NbSe_2$ and in the four-layered $NbSe_2$. In this report, which presents the influence of stoichiometry on the superconductivity, they emphasize the fact that equilibrium and reproducibility in the Nb- and Ta-Se systems were particularly difficult to achieve. Thus the number of phases and their respective homogeneity ranges in these systems should be regarded with some reserve.

Up to the present time it seems that only the following polytypes of $NbSe_2$ are known [11, 12, 72]: the $2s$- and $4s$-forms – alternatively denoted as 2H and 4H(a) $NbSe_2$ – which are stable at lower temperatures, and the two high temperature forms $1s$ and $4s(d)$ $NbSe_2$ respectively denoted as 1T- and 4H(d) $NbSe_2$. The form announced as 3R-$NbSe_2$ is also reported to be 3R $Nb_{1+x}Se_2$. Trigonal-prismatic coordination occurs in the 2H and 4H(a)-form, the 1T-form has octahedral coordination, while in the 4H(d)-form layers with two types of coordination alternate. In the 3R-form only a prismatic coordination occurs.

3.3.1. Compound Preparation

In all reports, the compounds are prepared by direct synthesis in quartz ampoules in the temperature range 600°–1000 °C [2, 67, 71]. Periods of 72–300 h are needed when elements in powder form are employed and 250–300 h when elements in sheet and pellet form are used [71]. In Revolinsky's work [71] a two-layered $NbSe_2$ type is obtained after firing between 600°–800 °C, while a four-layered type is the result after firing between 800°–1000 °C. Brixner [65] heated the elements at 600°–700 °C for 15 h, and thereafter for another 10–15 h at 1000°–1200 °C and obtained a grey looking homogeneous polycrystalline product. The preparation of the $Nb_{1+x}Se_2$ – with $0 < x < 0.3$ – and analogous Ta-Se phases have been discussed by Huisman et al. [70]. It proceeds along the same lines. Depending on the composition and preparation temperature, various forms of Nb-rich diselenides are obtained which differ in stacking sequence.

Samples prepared at 800 °C are of the $3s$- and $2s(b)$-form with $0.05 \leqslant x \leqslant 0.15$, while more metal-rich samples are of the $2s(a)$-form. Lower temperatures demand longer reaction times, and annealing periods of 3 weeks at 600°–700 °C were not sufficient to obtain well-crystallized samples of the Nb- and Ta-rich substances. Huisman also emphasized the fact that the metals used should be free from hydrides, otherwise the quartz tubes were attacked – a fact apparently observed by Schäfer and Dohman [73].

Investigating the different polytypes, Kadijk and Jellinek prepared the low-temperature forms of $NbSe_2$ ($2s(a)$ and $4s(a)$) by heating the elements between 600° and 1100 °C followed by quenching to room temperature [12]. Their X-ray analyses indicated that the majority of the samples thus prepared were $2s(a)$-NbSe$_2$; $4s(a)$-NbSe$_2$ was only obtained in slightly Se-rich material that had been quenched from 850 °C. Heating $2s(a)$-NbSe$_2$ with additional Se at 850 °C followed by quenching also resulted in $4s(a)$-NbSe$_2$ formation. On the other hand, if $4s(a)$-NbSe$_2$ was heated in an evacuated and sealed quartz ampoule at 850 °C and subsequently quenched to room temperature, the $2s(a)$-form was obtained, probably due to loss of selenium from the substance. The high-temperature behaviour of NbSe$_2$ – investigated by D.T.A. and high temperature X-ray techniques – showed that the irreversible exothermal conversion of $4s(a)$-NbSe$_2$ into $2s(a)$-NbSe$_2$ occurs between 600° and 700 °C. Further heating of the $2s$-form resulted in two more transitions, occurring at 910° and 980 °C respectively. It was, however, impossible to retain the high temperature forms by quenching to room temperature.

The type stable above 980 °C has a trigonal structure (Cd(OH)$_2$ type) and was denoted $1s$-NbSe$_2$; the type stable in the small temperature range 910°–980 °C is hexagonal and was denoted $4s(d)$-NbSe$_2$. It is assumed that – as in the $4s(d)$-TaSe$_2$ case – the SeNbSe slabs with octahedral coordination of the metal alternate with slabs having trigonal prismatic coordination of the metal.

Kershaw [74] reacts the elements at 600 °C; this temperature is reached by heating at 15° h^{-1}. The reaction period is 4 days. The tubes are then opened in a drybox, ground and resealed in a silica tube and refired for 4 days at 600 °C. In this way care is taken to insure that the starting material is pure and never exposed to air during manipulation.

3.3.2. Crystal Growth

Growth experiments have been performed with iodine as the carrier in all reports [11, 65, 67, 74] but experimental conditions are not always fully specified. Iodine concentrations of about 5 mg cm^{-3} tube volume are common, but ampoule dimensions, temperature gradients, growth times and crystal dimensions and appearances are often not presented.

Brixner [65] uses 1–3 mg iodine and transports in a gradient $T_H = 900° \rightarrow T_L = 700\,°C$ in a period of 10–15 h an amount of approximately 10 g. The reaction suggested is probably:
$$NbCh_{2(s)} + \tfrac{5}{2} I_{2(g)} \rightleftharpoons NbI_{5(g)} + 2Ch_{(g)}$$
Revolinsky et al. [67] employs tubes of 380 mm × 17.5 mm with 200 to 400 mg. of iodine and observes transport in 75 to 100 h in a gradient
$$T_H = 900°-1000° \rightarrow T_L = 600°-900\,°C.$$
Kershaw [74] presents in detail his refined technique – in a modification of Brixner's procedure – to produce large homogeneous well-formed crystals. A three-zone furnace having an auxiliary heater placed after the center (= growth) zone is used. The temperature profile in the furnace is adjusted by varying the temperature of this extra heater. The best results are obtained when a smooth decrease of the temperature along the ampoule – from charge end to growth zone – can be achieved. The growth zone temperature is constant over a distance of 7 to 8 cm and then gradually increases

towards the empty tip of the ampoule due to the effect of the auxiliary heater. This gradient causes the coolest part of the reaction ampoule – where the crystals grow – to be at the centre of the tube, rather then at the end. As a result the growing crystals have ample space and are well formed instead of being crowded into the tip of the ampoule.

Transport is performed with iodine – 5 mg cm^{-3} tube volume – in 270 mm × 15 mm tubes in a gradient $T_H = 800° \rightarrow T_L = 725$ °C. The growth of a small number of large crystals is optimized by keeping the charge end of the tube at room temperature, while the other end, including the growth zone, is heated to 875 °C. This gradient minimizes the number of nuclei in the center (growth) zone. After 24 h the charge zone is raised to 800 °C, while the other end is maintained at the above-mentioned 875 °C. Thereafter the temperature of the growth zone is gradually reduced to 725 °C at a rate of 1° h^{-1} and 5 days after the growth zone has reached this desired temperature, the whole furnace is shut off and the ampoule with its contents is allowed to cool. The crystals are afterwards washed with carbon tetrachloride to remove any excess of iodine and according to the analyses there appears to be little substitution of selenium by iodine in these samples.

Brown *et al.* [11] mention the difference in habit between the two-layer NbSe$_2$ forms and those having more than two layers. The two-layer forms tend to be thin hexagonal plates, with prominent (0001) faces; the three- and four-layer varieties have more strongly developed rhombohedral and prismatic pyramidal faces, so that they form thicker plates than the two-layer variety. The crystallization temperature of the 2H-type is significantly lower (~750 °C) than that of the 4H-type (~950 °C).

Table I contains data on the structure and lattice parameters and Table V the data on preparation procedures.

3.4. NbTe$_2$

According to Brixner [65], niobium ditelluride is a grey substance with a metal-like conductivity while Schäfer *et al.* [64] report black coloured crystals. The structure shows a distorted octahedral coordination of the metal atoms by tellurium atoms [75]. This distortion is a response to a pattern of metal–metal bonding. in which the metals are grouped into triple rows. This grouping puts two-thirds of the metal atoms in distorted octahedra and one-third in relatively regular octahedra.

In the available reports on Nb-Te systems several metal-rich and several tellurium-rich phases are discussed. Revolinsky [67], who studied the composition range between TCh and TCh$_2$, found three phases and a small homogeneity range for NbTe$_2$. Selte *et al.* [68] reported four other phases, both Te-rich and Nb-rich.

From Brixner's measurements on the electrical resistivity, the thermal conductivity and the Seebeck coefficient, it follows that NbTe$_2$ is a metal-like conductor. According to Wilson and Yoffe [1], the substance is diamagnetic, and the investigations of van Maaren *et al.* [57] have shown a superconductivity with a transition temperature of 0.74 K. So far no polytypism has been mentioned in the current literature.

3.4.1. Compound Preparation

In the preparation of NbTe$_2$ outlined by Brixner [65], the polycrystalline powder is obtained by heating the combined elements in quartz ampoules – 200 mm × 15 mm – in a vacuum of ∼10^{-5} torr. In the first stage of the reaction procedure, the material is heated to 600°–700 °C for 10–15 h, after which time usually a black-looking free flowing reaction product is obtained that can be remixed inside the tube by shaking. A second firing at 1000°–1200 °C for another 10–15 h assures completeness of reaction and a homogeneous product is obtained. The procedure followed by Revolinsky et al. [67] has been discussed in the section on NbSe$_2$.

3.4.2. Crystal Growth

Single crystals of NbTe$_2$ may be prepared by vapour phase transport with the aid of iodine or bromine [64, 65]. Schäfer and Fuhr [64] used ampoules with 48 cm^3 volume and a total charge of 51 mg iodine. In 5 days complete transport was observed from $T_H = 730° \rightarrow T_L = 490$ °C.

Brixner's experiments [65] have been discussed for NbSe$_2$; also bromine was useful as a carrier for transport from $T_H = 900° \rightarrow T_L = 700$ °C in 10–15 h. Revolinsky and co-workers [67] utilized 380 mm × 17.5 mm vitreous silica ampoules, filled with ∼10 g of polycrystalline material and added 200–400 mg iodine. Temperature gradients in the range $T_H = 900$–1000 °C and $T_L - 600°$–900 °C were used. In 75 to 100 h thin platelets grew in the colder zone.

The various data on structure type and lattice parameters are summarized in Table I those on preparation techniques in Table V.

3.5. TaS$_2$

Tantalum disulfide is a substance which has aroused much interest on account of its varying properties observed in the various forms: from semiconducting in the 1T-structure and superconducting in the 2H-form to a 'mix' of both in the recently discovered 4H(b)-TaS$_2$ type. The colours reported are yellow for 1T-TaS$_2$ and grey metallic for 2H-TaS$_2$.

Systematic investigations of the Ta-S system have been reported by Biltz and Köcher in 1938 [76], by Hägg and Schönberg in 1954 [59] and by Jellinek [77], and various Ta$_x$S$_y$ compounds have been found in these studies.

In view of the almost complete disagreement of the results presented in [59] and [76], Jellinek reinvestigated the Ta-S system, mainly by X-ray powder methods [77], and observed the following polytypes.

1s-TaS$_2$, 2s-TaS$_2$, 3s-TaS$_2$ and 6s-TaS$_2$ indicating that the repeat units in the c-direction have a thickness of one, two, three and six TaS$_2$ slabs respectively. They are also alternatively denoted 1T-, 2H-, 3R- and 6R-TaS$_2$. Further investigations have recently revealed the existence of a 4H(b)-form [78] and a superstructure of 1T-TaS$_2$ [79].

In view of the complexity of TaS$_2$, where such a rich variety of different structures occur, the preparation techniques of the different types will be discussed separately. It will become quite clear that stringent precautions have to be exercised in the preparation of the various materials needed for further investigation and great attention is required for preparative details to obtain single-phase stoichiometric TaS_2.

3.5.1. Compound Preparation

1T-TaS_2. This is the high temperature structure, stable above 780 °C, which can be retained as a metastable state by quenching to room temperature. It is a diamagnetic semiconducting substance below 80 K, metallic above 80 K and not superconducting [80] as has been previously reported (see [1]). The preparation method referred to in [78] is heating of stoichiometric 2H-TaS_2 powder at 950 °C with a small excess of sulfur ($\sim$1 atm. pressure) for 24 h and then quenching in air. The small amount of extra sulfur used to obtain the 1T-form, will condense out on the tube wall upon quenching. When TaS_2 powder is quenched from 950 °C without excess sulfur, the resulting product is a mixture of 4H(b), 1T, and 6R TaS_2 [57, 77, 78]. At higher temperatures TaS_2 starts to develop its own sulfur vapour pressure and the resulting material becomes non-stoichiometric.

In Gamble's procedure [29], the preparation is along the same lines. Ta-wire – the low surface area is preferred to avoid the problem of absorbed gases on the metal – and sulfur are first heated to 400 °C. The temperature is then raised to 500 °C in a few hours and kept there for 48 h until most of the sulfur has reacted. Thereafter it is slowly raised to 950 °C in a period of 3–4 days and held at that temperature for one week. The tube is subsequently water quenched. The use of $\sim$100 ml of argon in the tube may increase the quenching rate.

The product is a golden metallic free-flowing powder.

2H-TaS_2. Gamble and co-workers [81] prepare 2H-TaS_2 – the phase stable at room temperature – by heating stoichiometric ratios of the elements in evacuated quartz ampoules at 900 °C for several days, followed by slowly cooling the reaction ampoule and its contents.

The material is a black, free-flowing powder. Pure stoichiometric 2H-TaS_2 has a superconducting critical temperature of 0.75 K, material mixed with 6R, or 4H(b) will have a higher critical temperature, above 1 K.

3R-TaS_2. Jellinek [77] reported a 3R-phase together with four other TaS_2 forms, with TaS_3 and with three lower sulfides of composition $Ta_{1+x}S_2$ as a product of direct reaction between elements at 600°–800° and 1000 °C. In some cases the reaction products were homogenized and annealed again at 800°–1000 °C and some samples were tempered at 200°–300 °C before exposure to X-rays, in order to remove the effect of cold work from unreacted metal present in the samples.

The 3R-form was less frequently obtained than the other TaS_2 forms. Futhermore this 3R-form showed a considerable degree of stacking disorder in the event it had been heated for a short time only, but this disorder could be removed by prolonged heating at 1000 °C. So far no other workers have reported the presence of 3R-TaS$_2$.

4H(b)-TaS$_2$. This polytype was prepared for the first time by di Salvo and co-workers [78] in 1973. In this form, alternate layers contain Ta in either octahedral or trigonal prismatic coordination. Every other layer is like 1T-TaS$_2$ and the remaining are like 2H-TaS$_2$; the properties are described as a mix of 1T-TaS$_2$ and 2H-TaS$_2$. This particular form has only been obtained by chemical vapour transport with iodine as a carrier.

6R-TaS$_2$. The 6R-type was first obtained by Hägg and Schönberg [59] and called δ-TaS$_2$. Jellinek produced the same type in 1962 [77] and determined it to be the 6R-form. It always coexists with the 1T- and 2H-forms but there seem to be indications that the proportion of 6R increases after prolonged heating at 800 °C. Jellinek assumes that the 6R-form is the stable phase at this temperature. As in the 2H-form, stacking disorder was found in the 6R-type, although less pronounced in samples that had been annealed for long times.

Random TaS$_2$. In Jellinek's batch, one sample which had been heated for a short time gave a diffraction pattern that could be indexed with a unit cell containing only 1/3 TaS$_2$, and this had evidently a random layer structure (probably similar to that of NiBr$_2$ or CdBr$_2$).

In the report of van Maaren and Harland an energy band model for Nb- and Ta-dichalcogenide superconductors is presented. Those interested in this topic are referred to [82] which presents data on the specific heat C_v between 1.5 and 10 K, the resistivity ρ, the Hall constant R_H and the transition temperature of the NbCh$_2$ and TaCh$_2$.

3.5.2. Crystal Growth

Crystals of 1T-TaS$_2$ were prepared with iodine as the transporter by Brouwer et al. [79], Revelli et al. [83], Conroy et al. [84], and by Schäfer et al. [32] with the aid of sulfur vapour. Temperature gradients differ slightly, $T_H=950° \rightarrow T_L=900\,°C$ in [79] and $T_H=870 \rightarrow T_L=750\,°C$ in [83]. In [84] transport is from $T_H=1000° \rightarrow T_L=800\,°C$ and to induce the growth of 1T-TaS$_2$ a small quantity (0.2 mg ml^{-1}) of the isomorphous SnS$_2$ is introduced into the transport ampoule.

Various ampoule sizes are reported, ranging from 150 to 250 mm in length and 25 mm in diameter. The iodine concentration was 4 mg cm^{-3} tube volume, the reaction time about one week and yellow 1T-TaS$_2$ plates the product. Schäfer and co-workers [32] reported the transport of TaS$_2$ with sulfur vapour as a carrier in the range $T_H=800° \rightarrow T_L=700\,°C$ as well as from $T_L=800° \rightarrow T_H=1000\,°C$.

Sulfur pressures employed at these temperatures are about 5–10 atm. Ampoules were 110 mm × 17 mm and the product contained black crystals with yellow metallic lustre.

Van Maaren [57] reports the use of bromine in the gradient $T_H=925° \to T_L=725\,°C$ and of iodine in the gradient $T_H=900° \to T_L=700\,°C$.

Crystals of the 2H-variety – wrinkled grey-black – were obtained in the experiments of Revelli and Phillips [83]. They prereacted the elements in the presence of iodine at 950 °C for several days followed by transport from $T_H=870 \to T_L=750\,°C$ in 10 days' time. No tube dimensions or halogen concentrations are mentioned. Conroy et al. [84] also obtained the 2H-form when SnS_2 was not added to the starting material. Whereas the 1T-material is always highly ordered, giving sharp X-ray diffraction patterns, the 2H-form is strongly faulted material; only reflections for which $h-k=3n$ are sharp. Thompson et al. [80] transport prereacted material in the course of one week with iodine in the gradient $T_H=950 \to T_L=850\,°C$.

Crystals of the 3R-variety have not been reported in growth experiments.

Crystals of 4H(b)-TaS_2 have been produced by di Salvo and co-workers [78] by chemical transport with iodine as the transporter. Prereacted starting material is employed for the reason that complete reaction of the elements may take several weeks even at 950 °C and incomplete reactions leave behind some sulfur. This excess sulfur present in the tube during growth would then promote the growth of only the 1T-form. It is on these grounds that prereacted charge material is preferred. The source temperature and growth times are not specified but the growth temperature is given as 700 °C and crystals of $4 \times 4 \times 2$ mm^3 are obtained. The plate thickness of 2 mm is exceptional, according to these investigators, as the 1T- and 2H-forms always grow in thin plates of $<100\,\mu$ thick.

Mention of crystals of the 6R-variety and the random type of TaS_2 have so far not been found in literature.

Data on structure type and lattice constants are presented in Table I, preparation and growth techniques are summarized in Table V.

3.6. $TaSe_2$

Tantalum diselenide is a substance that resembles the disulfide in many ways. This is again a rather complex system, with many polytypic forms, also in the non-stoichiometric range $Ta_{1+x}Se_2$. In other words, great attention should be given to details in preparative work when dealing with the different compounds. As mentioned for TaS_2, different polytypes have different colours; 1T-$TaSe_2$ is golden metallic, 2H(a)-$TaSe_2$ crystals are silvery [83], the other $TaSe_2$ forms are black [85].

A great deal of work has been performed and reported in the 1960's and some rather complete, systematic, studies have been published in those years.

In 1964 Kadijk et al. [72] reported the presence of the $2s$, $3s$, $4s$ and $6s$-forms of $TaSe_2$.

In their report on intermediate phases in the Ta-Se system in 1966, Selte et al. [68] described the system roughly as having only two intermediate phases, i.e. $Ta_{1+x}Se_2$ and $TaSe_3$. In the metal-rich part they described their results as less satisfactory as compared with the results in the Nb-Se system. Most of the Ta-Se samples seem to contain impurities, some in considerable concentrations, probably due to the (unavoidable) reaction between the metal and the tube material. For the $Ta_{1+x}Se_2$ family they report 6 forms, i.e. the 1T, the 2H, the 3R, the 4H(a) and 4H(b) and the 6R-types, and they discuss the various homogeneity ranges. In 1967 Bjerkelund and Kjekshus [86] followed with a review of the $TaSe_2$ and $Ta_{1+x}Se_2$ polytypes. In this rather extensive report crystallographic data, density measurements and magnetic susceptibility data are presented.

In 1969, Huisman and Jellinek [9] announced a new polytype in the 4H(c)-series of $TaSe_2$ and proposed some reversible and irreversible transitions on account of investigations of the thermal behaviour of the various forms. Furthermore, in their report Bjerkelund and Kjekshus [86] also presented polytypic modifications of the $Ta_{1+x}Se_2$ phase. The composition range extends from ~ 55 to 66.7 at % Se for samples cooled slowly from 900 °C. According to these investigators, it was difficult to obtain consistent, reproducible results within the homogeneity range, i.e. when $x > 0$, due to the close structural relationships between the possible polytypes. A large number of samples within the concentration range of the $Ta_{1+x}Se_2$ phase contained a mixture of two or three of the polytypes.

Consistent results were obtained after relatively short heat treatments and re-annealing in general did not improve the material but occasionally worsened the homogeneity. However, some trends have been noticed for the occurrence of polytypism in $Ta_{1+x}Se_2$. The 1T-, the 4H(a)-, the 4H(b)-, and the 6R-form could only be obtained for $x = 0.0$, the form $2H-Ta_{1+x}Se_2$ only for $0.00 \leqslant x < 0.03$ and $0.25 < x < 0.64$, and the 3R-form for $0.03 < x < 0.25$ but almost invariably mixed with traces of $2H-Ta_{1+x}Se_2$. (See also [70].)

The relationships between the different $Ta_{1+x}Se_2$ and those between $TaSe_2$ polytypes are pointed out by Kadijk et al. [72] and Selte et al. [68], who also discuss the theory of polytypism resulting from spiral growth initiated by screw dislocations.

The physical properties of $TaSe_2$ depend on the coordination of the metal [85]; $1T-TaSe_2$ with octahedral coordination is gold-like and diamagnetic, the other forms are black and metallic and show nearly temperature independent paramagnetism.

As for the disulfide, the preparation of the different types will be discussed separately. It will be clear that – due to the occurrence of a rich variety of different structures with different properties – the preparation of the various substances needs great attention.

3.6.1. Compound Preparation

1T-TaSe₂. This is the form stable above about 800 °C. Its structure can be retained at room temperature by quenching from 900°–950 °C [9, 86] using a high quenching rate. The preparation technique referred to in [9] employs direct synthesis from the elements in evacuated quartz ampoules, for times varying between 15 min and 36 h followed by quenching to room temperature.

Reaction temperatures are above 880 °C. (See the transition scheme as presented by Huisman et al. [9].) Bjerkelund et al. [86] heated the constituent elements for a period of 14 days at 900 °C, followed by slow cooling to room temperature, and thereafter gave all their samples a reannealing between 200° and 900 °C. 1T-TaSe$_2$ was only obtained in samples which were subsequently quenched from 900 °C. Aslanov et al. [87] were probably the first to prepare the 1T-phase. Where in the disulfide the 1T-form is stabilized by an overpressure of sulfur, it seems likely that Aslanov and co-workers were able to prepare the 1T-form at 750 °C because of a slight excess of selenium. Slow cooling of their samples yielded the triselenide. In a later report [88], the formation of TaSe$_2$ by decomposition of the triselenide at 780 °C is mentioned.

2H-TaSe$_2$. It was Brixner [65] in 1962 who first prepared this form. The constituent elements, weighed to the nearest milligram, were heated in quartz reaction ampoules (200 mm × 15 mm; vacuum ~10^{-5} torr) to 600°–700 °C for 10 to 15 h. The resulting black, free flowing powder which could be easily remixed by shaking the ampoule was then reheated at 1000° to 1200 °C for another 10 to 15 h to assure completeness of reaction. The cooling rate is not specified, the end product is homogeneous and of grey-metallic lustre. Kadijk et al. [72] observed the 2H-form as the only or as the main product when the elements reacted at 900° to 1200 °C. His cooling rate was not specified but it seems that it must be slow, as quenching seems to be reported for the 1T-phase only.

From the phase diagram of the Ta-Se system as presented by Revolinsky and co-workers [67] in 1965, it appears that the 2H-form of TaSe$_2$ is the low temperature phase. In their preparative procedure these investigators reacted stoichiometric proportions of Ta and Se in powdered form in quartz ampoules; portions of 5 g of 150 mesh material were heated to 600°–1000 °C for 72 h and then air quenched. Temperatures below 600 °C resulted in incomplete reaction products, while temperatures above 1000 °C approached the limitations of the ampoule. Within analytical error, the chemical analyses of their samples near the composition TCh$_2$ were correct to within 1% of the intended composition.

According to Bjerkelund and Kjekshus [86], the best condition for the preparation of the 2H-polytype may be a heat treatment at ~800 °C followed by slow cooling to room temperature.

3R-TaSe$_2$. The first report on this polytype goes back to 1962 when Brixner [65] observed this form. He was not quite sure whether this modification – found on basis of powder X-ray analysis – was a metastable high temperature modification obtained by quenching or a new polytype. In 1964 Kadijk and co-workers [72] observed that the 3R-form could always be obtained when the preparation was carried out between 500° and 600 °C and that a mixture of 2H and 3R was the result of heating at intermediate temperatures, i.e. between 500° and 1200 °C. The purest 3R-TaSe$_2$ samples – still containing small amounts of 2H-TaSe$_2$ as an impurity – were obtained by Bjerkelund and Kjekshus [86] in experiments when the metastable form 1T-TaSe$_2$ was heated at 300°–500 °C. Huisman et al. [9] confirmed this and determined the irreversible transformation of 1T-TaSe$_2$ to 3R-TaSe$_2$ on heating to about 200 °C.

4H(a)-TaSe₂. Of the 4H-forms, three different modifications are reported in literature. The first form was found in 1964 by Kadijk *et al.* [72] as a small portion of a mixture of a 2H- and a 3R-sample. In the same case the 6R-form was also found. The accuracy of the unit-cell determination of this polytype was not as high as in the case of the other types.

Bjerkelund *et al.* [86] prepared the four-slab $TaSe_2$ form by protracted annealing of their samples followed by slow cooling to room temperature, but Huisman *et al.* [9] did not observe this 4H(a)-form in their experiments.

4H(b)-TaSe₂. This form of the four-slab family was first prepared in 1965 by Revolinsky *et al.* [67]. It has a shuffled deck structure with alternate layers, having trigonal prismatic and octahedral coordination.

As in the 2H-NbS_2 structure, all the cations are aligned in the *c*-direction (see also [60] and [78]).

This $TaSe_2$ phase, the structure of which was determined by Brown *et al.* [11], could not be reproducibly prepared.

According to Bjerkelund *et al.* [86] a reproducible preparative procedure for obtaining pure 4H(b) $TaSe_2$ is to quench samples from $\sim 800\,°C$. It seems, however, that this method also yields 6R-$TaSe_2$, a fact confirmed in [9] where the preparation of the 4H(b)-form along two possible lines is presented, i.e. either by heating the 4H(c)-form above 770 °C or alternatively by cooling the 1T-form below 780 °C. From the transition scheme it can be seen that the 4H(b) stability range lies between $\sim 530°$ and $\sim 800\,°C$.

4H(c)-TaSe₂. This polytype was discovered by Huisman *et al.* [9] in 1969. They obtained it by cooling the 4H(b)-form below 530 °C. Here the Ta atoms are in trigonal prismatic coordination.

6R-TaSe₂. This form is first described in [87], where it is mentioned as being mixed with the 2H(a)-form.

Furthermore, this form also occurred in [72], where it is reported as a sublimate at the top of the reaction ampoule. In their experiments at intermediate temperatures, Kadijk *et al.* usually found a mixture of the 2H- and the 3R-forms, sometimes also the 6R-form was observed. It seems reasonable to assume that this deposit is the result of sublimation of $TaSe_2$ under its own vapour pressure during the process of heat treatment given to the various samples and not a result of a chemical vapour transport with a deliberately added transporting agent.

In [86] the 6R-form is reported as the result of quenching samples from $\sim 800\,°C$. However, this procedure also yields the 4H(b)-form. From the experiments discussed in [9], the 6R-form can be prepared by heating 2H-$TaSe_2$ above 800 °C but below 870 °C.

The following transition scheme is presented in [9]:

Reversible transitions are

$$4H(c)-TaSe_2 \underset{770\,°C}{\overset{530\,°C}{\rightleftarrows}} 4H(b)-TaSe_2 \underset{880\,°C}{\overset{780\,°C}{\rightleftarrows}} 1T-TaSe_2 \qquad (1)$$

$$2H-TaSe_2 \underset{800\,°C}{\overset{580\,°C}{\rightleftarrows}} 6R-TaSe_2 \underset{870\,°C}{\overset{780\,°C}{\rightleftarrows}} 1T-TaSe_2 \qquad (2)$$

Irreversible transitions are

$$1T-TaSe_2 \xrightarrow{300\,°C} 3R-TaSe_2 \qquad (3)$$

$$3R-TaSe_2 \xrightarrow{800\,°C} 6R-TaSe_2 \qquad (4)$$

$$6R-TaSe_2 \xrightarrow{225\,°C} 3R-TaSe_2 + 2H-TaSe_2 \qquad (5)$$

$$4H(b)-TaSe_2 \xrightarrow{220\,°C} 4H(c)-TaSe_2 \qquad (6)$$

Non-stoichiometric phases like $Ta_{1+x}Se_2$ have been investigated by Huisman and co-workers [70]. In the range $Ta_{1.05}Se_2$ to $Ta_{1.30}Se_2$ they find a kind of dependence of the polytypes on the temperature of preparation which resembles the $Nb_{1+x}Se_2$ polytypes. At a preparation temperature of 600 °C, all material seems to be of the 3R-modification; although for $0.20 \leqslant x \leqslant 0.30$ the samples probably contain unreacted metal.

When prepared at 700 °C the resulting product contains 2H(b), 2H(a), 3R- and 6R $Ta_{1+x}Se_2$, and the same is observed when a temperature of 800 °C is used. When heated at 900 °C most of the material is in the 2H(a) or 2H(b) form with occasionally some 3R. The same result is obtained when a temperature of 1000 °C is utilized.

3.6.2. Crystal Growth

Reports on crystal preparation techniques have been presented by Brixner [65], Brown *et al.* [11], Revolinsky *et al.* [67] and Revelli and Phillips [83].

In all reports chemical vapour transport is employed with iodine (or some times bromine) as the carrier.

Brixner's method [65] is discussed in the section on $NbSe_2$ (Section 3.3.2), in [11] crystal growth by vapour transport is only mentioned, the specific conditions are those discussed by Revolinsky and co-workers [67] and the reader is referred to the discussion on $NbSe_2$ growth where Revolinsky's procedure is presented. The experiments of Revelli and Phillips [83] are presented in the section on TaS_2-crystal growth; 2H-$TaSe_2$ crystals obtained were silver coloured, those of the 1T-modification were shiny gold hexagons. Structure data are presented in Table I, preparation techniques in Table V.

3.7. $TaTe_2$

Tantalum ditelluride has been discussed in 1959 by Ukrainskii *et al.* [89, 90] who observed four phases in their powder X-ray analysis of the Ta-Te system; one of these was the tritelluride, there were two modifications of a subtelluride $TaTe_{0.35-1.12}$ and the ditelluride $TaTe_{1.5-2.0}$. Using the CdI_2-type of structure as a model, the ditelluride was found to have an orthorhombic pseudohexagonal layer structure.

Brixner [65] analysed single crystal patterns and suggested the structure has a rhombohedral lattice, related to $CdCl_2$.

Revolinsky and co-workers [67], investigating the Ta-Te system, found three phases within the composition limit $TaTe-TaTe_3$. Their ditelluride was found to possess a monoclinic structure more closely related to the octahedrally coordinated CdI_2-type structure than to the trigonal prismatically coordinated MoS_2 and NbS_2 structure. The homogeneity range of this phase extends from TaTe to $TaTe_2$ in material sintered below 800 °C and from $Ta_{1.16}Te_2$ to $TaTe_2$ in material sintered above 800 °C.

In 1966, Brown [75] studied the crystal structures of the isostructural compounds $NbTe_2$ and $TaTe_2$ by Patterson analysis and reported a distorted version of the CdI_2-type with metal–metal bonding. Selte and co-workers [68] reported in 1966 on their study of the Ta-Te system; they found a narrow homogeneity range, in disagreement with the earlier mentioned results [67, 75] of Revolinsky and of Brown. The density of their material agrees exactly with the theoretical density of Brown.

3.7.1. Compound Preparation

Ukrainskii et al. [89, 90] prepared his substance by direct reaction of the elements at 900 °C for a period of 18 days, followed by cooling over 7 days to room temperature.

In Brixner's procedure [65] – as previously discussed for the $NbCh_2$ – the elements are heated for 15 h at 600° to 700 °C, cooled, remixed by shaking the tube, reheated for the same length of time at 1000° to 1200 °C and again cooled. No specific cooling rate is mentioned. Revolinsky and co-workers [67] prepared the polycrystalline material under consideration by direct synthesis from powdered elements at 600° to 1000 °C in a period of 72 h, followed by air quenching. If temperatures lower than 600 °C were used, the resulting substance showed incomplete reactions.

3.7.2. Crystal Growth

Brixner's method [65] of growing crystals and the experiments of Revolinsky and co-workers [67] have been discussed in the section devoted to $NbSe_2$; the preparation of $TaTe_2$ single crystals goes along those lines.

The structural data are summarized in Table I, the preparation techniques and crystal growth procedures are presented in Table V.

TABLE V
Compound preparation and Crystal growth techniques

Compound	Compound preparation procedures	Crystal growth techniques
VCh_2	a. Reaction between V_2O_3 and H_2Ch, at 850 °C for VS_2, at 600°–900 °C for VSe_2 [52], at 1000 °C for VTe_2 b. Direct synthesis from elements at 1000°–1500 °C. [53–55]	I_2 transport $T_H = 900° \to T_L = 850$ °C for VS_2, $T_H = 850° \to T_L = 800$ °C for VSe_2, with an iodine concentration of 0.05–0.1 millimoles cm^{-3} [48] $T_H = 900° \to T_L = 700$ °C for $V_{1.01}Te_2$ [57] See also [32]
NbS_2	Direct synthesis from elements, at >850 °C yields the 2H-form, stable up to 1000 °C. at <800 °C yields the 3R-form [60, 63] 2-weeks annealing at 750 °C, then at 600 °C, for 1-week gives the 3R-type, 1-week at 900 °C, then quenched, yields the 2H-type. [18]	I_2 transport $T_H = 900° \to T_L = 700$ °C; $21 mgI_2$; 20.5 ml tubes; 10 days, gives 3R-Type [64] See also [32] for sulfur transport
$Nb_{1+x}S_2$	Synthesis at 1100 °C gives 3R-type, with $x \approx 0.12$ to ≈ 0.25 at 800 °C gives 3R-type, with $x \approx 0.12$ to ≈ 0.5 The 2H-type is stable above 1000 °C in range $x \approx 0.30$ to $x \approx 0.43$ [60, 63]	I_2 transport $T_L = 820° \to T_H = 1040$ °C; $2mgI_2 \, cm^{-3}$ gives $2H$-$Nb_{1+x}S_2$ [2, 63]
$NbSe_2$	Direct synthesis at 600° to 1100 °C, gives $2Ha$-$NbSe_2$ and $2Ha$-$Nb_{1+x}Se_2$. [2] at 850 °C and then quenched it yields $4Ha$-$NnSe_2$, and at 800°–1100 °C, it yields $2Hb$-$Nb_{1+x}Se_2$ Heating $4Ha$-$NbSe_2$ at 970 °C gives $1T$-$NbSe_2$ Heating at 600°–700 °C, then at 1000°–1200 °C., gives $2H$- and $4H$-$NbSe_2$ [65] Heating at 600°–1000 °C, for 72 h then air quenching, source material is 150 mesh powder, yields 2H-, 4H- and 3R-$NbSe_2$ [67]	I_2 transport (or bromine transport) $T_H = 900° \to T_L = 700$ °C; 1–3 mg I_2 cm^{-3}; tubes 200 mm × 15 mm; 15 h, gives $2H$-and$4H$-$NbSe_2$ [65] $T_H = 900°$–$1000° \to T_L = 600°$–900 °C; 200–400 mg I_2; tubes 380 mm × 17.5 mm; time 75–100 h gives 2H-, 3R- and 4H-$NbSe_2$ [67] See also [11] $T_H = 800° \to T_L = 725$ °C; 5 mg cm^{-3} I_2; tubes 270 mm × 15 mm; 5 days period, gives $2H$-$NbSe_2$ [74]

Table V (Continued)

Compound	Compound preparation procedures		Crystal growth techniques
$Nb_{1+x}Se_2$	Direct synthesis at 800 °C for 8 days, then slow cooling over 2 days	[66]	No growth experiments to obtain $Nb_{1+x}Se_2$ crystals are mentioned
	Reaction at temperatures between 600° and 1150 °C, followed by quenching results		
	for $T=880$ °C in 2 Hb $T=1000$ °C in 2 Hb+3 R		
	$T=900$ °C in 2 Hb+3 R $T=1100$ °C in 2 H } with $x=0.5$		
	for $T=800$ °C in 2 Hb+3 R $T=1000$ °C in 3 R		
	$T=900$ °C in 3 R $T=1100$ °C in 2 Ha } with $x=0.10$ [70]		
	for $T=800$ °C in 2 Hb+3 R $T=1100$ °C in 2 Ha		
	$T=900$ °C in 3 R } with $x=0.15$		
	for $T=800$°–1100 °C in 2 Ha with $x=0.20$		
	for $T=800$° and 900 °C in 2 Ha with $x=0.30$ this substance may contain unreacted Nb.		
$NbTe_2$	Direct reaction of elements at 600°–700 °C, with second firing at 1000°–1200 °C.	[65]	I_2 transport
			$T_H=900°\to T_L=700$ °C; 1–3 mg cm^{-3} I_2; tubes 200 mm × 15 mm; 10–15 h [65]
	Synthesis at 600°–1000 °C for 72 h, air quenching, powder is 150 mesh material.	[67]	$T_H=900°$–1000 °C $\to T_L=600°$–900 °C; 200–400 mg I_2; 72–100 h; tubes are 380 mm × 17.5 mm [67]
			$T_H=925°\to T_L=650$ °C (Other conditions probably same as given in [67]) [75]
			$T_H=730°\to T_L=490$ °C; tubes of 48 ml.; 51 mg I_2; 5 days [64]
TaS_2	Direct synthesis at 600°, 800° or 1000 °C, either annealed at 800°–1000 °C or homogenized, gives 2H-, 3R-, 6R- and 1T-TaS_2. The 6R always mixed with 1T-form. Prolonged heating at 800 °C yields increasing amount of 6R-TaS_2.	[77]	I_2 transport
			$T_H=1000°\to T_L=800$ °C; 4mg I_2 cm^{-3}; tubes 250 mm × 25 mm or 150 mm × 25 mm period of 1-week, gives 2H-TaS_2, if SnS_2 is added, product is 2H- and 1T-TaS_2 [84]
	Direct synthesis at 900 °C for 1-week	[84]	$T_H=?\to T_L=700$ °C, gives 4 Hb-TaS_2. nothing else specified [78]
			$T_H=950°\to T_L=900$ °C, rapid quenching, nothing else specified [79]

Table V (Continued)

Compound	Compound preparation procedures		Crystal growth techniques	
TaS_2	Heating stoichiom. $2H-TaS_2$ at 950 °C under ~1 atm. S-press. for 24 h, air quenched, gives $1T-TaS_2$, quenched from 950 °C, without excess S, gives $4HB + 1T + 6R$	[78]	I_2 transport $T_H = 950° \rightarrow T_L = 850$ °C, 1-week, iodine. $T_H = 870° \rightarrow T_L = 750$ °C, in 10 days, nothing else specified. $T_H = 900° \rightarrow T_L = 700$ °C, with iodine, $T_H = 925° \rightarrow T_L = 725$ °C, with bromine, nothing else specified $T_H = 800° \rightarrow T_L = 700$ °C, with iodine, $T_L = 800° \rightarrow T_H = 1000$ °C, with sulfur,tubes: 110 mm $\times$ 17 mm	[80] [83] [57] [32]
	Reaction of stoichiom. amounts of Ta and S at 900 °C for days, slow cooling, gives $2H-TaS_2$, with slight excess S, and then water quenched, it yields $1T-TaS_2$	[29, 81]		
$Ta_{1+x}S_2$	Procedure reported in reference [77] for $2H-$, $3R-$, $6R-$ and $1T-TaS_2$ are also valid.	[9, 86]		
$TaSe_2$	Direct synthesis at temperatures > 950 °C, followed by quenching, yields $1T$-form, stable > 880 °C	[65]	I_2 transport $T_H = 870° \rightarrow T_L = 750$ °C, 10 days, quenching with excess Se, from 900 °C gives $1T-TaSe_2$ $T_H = 900° \rightarrow T_L = 700$ °C, tubes 200 mm $\times$ 15 mm; 1–3 mg I_2 cm^{-3}, gives $2H-TaSe_2$ and $3R-TaSe_2$, see $NbSe_2$	[83] [65]
	Reaction at 600°–700 °C, gives the $2H$-form, see $NbSe_2$.	[65]		
	Reaction at 600°–1000 °C in 72 h, air quenched yields $2H-$ and $4Hb-TaSe_2$, see $NbSe_2$	[11, 67]		
	Reaction between 900° and 1200 °C for periods between 30 minutes and several weeks, slow cooled or quenched gives different types of $TaSe_2$.	[72]	$T_H = 900°$–1000 °C $\rightarrow T_L = 600°$–900 °C, tubes 380 mm $\times$ 17.5 mm, 75–100 h, gives $2H-TaSe_2$, see $NbSe_2$.	[67]
	Heat treatment of samples at ~800 °C and slow cooling gives $2H-TaSe_2$.	[86]	Same conditions as for the $2H-TaSe_2$ form gives also $4Hb-TaSe_2$.	[67]
	Heat treatment of $1T$-form at 300°–500 °C gives $3R$-form.	[9]		
	Heating $1T-TaSe_2$ slowly to 200 °C gives $3R-TaSe_2$			
	Protracted annealing of samples at 900°, and slow cooling yields $4Ha-TaSe_2$	[86]		
	Quenching from 800 °C gives $4 Hb-$ and $6R-TaSe_2$			

Table V (Continued)

Compound	Compound preparation procedures		Crystal growth techniques
TaSe$_2$ (Continued)	Heating the elements at 850 °C for 36 h, then quenching gives 4Hb-form Heating 4Hb-TaSe$_2$ at 220 °C, gives 4Hc-TaSe$_2$ Heating Ta. and Se. between 500°–1200 °C, gives 6R-type. The 6R-form is stable between 780° and 580 °C	[9] [72] [9]	
Ta$_{1+x}$Se$_2$	Prepared at 600 °C, it yields 3R-Ta$_{1.05}$Se$_2$ and 3R-Ta$_{1.10}$Se$_2$. at 700 °C it yields 2Hb-Ta$_{1.09}$Se$_2$ and 2Hb-Ta$_{1.12}$Se$_2$. at 1000 °C it yields 2Ha-Ta$_{1.15}$Se$_2$ and 2Ha-Ta$_{1.20}$Se$_2$.	[70]	I_2 transport $T_H = 1092° \to T_L = 1041$ °C gives 2Ha-Ta$_{1.19}$Se$_2$. $T_L = 700° \to T_H = 1050$ °C gives 2Ha-Ta$_{1.20}$Se$_2$. [70]
TaTe$_2$	See preparation of NbTe$_2$, in ref. [65 and 67] analogous procedures, are used for TaTe$_2$		Analogous procedures are utilized to grow single crystals as described for NbTe$_2$. See ref. [32, 65 and 75]

3.8. SOLID SOLUTIONS

Reports of solid solution formation have been presented by Kadijk [2] and Revelli and Phillips [83] while the results of the work of Thompson and coworkers in the Ti-Ta-S system are discussed in Section 2.10 together with the other solid solutions of the group IV compounds. Kadijk's experiments were directed to mixed niobium-molybdenum sulfides, those of Revelli and Phillips to the Ta-S-Se system.

3.8.1. Compound Preparations and Crystal Growth

Kadijk [2] prepared two series of samples with initial compositions $Nb_xMo_{1-x}S_{2.00}$ and $Nb_xMo_{1-x}S_{2.25}$ respectively, with $x=0.20$, 0.40, 0.60 and 0.80. The mixed elements were heated to 900 °C for some days and both series gave essentially the same results. The Mo-rich samples appeared to be very badly crystallized; reheating the material at 1000 °C for some days did not improve their crystallinity,

The X-ray diffraction patterns are very similar to those of badly crystallized $2H-MoS_2$. The unit-cell parameters, though not accurate, suggest that solid solutions are formed. The Nb-rich material shows X-ray diffraction patterns of the rhombohedral 3R-structure type. Unit cell parameters are intermediate between those of $3R-NbS_2$ and $3R-MoS_2$. In Table VI these data are summarized. No specific crystal experiments with halogen or sulfur as a carrier have been mentioned.

Revelli and Phillips [83] prepared material $TaS_{2-x}Se_x$ with x from 0 to 2.0. It follows from their experiments that mixed anion samples exhibit a series of polytypes similar to those found in TaS_2 and $TaSe_2$ with a- and c-lattice parameters increasing monotonically from TaS_2 to $TaSe_2$. The measurements included electrical and superconducting behaviour and the degree of intercalation of pyridine and collidine in these mixed anion substances.

Mixed anion material was prepared by heating stoichiometric mixtures of the elements at 950 °C for a period of one week followed by cooling slowly over one week to 300 °C and then annealing the substance at this temperature for another week. Single crystals were prepared with the iodine transport process, usually by prereacting the elements with iodine at 950 °C for several days, followed by transport in the gradient $T_H=870° \rightarrow T_L=750$ °C in 10 days' time.

These samples were then annealed at 500 °C for several days and cooled to room temperature in 4–5 h.

An alternative procedure was to keep the elements with the halogen at a uniform temperature of 950 °C for about 10 days and then to give it a similar annealing treatment. In a discussion of the phase diagram the characteristics of their samples as a function of composition are summarized as follows. For all x-values there are in general low, medium and high temperature forms. The purely trigonal prismatic phases 2H(a), 3R- and 4H(c) occur in the low temperature region, the 4H(b) and 6R phases in an intermediate range and the octahedral 1T-phase occurs at highest temperatures. The temperature at which the 1T stabilizes seems to increase with Se content.

In the low temperature region for $x > 0.4$ the 3R-phase predominates but for $x < 0.2$ the 2H(a) phase is stable. Annealing experiments (300 °C, 3-weeks) for $x > 0.4$ showed that the 3R-phase remained while the 2H(a)-phase seemed to diminish or to disappear. For $0.2 < x < 0.4$ the 4H(c) exists at low temperatures.

The intermediate temperature range $- 500\ °C \leqslant T \leqslant 900\ °C$ – is complex. For $x > 0.4$ a pure 1T phase could not be obtained by quenching; only in the cases of TaS_2 and $TaSe_2$ was this possible. The mixed anion samples separated into two 1T phases, corresponding to slightly different compositions.

Discussing the correlation between slab thickness and stability of the phase, Revelli et al. indicate that the layer slab thickness is associated with the most stable phase. Another observation made is that the phases with more octahedral character occur at higher temperatures. For more details and for the plot of a-axis variation as a function of x for the various $TaS_{2-x}Se_x$ phases, the reader is referred to [83].

TABLE VI
Solid solutions of $NbCh_2$ and $TaCh_2$

Compound	Transport. agent	Lattice	Parameters (Å)		T_H (°C)	T_L (°C)	Reaction (days)	Colour	Ref.
			a	c					
NbS_2	N.T.	2H-form	3.31	11.89	N.T.	N.T.	N.T.	N.S.	[2]
NbS_2	,,	3R-form	3.33	17.81	,,	,,	,,	,,	,,
$Nb_{0.8}Mo_{0.2}S_2$	,,	3R-form	3.31	17.97	,,	,,	,,	,,	,,
$Nb_{0.6}Mo_{0.4}S_2$	,,	3R-form	3.25	18.01	,,	,,	,,	,,	,,
$Nb_{0.4}Mo_{0.6}S_2$	,,	2H-form	3.25	12.01	,,	,,	,,	,,	,,
$Nb_{0.2}Mo_{0.8}S_2$	,,	2H-form	3.20	12.05	,,	,,	,,	,,	,,
MoS_2	,,	2H-form	3.16	12.29	,,	,,	,,	,,	,,
MoS_2	,,	3R-form	3.16	18.37	,,	,,	,,	,,	,,
$TaS_{2-x}Se_x$	I_2 for all x-values	2H-TaS_2	see ref. [83]		870	750	10	grey-bl.	[83]
		2H-$TaSe_2$	,,		,,	,,	,,	silver	,,
		3R-forms	,,		,,	,,	,,	shiny-black.	,,
		4Hc-forms	,,		,,	,,	,,		,,
		4Hb-forms	,,		,,	,,	,,	,,	,,
		6R-forms	,,		,,	,,	,,	,,	,,
		1T-forms	,,		,,	,,	,,	golden	,,

N.S. = not specified, N.T. = not transported according to data.

4. Group VI compounds

4.1. MoS_2

Molybdenum disulfide, or molybdenite, is a grey black substance and undoubtly the best-known layered transition metal dichalcogenide. It is one of the few naturally occurring layered dichalcogenides and the primary source for elemental molybdenum. The lubricating properties – although known for many years – did not receive attention before the second world war [91].

Furthermore, this disulfide and also those of tungsten and rhenium are extensively applied as catalysts in industrial processess such as hydrogenation, hydrocracking and desulfurization reactions. In the system Mo-S various compounds exist, i.e. Mo_2S_3, MoS_2, Mo_2S_5, MoS_3, and MoS_4 [92]. Of the disulfide, today only two polytypes have been detected in synthesized material; these are the two-layer, hexagonal 2H-modification and the three-layered rhombohedral 3R-modification [60, 93]. In the mineral molybdenite the 2H-form was detected in 1923 by Dickinson and Pauling [94] and according to the report of Traill of 1963 [95] the 3R-form also occurs in nature.

4.1.1. Compound Preparation

The synthesis of MoS_2 has been described by a number of early investigators. Such various methods of preparations as that proposed by Berzelius [96] in 1826 (decomposition of MoS_3 in a vacuum) or the procedure proposed by Svanberg et al. [97] (reaction of $MoO_3 + H_2S$ at red heat) in 1848, the method used by De Schulten [98] in 1889 and the procedure of heating Mo and S in an iron tube as proposed by van Arkel [99] in 1926, are only a few of those found in literature. The preparative procedures can be divided into those yielding the 2H-modification and those yielding the 3R-form [91]. The hexagonal 2H-modification is formed by direct synthesis from the elements at temperatures higher than 570 °C (see also [100]), although, to obtain a well-crystallized product without many stacking faults, a temperature of 900° to 1000 °C is preferred [91]. Another procedure is by thermal decomposition of the higher sulfide at 1100 °C or of a thiometalate above 200 °C in vacuum or in an inert atmosphere [91, 92]. The rhombohedral 3R-modification, on the other hand, is obtained when the trioxide or a molybdate is reacted with sulfur in a flux of an alkali carbonate at about 900 °C. This last method is an old one, first applied by De Schulten in 1889 [98]. Silverman [101] was able to prepare the rhombohedral MoS_2 modification from the elements at high temperatures (~ 2000 °C) and ultra high pressures (~ 70 kbar) and also succeeded in transforming normal 2H-molybdenite completely into the 3R-form at 40–75 kbar and 1900°–2000 °C in 1 to 5 min. The temperatures required in this transformation process are much higher than those needed for the synthesis of the 3R-form from the elements. The author concludes that the direct synthesis of this 3R-form is mainly a temperature dependent procedure; a Mo-S mixture (ratio 1:2) hot pressed at 27 kbar and 900 °C and also at 47 kbar and 800 °C only yielded the 2H-form after 5 min. At 47 kbar and 1000 °C, 74 kbar and 1100 °C and up to 70 kbar and 2000 °C, Silverman found quantitative yields of the 3R-form [101].

In the flux methods, as proposed by De Schulten [98] and applied by Dickinson et al. [94] and Bell et al. [93], the 3R-modification is formed by either heating molybdenum trioxyde with sulfur in a potassium carbonate melt [98] or by fusing together ammonium molybdate with sulfur and potassium carbonate [94]. Wildervanck followed this flux procedure and heated 4 g of sulfur and 6 g of anhydrous sodium carbonate at 675 °C in an open vitreous silica tube for a few hours. After cooling, one third part of a mixture of 5 g Mo.O$_3$, 1 g S and 2 g. Na$_2$CO$_3$ was added to the cooled material and the batch was heated again to ~900 °C for 1.5h. After cooling, another part of the MoO$_3$—S—Na$_2$CO$_3$ mixture was added and so on. After cooling, the reaction product was extracted with boiling ammonia, water and hydrochloric acid and the remaining black powder washed with carbon disulfide, alcohol and ether. In later experiments sodium carbonate was replaced by potassium carbonate, since the former severely attacks the quartz tube wall. Analyses gave a S/Mo ratio of 1.94 [91, 92]. Prolonged heating of the 3R-form at 1000 °C in vacuo resulted in the conversion into the 2H-form indicating that at least at elevated temperatures the 2H-form is the stable form [92]. This fact could not be confirmed by Zelikman et al. [102] who found a large sulfur pressure (4–18 atm.) necessary during annealing attemts. According to them, the kinetics of transformation depended on the deviation from stoichiometry of MoS$_2$. The stability of the hexagonal form was beyond doubt.

In repeating these experiments Wildervanck could confirm these results. In [103] a stoichiometric mixture of Mo and S is heated at 600°–700 °C for 24 h and then cooled. The furnace is heated up in steps of 50 °C. A second heat treatment is given at 1000°–1050 °C for a period of one week.

4.1.2. Crystal Growth

The preparation of single crystals from the vapour phase have been described by Wildervanck [91], by Schäfer and co-workers [104] and Nitsche [48]. Growth in a flux has been discussed in [91] and [93], while Al-Hilli and Evans [103] devised a method of direct vapour transport without using halogen transport. Wildervanck attempted to develop a method of preparation of pure rhombohedral MoS$_2$ and to study the conditions of the formation of the two polytypes from the gas phase. Unfortunately he was unsuccessful and could not predict which of the modifications should be formed in a particular experiment. Chlorine, bromine as well as iodine and in a few cases ammonium chloride were tried. Hexachloroplumbate was used as a chlorine donor in many cases. Iodine concentrations were usually about 5 mg cm^{-3} tube volume or an equivalent of chlorine or bromine; higher concentrations resulted in more rapid transport, giving lumps of small crystals, while smaller concentrations enhanced the growth of large isolated crystals. The temperature gradient also affected the crystal size; normally gradients of 50–100 °C over 20 cm were used. A further observation was that excess sulfur stimulated transport and yielded better developed crystals although with a composition at the sulfur-rich limit of the homogeneity range (which is very small in MoS$_2$).

Experiments with Cl$_2$ were carried out with 2H-MoS$_2$ source material in gradients $T_H = 1000°–1120 °C \rightarrow T_L = 810°–960 °C$. Reaction times were from 3 days to 3.5 weeks. The resulting crystals have sometimes been marked as being large. With 3R-MoS$_2$ source material, chlorine transport was tried between $T_H = 940°–1030 °C \rightarrow T_L = 830°–970 °C$ in 1–3 weeks. Bromine transport experiments were performed in gradients ranging from $T_H = 1050°–1080 °C \rightarrow T_L = 900°–1300 °C$ with 2H-material in 1–2 weeks time but the size of the plate-like crystals was small. With iodine the experiments occurred between $T_H = 830°-1000\ °C \rightarrow T_L = 795°-875\ °C$ in 4–5 weeks also with 2H-material in the source zone. With NH_4Cl no transport was observed. Nitsche [48] transported with 0.05–0.1 millimoles $Br_2\ cm^{-3}$ tube volume in 120 mm × 10 mm tubes in a period of ≈ 48 h from $T_H = 900° \rightarrow T_L = 800\ °C$ and obtained hexagonal, black coloured plates of $5 \times 5 \times 0.1\ mm^3$. Schäfer and co-workers investigated the chemical transport process of MoS_2 and WS_2 in the presence of iodine. In this extensive and complete report they showed that this transport with iodine as transporting agent proceeds via the gaseous oxide iodines MoO_2I_2 and WO_2I_2, which are produced by the presence of small amounts of water, an influence which has also been discussed by Rimmington and Balchin for the case of mixed crystal formation in the system Ti-S-Se [51]. It is obvious that the influence of water deserves more attention. Al-Hilli and Evans [103] directed their effort to transport of some TCh_2 compounds without using a halogen and obtained crystals superior to those grown with bromine or iodine as an agent. In their bromine transport experiments a halogen concentration of 0.428 mg cm^{-3} was used, tubes of 190 mm × 22.5 mm were used in a temperature gradient of $T_H = 947° \rightarrow T_L = 890\ °C$. Reaction time was 5 days.

In direct vapour transport the free flowing black powder – which was obtained in the synthesis from the elements – was distributed along the length of the tube (190 mm × 22.5 mm) to within 2 cm of each end and then heat treated for about two weeks. The reaction tube was situated in the furnace in such a way that both tube ends

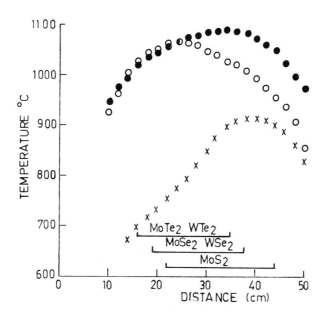

Fig. 3. The furnace temperature profile and ampoule position employed by Al-Hilli and Evans [103] in their growth experiments with and without the aid of a halogen carrier.
(●) = MoS_2, (○) = $MoSe_2$ and WSe_2, (X) = $MoTe_2$ and WTe_2.
MoS_2, $MoSe_2$ and WSe_2 are grown without the aid of halogen, $MoTe_2$ and WTe_2 are grown with the aid of bromine as a transporter.

were at ~1050 °C, while the maximum temperature was ~1090 °C. The authors emphasize the fact that the position of the ampoule in the temperature profile and the profile itself are very important. It is necessary to keep the temperature to within 10 °C of the value presented, otherwise there will be no yield. Profile and ampoule position are shown in Figure 3.

At the end of the two weeks' reaction time the furnace was allowed to cool to about 300 °C before the reaction tube was removed. Removal of the ampoule at higher temperatures (~1000 °C) brought the risk of an explosion of the ampoule.

Large single crystals were obtained with this technique; they had a hexagonal structure whereas the bromine transported crystals had a rhombohedral structure showing triangular growth features.

Information about the structure type and lattice parameters are given in Table I, preparation and crystal growth data are presented in Table VII.

4.2. $MoSe_2$

Molybdenum diselenide is similar to but has not proven to be as useful as the disulfide. The material is not known to occur naturally and the cost of preparing it makes it noncompetitive with MoS_2 as a lubricant.

In the system Mo-Se three phases seem to be well-established, i.e. Mo_3Se_4, $MoSe_2$ and $MoSe_3$ [91], and at present only two polytypes of $MoSe_2$ are known – the same modifications 2H and 3R as found for the disulfide.

4.2.1. *Compound Preparation*

The current preparation techniques resemble those used for MoS_2. The 3R-modification has only been prepared by the high temperature – high pressure method of Towle *et al.* [105] and Silverman [101]. With elemental Mo and Se, in 1:2 atomic ratio mixtures, below 800 °C and pressures between 17 and 78 kbar, only the hexagonal form was obtained. From 1100° to 1400 °C, with the same pressures, the resulting product was a mixture of 2H- and 3R-$MoSe_2$. At 1700 °C under 47 kbar and at 2000 °C under 70 kbar the complete reaction occurred. Heating the 3R-modification, obtained under such conditions of T and p, in vacuo at 1000 °C in quartz tubes resulted in a complete conversion into the 2H-modification.

Attempts by Wildervanck [91] to utilize the flux technique of De Schulten [98] did not lead to the preparation of the diselenide at all. Synthesis from the elements at temperatures above 340 °C seems to be the most convenient preparation procedure which always leads to 2H-production [91, 100]; to obtain the compound in a well-crystallized form, however, higher temperatures – in the range 900° to 1100 °C – have to be applied for periods between one day and two weeks [91], although Brixner [65] mentioned short periods of 10–15 h at 600°–700 °C as sufficient. For a detailed discussion of Brixner's method the reader is referred to the section on $NbCh_2$ compounds. The procedure followed by Al-Hilli and Evans [103] has been described in the section on MoS_2.

4.2.2. Crystal Growth

Growth experiments have been described by Brixner [65], by Wildervanck [91] and Al-Hilli and Evans [103]. Brixner's growth conditions with bromine are analogous to those described extensively in the section on $NbCh_2$ and Al-Hilli's work is described in the section on MoS_2. For $MoSe_2$ the characteristic temperatures in the bromine assisted transport process [103] were $T_H = 910 \rightarrow T_L = 730\,°C$ for a period of 3 days with 2 mg bromine cm^{-3} tube volume.

The conditions for vapour transport without halogen addition are analogous to those mentioned for MoS_2 and the situation of the reaction tube (200 mm × 22.5 mm) in the furnace is illustrated in Figure 3. Evans et al. [106] advise placing the ampoule in a uniform temperature of 1070 °C. According to Al-Hilli's report, their growth experiments with bromine resulted in a mixture of 2H- and 3R-crystals. The surfaces of such crystals showed six-sided growth spirals in which the ratio of the side lengths varied to give both hexagonal and triangular forms. Crystals grown without bromine were of the 2H-type only and showed well-defined hexagonal growth spirals.

In Wildervanck's experiments on $MoSe_2$ crystal growth [91], only bromine and iodine were used. Both Mo and Se in powder form as well as the prereacted compound were employed as source material; with bromine, transport took place from $T_H = 995°–1100\,°C \rightarrow T_L = 905°–1060\,°C$. Sometimes thick large crystals could be produced.

Reaction times varied between 4 days and 1 week. With iodine, reaction times of 2 days were needed to transport from $T_H = 1085° \rightarrow T_L = 1030\,°C$. In both cases halogen transport yielded 2H-$MoSe_2$ and Wildervanck was not able to grow 3R-$MoSe_2$ in halogen assisted experiments.

As with MoS_2, more work will be needed to specify the effects of water vapour and selenium vapour on transport. The reader is also referred to [32] for a describtion of transport possibilities with selenium as the transporting agent.

Table I contains information on crystal structure and lattice parameters, in Table VII the preparative data are summarized.

4.3. $MoTe_2$

Molybdenum ditelluride does not occur in nature and has not received as much attention as the disulfide. Previously $MoTe_2$ was considered to possess only a hexagonal layer type structure of the MoS_2-form [107], but recently a layered monoclinic structure closely related to the CdI_2-type has been found at high temperatures [108].

The MoS_2-type – also denoted α-$MoTe_2$ – is the polytype stable below ~815 °C; the so-called β-$MoTe_2$ form or high temperature modification, is stable above 900 °C. Room temperature resistivities of these two structures differ about 10^6 [109], a fact which is attributed to differences in metal-metal bonding in the two structures. The resistivity of the α-type material is affected by deviations from stoichiometry.

Vellinga et al. [110] showed that $MoTe_2$ undergoes a transition from the diamagnetic semiconducting α-phase to the metallic paulimagnetic β-phase, at a temperature which is 820 °C for Te-rich material and 880 °C for Mo-rich samples.

It seems that only two compounds exist in the Mo-Te system, i.e. $MoTe_2$ and Mo_3Te_4. At the present time no rhombohedral $MoTe_2$ modification has been reported.

4.3.1. Compound Preparation

In his 1962 report, Brixner [65] also discusses the preparation of $MoTe_2$ and for details the reader is referred to the section on the preparation and crystal growth of the $NbCh_2$ compounds.

The procedure reported by Revolinsky and Beerntsen [111] makes use of a homogeneous mixture of powdered elements which is heated to 700°–800 °C for 72 h and air quenched. In a later report these authors [109] describe extended preparation conditions in which temperatures between 700° and 1000° are used for periods of 75–100 h. They found the low temperature α-form to be stable below 815 °C, while the β-modification was found to be stable above 900 °C. To retain the β-form, the reaction ampoule had to be water quenched.

Attempts of Silverman [101] to prepare rhombohedral $MoTe_2$ by his high pressure –high temperature technique did not yield this desired form. In all cases the mixture either did not react or formed α-$MoTe_2$ and β-$MoTe_2$. This last form was obtained in partial yields at 45 kbar, and 1800 °C in 3 minutes' time and in quantitative amounts at 60 kbar and 2200 °C in 2 minutes' time.

Wildervanck [91] also prepared $MoTe_2$ from the elements at 900°–1100 °C; his temperature of transformation is given as 875 °C. According to D.T.A. measurements, the formation temperature of $MoTe_2$ lies at 430 °C [100]. Al-Hilli et al. [103] prepare $MoTe_2$ in powder form, following the procedure described for MoS_2, while Vellinga et al. [110] heated the oxygen-free elements in weighed amounts for one week at temperatures between 600° and 1000 °C. Compositions corresponding to the Te-rich and Mo-rich limits of the homogeneity region were prepared by bringing the samples in equilibrium with mixtures of $MoTe_2+Te$ and $MoTe_2+Mo_3Te_4$ respectively at temperatures between 750° and 950 °C. The compositions were then chemically analysed and the homogeneity region found to lie between $MoTe_{1.90}$ and $MoTe_{1.99}$ at the above-mentioned temperatures between 750° and 950 °C [110].

Vellinga's results indicate a narrower homogeneity range than that reported by Spiesser and Rouxel, i.e. from $MoTe_{1.75}$ to $MoTe_2$ [112].

According to the work of Vellinga and co-workers, the onset of the α-β transition lies at 820 °C for the Te-rich and at 880 °C for the Mo-rich material.

The β-phase can be retained at room temperature by quenching. In heating this β-modification it transforms at ∼500 °C to the α-phase.

4.3.2. Crystal Growth

Information on $MoTe_2$ crystal growth techniques have been presented by Brixner [65], Al-Hilli et al. [103] and Revolinsky et al. [109].

The procedure followed by Brixner – using 1–3 mg bromine for Mo-compounds – is described in detail for the NbCh$_2$ compounds. MoTe$_2$ was not transported in the event iodine was used as a carrier.

The method reported in [103] is similar to the one discussed for MoS$_2$ and MoSe$_2$. With bromine as transporter – 2.7 mg cm^{-3} tube volume – α-MoTe$_2$ crystals were obtained in 4 days, in a gradient $T_H = 900° \rightarrow T_L = 700$ °C with ampoules having a length of 190 mm and a diameter of 22.5 mm. Attempts to grow MoTe$_2$ crystals without the aid of halogen were unsuccessful. Revolinsky and Beerntsen [109] have grown single crystals of both α-and β-MoTe$_2$ using bromine as the transporter; the deposition temperatures ranged from 650°–800 °C for α-MoTe$_2$ and from 900°–1000 °C for β-MoTe$_2$. Crystals used by Brown [108] were bromine transported and were Mo$_{0.985}$Te$_2$.

Structure type and lattice parameters are given in Table I, data on preparation and growth are presented in Table VII.

4.4. WS$_2$

Tungsten disulfide is a grey black semiconducting compound. In nature WS$_2$ is found as the mineral tungstenite and, as mentioned for MoS$_2$, the disulfide of tungsten has also found application as a catalyst in industrial processes. In the system W-S, there seem to be two well-established compounds, WS$_3$ and WS$_2$, and no evidence has been found for the existence of W$_2$S$_3$ corresponding to Mo$_2$S$_3$ [92]. Today only two polytypes for WS$_2$ have been reported in literature; the 2H-type with the 2H-MoS$_2$ structure and the rhombohedral 3R-modification having the 3R-MoS$_2$ type of structure.

4.4.1. *Compound Preparation*

Preparation procedures for WS$_2$ can – analogous to MoS$_2$ – be divided in those yielding the hexagonal modification and those yielding the rhombohedral form [91]. Synthesis from the elements at 900°–1000 °C is a favourable method for obtaining a 2H product without too much stacking disorder. According to D.T.A. experiments described in [100], the temperature of formation for WS$_2$ lies at ≈ 400 °C.

Glemser *et al.* [113] report the synthesis at 800 °C for 24 h. As for MoS$_2$, the thermal decomposition of WS$_3$ will yield hexagonal WS$_2$; according to Wildervanck and Jellinek [92], tempering at 1100 °C gives a well-crystallized disulfide, isostructural with 2H-MoS$_2$.

The preparation of 3R-WS$_2$ via the carbonate flux method has been described briefly in [92]; the yield, however, was much lower than that of MoS$_2$ and attempts to improve this yield resulted in products having a larger degree of stacking disorder. Another procedure for preparing WS$_2$ is the reaction of tungsten trioxide and sulfur described by Tiede and Lemke [114]. The high pressure–high temperature technique of Silverman [101] has also been successfully applied to obtain 3R-WS$_2$. A W-S 1:2 mixture was compressed at 45 kbar and heated to 1800 °C for 2–3 min and the silver-grey soft product formed was completely rhombohedral.

4.4.2. Crystal Growth

Well-defined experiments are mentioned by Nitsche [48], Wildervanck [91] and Schäfer et al. [32, 104].

The conditions for bromine assisted transport of WS_2 presented in [48] are the same as those given for MoS_2, and the reader is referred to section 4.1.1 for this discussion. Wildervanck [91] attempted to grow WS_2 by transport with Cl_2, Br_2, I_2 and NH_4Cl.

With $2H$-WS_2 and $3R$-WS_2 as starting material, the Cl_2 experiments were carried out between $T_H = 960°$–$1030\,°C$ and $T_L = 875°$–$945\,°C$ in a period varying between 10 days and 2.5 weeks. With 2H-source material and Br_2 as carrier, the experiments were carried out in 1–2 weeks in varying temperature gradients
$$T_H = 865°\text{--}1090\,°C \rightarrow T_L = 787°\text{--}1009\,°C.$$
With the aid of iodine, 2H-source material was transported from
$$T_H = 830°\text{--}930\,°C \rightarrow T_L = 800°\text{--}870\,°C \text{ in 2--4 weeks.}$$
Transport with NH_4Cl from $T_H = 900° \rightarrow T_L = 800\,°C$ was incomplete.

Pure 2H and 3R products were obtained in bromine assisted transport, however, most of the crystals were denoted as small.

Schäfer and co-workers indicated in 1968 the possibility of sulfur assisted transport of WS_2 [32]; in 1973 they reported on the influence of water on the iodine assisted transport of WS_2 [104]. For a detailed discussion the reader is referred to this report.

Table I contains the structure type and lattice parameters, and preparation and crystal growth data are presented in Table VII.

4.5. WSe_2

Tungsten diselenide is a semiconducting compound with a relatively low electrical conductivity increasing with temperature and a high Seebeck coefficient. In the system W–Se–extensively studied by Hicks [115]–no evidence was found for the existence of WSe_3, mentioned by Hansen [116], nor could selenides of tungsten lower than WSe_2 be detected. Investigations concerning the stoichiometry of WSe_2 have indicated that the homogeneity range of this compound must be extremely narrow. From the experiments described in [115], it is clear that WSe_{2+y} has a composition with y having values between -0.01 and $+0.01$.

The structure of the diselenide is like that of the disulfide, with the metal trigonal prismatically surrounded by selenium [113]. It seems that, unlike WS_2, the only known modification of WSe_2 is the 2H-type and that no rhombohedral diselenide exists.

4.5.1. Compound Preparation

Several reports exist on the preparative conditions; Silverman's high pressure–high temperature experiments (covering a very wide pressure–temperature profile, up to 70 kbar and 2400 °C) gave only the hexagonal form. Wildervanck's work [91] and the studies of Al-Hilli *et al.* [103], Brixner [65, 117], Kershaw *et al.* [74], Revolinsky *et al.* [111] and Hicks [115] report the preparations of the polycrystalline substance by direct reaction of the elements in stoichiometric ratio. D.T.A. experiments [100] give a temperature of formation of 480 °C for WSe_2, however, most of the described procedures utilize higher temperatures. For example, [91] heats at 900–1000 °C for one week, while in [103] a temperature of 600–700 °C for 24 h seems to be sufficient. Brixner [65, 117] heats at 600–700 °C for 10–15 h and gives a second firing at 1000–1200 °C for the same period of time. In [111], a temperature of 700–800 °C is used for 72 h, while in [115] a temperature of 550 °C is reported, followed by a 30 min mixing in a mechanical vibrator, whereafter a second heat treatment follows at 1000 °C. [74] mentions a temperature of 600 °C for 4 days, followed by a refiring period of 4 days at 800 °C. In all cases the yield consisted of $2H\text{-}WSe_2$.

4.5.2. *Crystal Growth*

Wildervanck's growth experiments on WSe_2 [91] followed the same line as discussed for MoS_2, $MoSe_2$ and WS_2; both bromine and iodine were used and reaction were usually between 2.5 and 3 weeks. With bromine transport took place from $T_H = 800\text{–}850\,°C \rightarrow T_L = 750\,°C$ and with iodine the gradient was $T_H = 1055° \rightarrow T_L = 890\,°C$. The produced samples are described as thin small crystals having the hexagonal $2H\text{-}MoS_2$ type lattice. Al-Hilli *et al.* [103] have grown WSe_2 with and without bromine; with halogen as the carrier, transport was from $T_H = 800° \rightarrow T_L = 700\,°C$ with 2.7 mg Br_2 cm^{-3} in 4 days. The produced material consisted of thin buckled crystals with no obvious surface features. Crystals grown without the aid of Br_2 were plates with well-defined hexagonal growth features. In both cases the lattice parameters were similar to those published by Glemser *et al.* [113].

Brixner's preparative technique [65] has been treated in detail in the section on $NbCh_2$. The same conditions have been applied in WSe_2 growth experiments. The method employed by Kershaw *et al.* [74] discussed for $NbSe_2$ has also been used for WSe_2; the reader is referred to the section on $NbSe_2$ for full details.

In Table I the structural data are given and the information on preparative procedures and growth conditions are presented in Table VII.

4.6. WTe_2

Tungsten ditelluride is a grey black substance with a character somewhat between that of the semiconducting $MoSe_2$ $MoTe_2$ and WSe_2 and the metal-like $NbTe_2$ and $TaTe_2$.

Structurally there is a difference between $MoTe_2$ and WTe_2; according to the X-ray investigations of Knop and Haraldsen [118] – based on a single crystal found in an otherwise polycrystalline reaction product – WTe_2 does not have the MoS_2 type of structure but a distorted octahedral coordination about the tungsten atom and is thus more closely related to the CdI_2-type of lattice. WTe_2 is then effectively isostructural with $\beta\text{-}MoTe_2$, the high temperature $MoTe_2$ phase. Also the physical properties of $\beta\text{-}MoTe_2$ and WTe_2 are similar [110]. In the system W-Te, the ditelluride is commonly accepted as the only compound formed.

4.6.1. Compound Preparation

In literature the preparation of WTe_2 follows the line given for the other WCh_2, i.e. reaction of the elements in stoichiometric ratio. The temperature of formation from the elements lies at 630 °C according to D.T.A. measurements [100]. Knop and Haraldsen [118] used a temperature of 700°–800 °C and Brixner's procedure [65] has been treated in the section on $NbCh_2$.

Al-Hilli's method [103] for preparing polycrystalline source material has also been discussed previously in the section on MoS_2. Silverman [101] has also applied his technique on the ditelluride; like in the $MoTe_2$ experiments his high pressure–high temperature attempts either did not yield any reaction product at all, or the product obtained was WTe_2 with the β-$MoTe_2$ structure.

Revolinsky and Beerntsen [111] prepared the polycrystalline substance by direct reaction at 700°–800 °C of the powdered elements.

4.6.2. Crystal Growth

Crystal growth has been reported by Brixner [65], Al-Hilli and Evans [103] and Brown [108]. The conditions of Brisner's experiments are discussed in the section on $NbCh_2$. The experiments of Al-Hilli *et al.* without Br_2 vapour failed to yield WTe_2; with 2.7 mg cm^{-3} Br_2 they could grow WTe_2 in the form of thin long sheets in 4 days in a gradient $T_H = 900° \rightarrow T_L = 700 °C$. Brown [108] did his structure determination work on bromine transported WTe_2 which turned out to have the formula $W_{0.94}Te_2$.

Information concerning the structure can be found in Table I, the preparative procedures can be found in Table VII.

4.7. SOLID SOLUTIONS

Information on solid solution formation can be found in reports from Brixner [117], Brixner and Teufer [119], Hicks [115] and Revolinsky and Beerntsen [111]. Brixner intended to see if $TaSe_2$ and WSe_2 could be combined advantageously in the form of single-phase solid solutions with improved or optimized overall thermoelectric properties. Brixner and Teufer's X-ray study devoted to the $W_{0.5\,x}\,Mo_{0.5\,x}\,Ta_{1-x}\,Se_2$ system had the same intention. The investigation of Hicks were focused on the semiconducting behaviour of substituted WSe_2, $MoSe_2$ and $MoTe_2$ while Revolinsky *et al.* studied the electrical properties of the $MoSe_2$–WSe_2 and the $MoTe_2$–WTe_2 systems.

4.7.1. Compound Preparation

Brixner [117] prepared the solid solutions by heating the constituent elements W, Ta and Se in evacuated sealed quartz ampoules for 10–15 h at 600°–700 °C after which the black looking reaction products were remixed inside the tube by shaking. Thereafter a second heat treatment followed at 1000°–1200 °C for another 10–15 h.

The resulting end product was a homogeneous polycrystalline grey-metallic substance. Structure data of these compounds are presented in Table VIII. Up to

TRANSITION METAL DICHALCOGENIDES   211

TABLE VII
Compound preparation and Crystal growth techniques

Compound	Compound preparation procedures		Crystal growth techniques	
MoS_2	a. Direct synthesis from the elements at 1100 °C, gives 2H-MoS_2. at 600°–700 °C, for 24 h, followed by second firing at 1000°–1050 °C for 1-week in 220 mm × 22.5 mm tubes. See also ref. [65].	[91] [103]	*Flux method* [91,93,94] *Br_2 transport* $T_H = 900° \to T_L = 800$ °C, tubes 120 mm × 10 mm; 0.05–0.1 milli-moles cm^{-3}, (see also $NbCh_2$, ref [65].) in 48 h	[48]
	b. Thermal degradation of higher sulfide, above 200 °C in vacuo or inert gas, yields 2H-type	[91, 92]	*Cl_2 transport* $T_H = 1000°$–1120 °C $\to T_L = 810$–960 °C; 5 mg Cl_2 cm^{-3}; 3 days–3.5 weeks; 2H-as source material. With 3R-source in 1–3 weeks, gradient is $T_H = 940°$–1030 °C $\to T_L = 830°$–970 °C, yield is 2H + 3R – MoS_2.	[91]
	c. Flux method, carbonate melt at 900 °C, gives 3R-modification	[91, 93, 94]	*Br_2 transport* $T_H = 1050°$–$1080° \to T_L = 900°$–1030 °C; 5mg Br_2 cm^{-3}; in 1–2 weeks with 2H-source. Yield is 2H- and 3R-MoS_2	
	d. Ultra high pressure-High temperature (70 kbar and ~2000 °C) gives the 3R-type of MoS_2	[101]	*I_2 transport* $T_H = 830°$–$1000° \to T_L = 795°$–875 °C; 5 mg I_2 cm^{-3}; in 4–5 weeks with 2H-source pure 2H-MoS_2. See also ref. [32] and [104].	
			Br_2 transport $T_H = 947° \to T_L = 890$ °C; 0.428 mg cm^{-3} in 190 mm × 22.5 tubes. Direct transport without halogen in temp. grad., see Figure 3	[103]
$MoSe_2$	a. Same procedure as MoS_2, at 900°–1100 °C gives 2H-type. at 600°–700 °C, then at 1000°–1050 °C in 190 mm × 22.5 mm tubes c.f. MoS_2 and ref. [65]	[91] [103]	*Br_2 transport* Conditions are analogous to those for MoS_2, ref. [65]. $T_H = 910° \to T_L = 730$ °C; 2 mg Br_2 cm^{-3}; 2 days [103].	
	b. Ultra high pressure- high temperature, see MoS_2	[101, 105]	Transport without halogen in temp. of 1070 °C [106]. See also Figure 3 and ref. [103] for $NbCh_2$ growth procedure. $T_H = 995°$–$1100° \to T_L = 905°$–1060 °C; 4 days – 1 week, yields 2H-$MoSe_2$ [91].	
	c. Flux method is not valid.	[91]	*I_2 transport* $T_H = 1085° \to T_L = 1030$ °C; 2 days, gives 2H-$MoSe_2$ See also ref. [32]	[91]

Table VII (Continued)

Compound	Compound preparation		Crystal growth techniques	
MoTe$_2$	a. Synthesis from elements at 700°–800°C, 72 h and air-quenching, yields MoTe$_2$ also at 700°–1000°C for 75–500 h MoTe$_2$ is $\alpha+\beta$ α-type stable <850°C, β stable >900°C.	[109, 111]	Br$_2$ transport $T_H=900° \to T_L=700°C$; 2.7 mg cm^{-3} Br$_2$; 4 days; 190 mm × 22.5 mm tubes. Growth without halogen not successful. Similar conditions as for NbCh$_2$, 1–3 mg Br$_2$ cm^{-3}. With iodine no MoTe$_2$ crystal growth.	[103]
	Synthesis at 900°–1100°C, transformation temp. lies at 875°C Synthesis at 600°–1000°C, for 1 week, annealing at 750° and 950°C of mixtures of MoTe$_2$+Mo$_3$Te$_4$ and MoTe$_2$+Te to affect composition.	[91] [110]		[65]
	b. Ultra High pressure- High temp. methode, yields only $\alpha+\beta$ mixture, or β-MoTe$_2$ (60 kbar, 2200 °C) no rhombohedral MoTe$_2$ could be obtained.	[101]	α-MoTe$_2$ and β-MoTe$_2$ were grown with Br$_2$, deposition temp. are: 650°–800 °C for α-MoTe$_2$ and 900°–1000 °C for β-MoTe$_2$	[109]
	See also procedures used for NbCh$_2$, ref. [65], and ref. [103] for conditions of MoS$_2$ synthesis; they are also valid.			
WS$_2$	a. Reaction of elements at 900°–1000 °C gives 2H-WS$_2$ (see MoS$_2$).	[91]	Br$_2$ transport $T_H=900° \to T_L=800°C$; further same conditions as for MoS$_2$ see also ref. [32] and [104].	[48]
	b. Thermal decomposition of WS$_3$, yields 2H-WS$_2$	[91, 92]	Cl$_2$ transport $T_H=960°–1030° \to T_L=875°–945°C$; 10 days to 2.5 weeks, with 2H-source material yields 2H- and 3R-WS$_2$.	
	c. Flux method (see MoS$_2$) gives 3R-WS$_2$	[92]	Br$_2$ transport $T_H=865°–1090° \to T_L=787°–1009°C$; 1–2 weeks, 2H- source, yields 2H- and 3R-type WS$_2$.	[91]
	d. Reaction of WO$_3$ and sulfur	[114]	I$_2$ transport $T_H=830°–930° \to T_L=800°–870°C$; 2–4 weeks, 2H-source. This yields 2H+3R-WS$_2$.	
	e. Ultra high pressure- High temp. (45 kbar and 1800 °C) in 2-3 min	[101]	NH$_4$Cl transport $T_H=900° \to T_L=800°C$. Transport is incomplete, yields 2H-WS$_2$.	

Table VII (Continued)

Compound	Compound preparation		Crystal Growth techniques	
WSe$_2$	a. Reaction of elements at 900°–1000 °C for 1-week [91]. at 600°–700 °C, 24 h then at 1000°–1050 °C for 1-week in 190 mm × 22.5 mm tubes.	[103]	Br$_2$ transport $T_H = 800°$–$850° \rightarrow T_L = 750 °C$, in 2.5-3-weeks	[91]
	at 600°–700 °C for 10-15 h then second firing at 1000°–1200 °C for 10–15 h at 700°–800 °C for 72 h [111].	[65, 117]	I$_2$ transport $T_H = 1055° \rightarrow T_L = 890 °C$, in 2.5-3-weeks The yield is in these cases 2H-WSe$_2$.	
	at 550 °C, 30 min mixing and second heating at 1000 °C. at 600 °C for 4 days, then refiring at 800 °C for 4 days. In all these cases the yield is 2H-WSe$_2$.	[115] [65]	Br$_2$ transport $T_H = 800° \rightarrow T_L = 700 °C$; 2.7 mg cm^{-3}; in 4 days; 190 mm × 22.5 mm tubes. Also transport without halogen. Conditions reported in ref. [65] and [74] are also valid for crystal growth of WSe$_2$.	[103]
	b. Ultrahigh pressure- high temp. technique, gives 2H-WSe$_2$, with 70 kbar, and 2400 °C.	[101]		
WTe$_2$	a. Reaction of elements at 700°–800 °C. at 600°–700 °C, then at 1000°–1200 °C. at 700°–800 °C. Same conditions as applied for MoS$_2$ are valid.	[118] [65] [111] [103]	Transport conditions reported for other MoCh$_2$ and WCh$_2$ in refs. [65, 103 and 108] are also valid. Without Br$_2$, no WTe$_2$ transport or growth; with bromine transport from $T_H = 900° \rightarrow T_L = 700 °C$; 2.7 mg cm^{-3} in 4 days in 190 mm × 22.5 mm tubes.	[103]
	b. Ultra high pressure- High temp. procedure yields no 3R-WTe$_2$.	[101]		

$x=0.55$ the system $W_x Ta_{1-x} Se_2$ retains the $CdCl_2$-type of structure, then a rather narrow two-phase region is observed and after $x=0.65$ the substances change over to the single phase MoS_2-type. In the narrow two-phase region, $W_{0.60} Ta_{0.40} Se_2$ contains both phases at about equal intensities. Attempts were also made to substitute other elements like Hf, Zr, Ti, V, Re, Nb and Sb to study their effects. In most cases there was a minor effect due to a lack of solubility in WSe_2; only Re, Nb and V could be built in and Nb and V incorporation has a similar effect as that of Ta incorporation.

All measurements of Brixner were performed on cold-pressed (40–60 tsi) and subsequently sintered bars (600°–800 °C).

In the report of Brixner and Teufer [119] information is given about the more complex $W_{0.5x} Mo_{0.5x} Ta_{1-x} Se_2$ system.

In the course of their X-ray study and thermoelectric measurements of these substances, the system $Mo_x W_{1-x} Se_2$ was investigated in some more detail. Routine preparative work follows the line mentioned above for the $W_x Ta_{1-x} Se_2$ compounds; all measurements were performed on powders and growth experiments have not been mentioned.

In the system $Mo_x W_{1-x} Se_2$ all compositions crystallize in the MoS_2-type; in the system $W_{0.5x} Mo_{0.5x} Ta_{1-x} Se_2$ up to $x=0.65$ the $CdCl_2$-type is retained (D_{3d}^5), then a narrow two-region phase where both types coexist in roughly equal intensities and from $x>0.8$ all compositions crystallize in the MoS_2-type of structure (D_{6h}^4).

All investigated samples of the five-component system $Mo_{0.5x} W_{0.5x} Ta_{1-x} Se_y Te_{2-y}$ have the MoS_2-type of lattice.

Hicks [115] substituted up to 5 at. % Ta in WSe_2, $MoSe_2$ and $MoTe_2$ and studied the Hall coefficient as a function of temperature on sintered bars to elucidate the conductivity mechanism. Attempts were also made to substitute V, Nb, Ti, Zr, Hf and Al in WSe_2 and Mn, Re, Ru and Pt in $MoSe_2$. Hicks derived a general rule for solubility and hence effective doping in WSe_2 and $MoSe_2$, which states that not only must the doping element have an ionic radius close to that of the host metal, but also it must come from an adjacent group of the periodic table. A detailed discussion of the solubility limits is presented in [115]. In his preparative method Hicks follows the technique of Brixner mentioned above, and crystal preparations are not discussed. All measurements were performed on sintered bars.

Revolinsky and Beerntsen [111] prepared polycrystalline samples of the $MoSe_2$-WSe_2 and the $MoTe_2$-WTe_2 systems, by heating the powdered elements, weighed in the proper proportions to 700°–800 °C for a period of 72 h followed by air quenching.

While complete solid solubility was observed between WSe_2 and $MoSe_2$ – the small difference in their cell parameters results in a linear variation of the crystallographic parameters of the ternary compounds with composition – it appears that a two-phase region exists in the $MoTe_2$–WTe_2 system. The structural difference between the binary members enables the accomodation of approximately 35 at % Mo in the orthorhombic WTe_2 structure, while on the other hand the MoS_2-type of structure of $MoTe_2$ only can accept 10–15 at % W.

The solubility limits indicated are only characteristic for the preparation temperature – 750 °C – and the detection limits of the X-ray technique.

Electrical and thermoelectrical measurements were performed on wafers prepared by hot-pressing the powders for about one hour at 4900 kg cm^{-2} (70 000 psi) and 300°–325 °C. Wafer densities varied from 78% to 85% of the theoretical value. No lattice parameters are given and no attempts to grow crystals have been mentioned.

Information is collected in Table VIII.

TABLE VIII
Solid solutions of MoCh$_2$, WCh$_2$ and TaCh$_2$

Compound	Lattice	a (Å)	c (Å)	Density g cm^{-3}	Ref.	Remarks[a]
TaSe$_2$	CdCl$_2$	3.446	19.101	8.593	117	Substances are
W$_{.10}$Ta$_{.90}$Se$_2$	,,	3.423	19.107	8.641	,,	prepared by direct
W$_{.25}$Ta$_{.75}$Se$_2$	,,	3.408	19.132	8.790	,,	reaction of the
W$_{.50}$Ta$_{.50}$Se$_2$	,,	3.367	19.186	8.999	,,	elements.
W$_{.51}$Ta$_{.49}$Se$_2$	,,	3.366	19.186	9.005	,,	
W$_{.52}$Ta$_{.48}$Se$_2$	,,	3.365	19.188	9.010	,,	Crystal growth
W$_{.53}$Ta$_{.47}$Se$_2$	,,	3.363	19.189	9.022	,,	experiments are not
W$_{.54}$Ta$_{.46}$Se$_2$	,,	3.362	19.189	9.027	,,	mentioned.
W$_{.55}$Ta$_{.45}$Se$_2$	,,	3.360	19.192	9.038	,,	
W$_{.60}$Ta$_{.40}$Se$_2$	CdCl$_2$+ MoS$_2$	3.353	19.194/ 12.796	9.084+ 9.078	,,	Lattice parameters are based on powder patterns.
W$_{.65}$Ta$_{.35}$Se$_2$	MoS$_2$	3.345	12.806	9.119	,,	
W$_{.70}$Ta$_{.30}$Se$_2$	,,	3.336	12.836	9.150	,,	
W$_{.75}$Ta$_{.25}$Se$_2$	,,	3.329	12.852	9.191	,,	
W$_{.90}$Ta$_{.10}$Se$_2$	,,	3.303	12.899	9.307	,,	
W$_{.91}$Ta$_{.09}$Se$_2$	,,	3.302	12.902	9.309	,,	
W$_{.92}$Ta$_{.08}$Se$_2$	,,	3.300	12.908	9.317	,,	
W$_{.93}$Ta$_{.07}$Se$_2$	,,	3.298	12.912	9.325	,,	
W$_{.94}$Ta$_{.06}$Se$_2$	,,	3.297	12.920	9.326	,,	
W$_{.95}$Ta$_{.05}$Se$_2$	,,	3.295	12.923	9.336	,,	
W$_{.96}$Ta$_{.04}$Se$_2$	,,	3.293	12.928	9.345	,,	
W$_{.97}$Ta$_{.03}$Se$_2$	,,	3.292	12.938	9.344	,,	
W$_{.98}$Ta$_{.02}$Se$_2$	,,	3.290	12.945	9.351	,,	
W$_{.99}$Ta$_{.01}$Se$_2$	,,	3.288	12.966	9.348	,,	
WSe$_2$	,,	3.286	12.976	9.353	,,	
W$_{.15}$Mo$_{.85}$Se$_2$	,,	3.289	12.935	7.318	119	
W$_{.25}$Mo$_{.75}$Se$_2$	,,	3.288	12.941	7.560	,,	
W$_{.35}$Mo$_{.65}$Se$_2$	,,	3.288	12.945	7.799	,,	
W$_{.75}$Mo$_{.25}$Se$_2$	,,	3.287	12.965	8.754	,,	
MoSe$_2$	,,	3.289	12.930	6.959	,,	
Mo$_{.47}$W$_{.47}$Ta$_{.06}$Se$_{.5}$Te$_{1.5}$	MoS$_2$	3.400	13.770	8.991	,,	
Mo$_{.47}$W$_{.47}$Ta$_{.06}$SeTe	,,	3.390	13.560	8.586	,,	
Mo$_{.47}$W$_{.47}$Ta$_{.06}$Se$_{1.5}$Te$_{.5}$	,,	3.345	13.273	8.381	,,	
Mo$_{.47}$W$_{.47}$Ta$_{.06}$Se$_{1.7}$Te$_{.3}$	,,	3.328	13.218	8.247	,,	

[a] These remarks are not valid for all compounds in this table.

5. Group VII compounds

5.1. TcCh$_2$

Technetium and technecium compounds do not occur in nature. The element Tc was first obtained in weighable amounts by neutron irradiation of molybdenum in a nuclear reactor in 1948. Now it is also obtained from fission product wastes [91]. Whereas Hulliger [120] assumed that all TcCh$_2$ crystallize in the ReS$_2$ structure or a closely related structure, Wildervanck and Jellinek [121] in their 1971 report on the TcCh$_2$ and ReCh$_2$ found the Tc-compounds to have structures different from the corresponding Re-compounds.

Their single crystal X-ray studies showed TcS$_2$ and TcSe$_2$ to have triclinic symmetry; the shape of the unit cell may be described as a distorted Cd(OH)$_2$-type of cell, with hexagonal close-packed anions. According to their information, the dimensions of the a, b plane of TcS$_2$ and ReS$_2$ are almost the same and they therefore assume that the structure of TcS$_2$ contains layers – parallel to the a, b plane – quite similar to those found in ReS$_2$ with only a difference in the stacking of the layers.

TcSe$_2$ has a diffraction pattern closely resembling that of the disulfide, indicating that both compounds are isostructural. However, the unit cell of TcSe$_2$ could not be determined by Wildervanck et al. due to the diffuse diffraction spots and the broad lines in the Guinier diagrams.

They found the ditelluride to have a monoclinic layer structure, the unit cell shape resembling somewhat that of β-MoTe$_2$ (the high temperature form of MoTe$_2$).

All Tc-compounds have been reported semiconducting [91, 120, 121].

5.1.1. *Compound Preparation and Crystal Growth*

Although conditions for preparation of polycrystalline Tc-compounds have not been specified, neither in [91] nor in [120] or [121], it can be assumed that the method of direct synthesis from the elements at elevated temperatures, e.g. between 900° and 1100 °C in evacuated quartz tubes, will be a convenient procedure.

Single crystal preparation has been described in [91]; as in the majority of the cases chemical transport with Cl$_2$, Br$_2$ and I$_2$ has been successfully applied; for the sulfide, transport was performed with Cl$_2$ [91, 121].

For the sulfide the temperatures employed were $T_H = 1150° \rightarrow T_L = 1080$ °C, for the selenide the temperatures were $T_H = 1080° \rightarrow T_L = 1000$ °C and for the telluride from $T_H = 980° \rightarrow T_L = 840$ °C. Crystals obtained in these experiments showed a pronounced plate-like habit.

Information on the structure is given in Table I, Table IX contains the preparative information.

5.2. ReCh$_2$

The rhenium dichalcogenides have been investigated by the Oslo group of Kjekshus and co-workers [122, 123] and by the group of Jellinek and co-workers [91, 121].

According to the information presented by both groups, ReS_2 and $ReSe_2$ have layered structures whereas $ReTe_2$ does not show a plate-like habit. For ReS_2 different structures have been proposed; in 1931 Meisel [124] suggested a layer structure of trigonal habit, in 1954 Lagrenaudie [125] proposed a hexagonal molybdenite type of structure and other authors [126, 127] also used the MoS_2-type of unit-cell as a basis for their X-ray studies.

Wildervanck [91] and Wildervanck and Jellinek [121] in 1971 reported X-ray studies performed on $ReCh_2$ single crystals and this resulted in the conclusion that ReS_2 is triclinic and isostructural with $ReSe_2$. In 1965 Alcock and Kjekshus [122] had already determined the $ReSe_2$ structure by single crystal X-ray techniques; they found a triclinic symmetry related to the $Cd(OH)_2$-type of structure. The experiments of Furuseth and Kjekshus [123] in 1966 on the phase-relationship in the system Re-Te indicated that no similarity existed between the ditelluride structure and that of the sulfide and the selenide.

Whereas in [122] and [123] the fact is mentioned that only one intermediate phase has been found in the systems Re-Se and Re-Te respectively, Wildervanck *et al.* [121] reports the preparation of both Re_2S_7 and ReS_2. The existence of a compound Re_2Se_7 prepared by wet chemical reaction is only very briefly mentioned in [122]. According to Hulliger, Wildervanck and the Oslo research group [120, 121, 122 and 123] the $ReCh_2$ are essentially diamagnetic semiconductors.

5.2.1. *Compound Preparation*

ReS_2 can, according to [91] and [121], be prepared by direct synthesis from the elements and also by thermal decomposition of Re_2S_7. In the first procedure, an intimate mixture of the elements is heated in an evacuated and sealed quartz ampoule to about 1100 °C. In the second procedure Re_2S_7 is first prepared by treating an acid perrhenate solution with gaseous H_2S or with a solution of thioacetamide followed by careful washing of the dark brown precipitate and extracting it afterwards with CS_2. In this sequence air was excluded to avoid oxidation of the material.

Thereafter the product is dried in vacuum over P_2O_5 but this Re_2S_7 still contains water and some sulfur in excess of the ideal composition and the substance is amorphous to X-rays, even after a long period of tempering at about 120 °C. The excess water and sulfur starts to be expelled and Re_2S_7 starts to decompose when the temperature exceeds 150 °C. The heating is continued up to 600 °C the decompositions is then completed.

Tempering at 1100 °C in sealed ampoules is necessary to obtain well-crystallized ReS_2. $ReSe_2$ is, according to Wildervanck [91, 21] prepared in much the same way by direct reaction at 1100 °C. Alcock and Kjekshus first heat the elemental constituents at 1000 °C for 2 days, subsequently cool the material to 700 °C and keep it at this temperature for about 3 weeks and finally cool it slowly to room temperature over a period of 2 weeks. The preparation of the ditelluride, reported by Fureseth *et al.* [123] follows the same procedure of heating at several temperatures.

5.2.2. *Crystal Growth*

Crystal growth experiments are discussed in [91] and [121]; ReS_2 and $ReSe_2$ crystals were grown with the aid of bromine and iodine and in some cases – for the sulfide – also chlorine. Excess chalcogen was found to stimulate transport and to improve the quality of the produced crystals.

Chlorine transport took place in 1.5 days from $T_H = 1190° \rightarrow T_L = 1100\,°C$ and resulted in large crystals. With bromine large crystals were obtained in 19 days in a gradient $T_H = 1105° \rightarrow T_L = 1070\,°C$. When iodine was used, thin large crystals were obtained in 8 days and transport occurred from $T_H = 1015° \rightarrow T_L = 975\,°C$ [91].

6. Group VIII compounds

6.1. $CoTe_2$, $RhTe_2$, $IrTe_2$, $NiTe_2$, $PdTe_2$ AND $PtCh_2$

As little synthetic work has been reported, the layered members of group VIII will be mentioned only briefly.

The compounds under consideration resemble each other [1] and the ditellurides of Ni, Pd and Pt display a metallic character ($PtTe_2$ is a superconductor) and β-$CoTe_2$, β-$RhTe_2$ and $IrTe_2$ are expected to be metallic.

6.1.1. *Preparation Procedures*

$CoTe_2$. The system Co-Te has been reinvestigated by Haraldsen et al. [128]. As a result of their X-ray and magnetic study, two phases – denoted β and γ – were reported. The materials studied by Haraldsen et al. were prepared by direct synthesis from the elements in evacuated and sealed quartz ampoules. Samples up to 61.5 at % Te were heated to 1000 °C and subsequently cooled to 700 °C, kept at that temperature for 24 h and then slowly cooled to room temperature. The molten products were silver white in color. The tellurium rich samples were heated at 700 °C for 24 h and slowly cooled; the sintered lumps had a grey color. In the X-ray diffraction patterns, lines of the β-phase were observed in material – annealed during 6 days at 600 °C – having a composition in the range CoTe to $CoTe_{1.8}$ (from 50 at % Te to 64.29 at % Te). The structure is that of the CdI_2-type. Samples annealed for 7 months at 335 °C indicated that the β-phase limit should lie at $CoTe_{1.7}$ (having 63 at % Te).

Samples with higher tellurium concentration like $CoTe_{1.9}$ show a marcasite structure. The homogeneity region of the β-phase therefore does not exceed the limit $CoTe_{1.8}$, which means that the ditelluride or γ-phase does not exhibit a layer structure. Experiments to grow single crystals are not mentioned.

$RhTe_2$. Geller [129] determined the crystal structures of three rhodium telluride phases with the aid of X-ray powder diffraction technique. The monotelluride RhTe has an NiAs structure, the high temperature $RhTe_2$ form has a $Cd(OH)_2$ type structure, while the low temperature $RhTe_2$ modification occurs in the pyrite structure. The latter is a superconductor at 1.5 K.

In older work of Wöhler et al. [13] and of Biltz [131] (mentioned by Geller) the $RhTe_2$ was prepared by heating $RhCl_3$ with excess Te in a CO_2 stream at 750 °C. According to Wöhler et al. it was $RhTe_2$, while Biltz claimed the product to be Rh_2Te_5. In Geller's study all compounds were prepared by direct heating of a mixture of the elements in the right proportions. The evacuated and sealed quartz ampoules containing the mixtures were kept at 1200 °C for a period of one to three days for the preparation of the layered – high temperature – modification of $RhTe_2$.

No crystal growth experiments have been found in literature.

$IrTe_2$. The system iridium-tellurium has been examined systematically by Hockings and White [132] in 1960 by X-ray powder methods. In the study of the tellurides of the platinum metals presented in [130] and [131], various compositions were reported which prompted Hockings and White to reinvestigate the Ir-Te system. Only two compounds were observed: $IrTe_2$ and Ir_3Te_8. The compound Ir_3Te_8 had a defect pyrite structure and the ditelluride had the $Cd(OH)_2$-type of structure. In their experiments Hockings et al. prepared the substances under consideration by heating weighed quantities of the elements in evacuated sealed quartz ampoules. In their procedure the mixtures of the elements were heated for six hours at 950 °C, then maintained for eighteen hours at 600 °C and subsequently cooled rapidly. No report on crystal growth experiments have been found yet.

$NiTe_2$, $PdTe_2$ and $PtCh_2$. The crystal structures of these compounds have been redetermined by Furuseth et al. [133] by powder X-ray techniques. The materials were synthesized following the procedure described in 1960 in the report by Grönvold et al. [134] on the platinum chalcogenides. These investigators studied the phase relationships by means of X-rays and confirmed the observation of Thomassen [135] dating back to 1929 that all $PtCh_2$ had $Cd(OH)_2$ structures. In their report Grönvold et al. discuss the earlier information on the systems Pt-S, Pt-Se and Pt-Te. According to their observations, only the PtS and PtS_2 phases were found and the existence of the compounds Pt_2S_3 and Pt_5S_6 – claimed by other workers – was not confirmed. In the system Pt-Se, only two intermediate phases, i.e. $PtSe_{0.80}$ and $PtSe_2$, were identified and the existence of $PtSe_3$ was not confirmed. Finally in the Pt-Te system the earlier reported PtTe-phase and the $PtTe_2$-phase could be confirmed while the phase Pt_2Te was not observed.

The materials were prepared by heating accurately weighed quantities of platinum and sulfur, selenium or tellurium respectively in evacuated silica tubes at 825 °C for a period of 30 days and then cooled over another 30 day period to room temperature. Mixtures with the composition corresponding to PtS_2 were then annealed for 50 days at 650 °C and those corresponding to $PtSe_2$ were annealed at 1000° and 600 °C for 20 to 30 days.

Experiments to grow crystals have not been mentioned. See also Tables I and IX.

TABLE IX
Compound preparation and crystal growth techniques

Compound	Compound preparation procedures		Crystal growth techniques	
TcCh$_2$	Direct reaction of elements at 900°–1100 °C	[91, 121]	Cl_2, Br_2 and I_2 transport TcS$_2$: $T_H = 1150° \to T_L = 1080$ °C. TcSe$_2$: $T_H = 1080° \to T_L = 1000$ °C. TcTe$_2$: $T_H = 980° \to T_L = 840$ °C. Tubes had 200 mm length.	[121]
ReS$_2$ and ReSe$_2$	a. Direct synthesis from the elements at 1100 °C gives ReS$_2$ and ReSe$_2$. b. Thermal decomposition of Re$_2$S$_7$ above 150 °C. Thermal decomposition of Re$_2$Se$_7$ above 320 °C is mentioned, probably doubtful.	[121] [122]	Cl_2, Br_2 and I_2 transport $T_H = 1150° \to T_L = 1080$ °C for ReS$_2$. $T_H = 1080° \to T_L = 1000$ °C for ReSe$_2$. Excess chalcogen stimulates transport and improves quality of the crystals. Chlorine in some cases used for sulfide transport. See also ref. [91] for additional information on crystal growth conditions.	[121]
CoTe$_2$	Direct synthesis at 1000 °C, then cooled to- and kept at 700 °C for 24 h then slowly cooled to room temp. followed by annealing (6 days) at 600 °C. CoTe$_{1.8}$ has layer structure, Te-richer material not.	[128]	No reports on crystal growth experiments have been found for the group VIII layered compounds.	
RhTe$_2$	Direct synthesis at 1200 °C (1 to 3 days), the high-temperature modification has layer structure.	[129]		
IrTe$_2$	Direct synthesis at 950 °C, kept at 600 °C for 18 h, then cooled rapidly.	[132]		
NiTe$_2$ + PdTe$_2$ + PtCh$_2$	Direct synthesis from elements at 825 °C, this temp. is maintained for 30 days, then over 30 days cooled to roomtemperature.	[134]		

References

1. J. A. Wilson and A. D. Yoffe: *Advances in Phys.* **18** (1969), 193.
2. F. Kadijk: Thesis State University Groningen, 1969.
3. H. Schäfer: in *Chemical Transport Reactions*, Academic Press, New York, 1964.
4. R. W. G. Wijckoff: in *Crystal Structures*, vol. 1, chapter 4, Interscience Publ., New York 1960.
5. A. R. Verma and P. Krishna: in *Polymorphism and Polytypism in Crystals*, John Wiley & Sons Inc., New York, London, 1966.
6. A. A. Balchin: in F. Lévy (ed.), *Crystallography and Crystal Chemistry of materials with Layered Structures*, D. Reidel Publ. Co., Dordrecht, Boston, 1976.
7. E. Tronc and M. Huber: *J. Phys. Chem. Sol.* **34** (1973), 2045.
8. R. Moret and M. Huber: *Acta Cryst.* **B32** (1976), 1302.
9. R. Huisman and F. Jellinek: *J. Less Comm. Metals* **17** (1969), 111.
10. J. A. Wilson, F. J. Di Salvo, and S. Mahajan: *Adv. Phys.* **24** (1975), 117.
11. B. E. Brown and D. J. Beerntsen: *Acta Cryst.* **18** (1965), 31.
12. F. Kadijk and F. Jellinek: *J. Less Comm. Metals* **23** (1971), 437.
13. R. M. White and G. Lucovsky: *Sol. Stat. Comm.* **11** (1972), 1369.
14. G. Lucovsky, R. M. White, J. A. Benda, and J. F. Revelli: *Phys. Rev.* **B7** (1973), 3859.
15. R. B. Murray, R. A. Bromley, and A. D. Yoffe: *J. Phys. C., Sol. Stat.* **5** (1972), 746.
16. R. A. Bromley, R. B. Murray, and A. D. Yoffe: *J. Phys. C., Sol. Stat.* **5** (1972), 759.
17. R. Huisman, R. de Jong, C. Haas and F. Jellinek: *J. Sol. Stat. Chem.* **3** (1971), 56.
18. F. R. Gamble: *J. Sol. Stat. Chem.* **9** (1974), 358.
19. I. Oftedal: *Z. Phys. Chem.* **134** (1928), 301.
20. J. Benard and Y. Jeannin: in *Non Stoichiometric Compounds*, Advanc. in Chem. Series No. 39, Wash. D. C., 1963.
21. A. H. Thompson, F. R. Gamble, and C. R. Symon: *Mat. Res. Bull.* **10** (1975), 915.
22. W. Biltz, P. Ehrlich, and K. Meisel: *Z. anorg. allg. Chem.* **234** (1937), 97.
23. H. Hahn and B. Harder: *Z. anorg. allg. Chem.* **288** (1956), 241.
24. H. Hahn and P. Ness: *Naturwiss.* **44** (1957), 581, and *Z. anorg. allg. Chem.* **302** (1959), 17.
25. Y. Jeannin: *Ann. Chim.* **7** (1962), 1.
26. G. A. Wiegers and F. Jellinek: *J. Sol. Stat. Chem.* **1** (1970), 519.
27. G. Brauer in *Handbuch der Präperativen Anorganische Chemie*, Ferd. Enke Verlag, Stuttgart, 1962.
28. K. F. McTaggart and A. D. Wadsley: *Austral. J. Chem.* **11** (1958), 445.
29. F. R. Gamble and A. H. Thompson: private communication.
30. R. R. Chianelli, J. J. Scanlon, and A. H. Thompson: *Mat. Res. Bull.* **10** (1975), 1379.
31. S. Takeuchi and H. Katsuda: *J. Japan Inst. Metals* **34** (1970), 758.
32. H. Schäfer, F. Wehmeier, and M. Trenkel: *J. Less Comm. Metals* **16** (1968), 290.
33. H. P. B. Rimmington, A. A. Balchin, and B. K. Tanner: *J. Crystal Growth* **15** (1972), 51.
34. D. L. Greenway and R. Nitsche: *J. Phys. Chem. Solids* **26** (1965), 1445.
35. J. Bear and K. F. McTaggert: *Austral. J. Chem.* **11** (1958), 458.
36. P. Ehrlich: *Z. anorg. allg. Chem.* **260** (1949), 1.
37. P. Bernusset and Y. Jeannin: *Compt. Rend.* **255** (1962), 934.
38. F. Raaum, F. Grönvold, A. Kjekshus, and H. Haraldsen: *Z. anorg. allg. Chem.* **317** (1962), 91.
39. E. F. Strotzer, W. Biltz, and K. Meisel: *Z. anorg. allg. Chem.* **242** (1939), 249.
40. H. Hahn, B. Harder, U. Mutschke, and P. Ness: *Z. anorg. allg. Chem.* **292** (1957), 82.
41. A. Gleizes and Y. Jeannin: *J. Sol. Stat. Chem.* **1** (1970), 180.
42. A. E. van Arkel: *Physica* **4** (1924), 286.
43. H. Hahn and P. Ness: *Naturwiss.* **44** (1957), 534 and *Z. anorg. allg. Chem.* **302** (1959), 37.
44. H. Hahn and P. Ness: *Naturwiss.* **44** (1957), 534 and *Z. anorg. allg. Chem.* **302** (1959), 136.
45. A. Gleizes and Y. Jeannin: *J. Solid Stat. Chem.* **5** (1972), 42.
46. L. Brattås and A. Kjekshus: *Acta Chem. Scand.* **25** (1971), 2783.
47. J. G. Smeggil and S. Bartram: *J. Sol. Stat. Chem.* **5** (1972), 391.
48. J. Nitsche: in H. S. Peiser (ed.), *Crystal Growth*, Pergamon, Oxford, 1967.
49. A. H. Thompson, K. R. Pisharody, and R. F. Koehler: *Phys. Rev. Lett.* **29** (1972), 163.
50. C. R. Whitehouse, H. P. B. Rimmington, and A. A. Balchin: *Phys. Stat. Sol.* (a) **18** (1973), 623.
51. H. P. B. Rimmington and A. A. Balchin: *J. Crystal Growth* **21** (1974), 171.
52. E. Hoschek and W. Klemm: *Z. anorg. allg. Chem.* **242** (1939), 49.

53. E. Röst and L. Gjertsen: *Z. anorg. allg. Chem.* **328** (1964), 299.
54. E. Röst, L. Gjertsen, and H. Haraldsen: *Z. anorg. allg. Chem.* **333** (1964), 301.
55. F. Grönvold, O. Hagberg, and H. Haraldsen: *Acta Chem. Scand.* **12** (1958), 971.
56. P. Ehrlich: *Z. anorg. allg. Chem.* **260** (1949), 30.
57. M. H. van Maaren and G. M. Schaeffer: *Phys. Lett.* **24A**, (1967), 645.
58. W. Biltz and A. Köcher: *Z. anorg. allg. Chem.* **237** (1938), 369.
59. G. Hägg and N. Schönberg: *Arkiv Kemi* **7** (1954), 371.
60. F. Jellinek, G. Brauer, and H. Müller: *Nature* **185** (1960), 376.
61. F. Jellinek: *Arkiv Kemi* **20** (1963), 447.
62. M. H. van Maaren and G. M. Schaeffer, *Phys. Lett.* **20** (1966), 131.
63. F. Kadijk and F. Jellinek: *J. Less Comm. Metals* **19** (1969), 421.
64. H. Schäfer and W. Fuhr: *J. Less Comm. Metals* **8** (1965), 375.
65. L. H. Brixner: *J. Inorg. Nucl. Chem.* **24** (1962), 257.
66. K. Selte and A. Kjekshus: *Acta Chem. Scand.* **18** (1964), 697.
67. E. Revolinsky, B. E. Brown, D. J. Beerntsen, and C. H. Armitage: *J. Less Comm. Metals* **8** (1965), 63.
68. K. Selte, E. Bjerkelund, and A. Kjekshus: *J. Less Comm. Metals* **11** (1966), 14.
69. R. Huisman, F. Kadijk, and F. Jellinek: *J. Less Comm. Metals* **12** (1967), 423.
70. R. Huisman, F. Kadijk, and F. Jellinek: *J. Less Comm. Metals* **21** (1970), 187.
71. E. Revolinsky, G. A. Spiering, and D. J. Beerntsen: *J. Phys. Chem. Sol.* **26** (1965), 1029.
72. F. Kadijk, R. Huisman, and F. Jellinek: *Rec. Trav. Chem.* **83** (1964), 768.
73. H. Schäfer and K. D. Dohmann: *Z. anorg. allg. Chem.* **299** (1959), 197.
74. R. Kershaw, M. Vlasse, and A. Wold: *Inorg. Chem.* **6** (1967), 1599.
75. B. E. Brown: *Acta Cryst.* **20** (1966), 264.
76. W. Biltz and A. Köcher: *Z. anorg. all. Chem.* **238** (1938), 81.
77. F. Jellinek: *J. Less Comm. Metals* **4** (1962), 9.
78. F. J. Di Salvo, B. G. Bagly, J. M. Voorhoeve, and J. V. Waszczak: *J. Phys. Chem. Sol.* **34** (1973), 1357.
79. R. Brouwer and F. Jellinek: *Mat. Res. Bull.* **9** (1974), 827.
80. A. H. Thompson, F. R. Gamble, and J. F. Revelli: *Sol. Stat. Comm.* **9** (1971), 981.
81. F. R. Gamble, F. J. Di Salvo, R. A. Klemm, and T. H. Geballe: *Science* **168** (1970), 568.
82. M. H. van Maaren and H. B. Harland: *Phys. Lett.* **29A** (1969), 571.
83. J. F. Revelli and W. A. Phillips: *J. Sol. Stat. Chem.* **9** (1974), 176.
84. L. E. Conrey and K. P. Pisharody: *J. Sol. Stat. Chem.* **4** (1972), 345.
85. W. Geertsma, C. Haas, R. Huisman, and F. Jellinek: *Sol. Stat. Comm.* **10** (1972), 75.
86. E. Bjerkelund and A. Kjekshus: *Acta Chem. Scand.* **21** (1967), 513.
87. L. A. Aslanov, Yu. P. Simanov, A. V. Novoselova, and Yu. M. Ukrainskii: *Russ. J. Inorg. Chem.* (Engl. transl.) **8** (1963), 937.
88. L. A. Aslanov, Yu. P. Simanov, A. V. Novoselova, and Yu. M. Ukrainskii: *Russ. J. Inorg. Chem.* (Engl. transl.) **8** (1963), 1381.
89. Yu. M. Ukrainskii, L. M. Kovba, Yu. P. Simanov, and A. V. Novoselova: *Russ. J. Inorg. Chem.* Engl. transl.) **4** (1959), 60.
90. Yu. M. Ukrainskii, L. M. Kovba, Yu. P. Simanov, and A. V. Novoselova: *Russ. J. Inorg. Chem.* (Engl. transl.) **4** (1959), 1305.
91. J. C. Wildervanck: Thesis State University Groningen, 1970.
92. J. C. Wildervanck and F. Jellinek: *Z. anorg. allg. Chem.* **328** (1964), 309.
93. R. E. Bell and R. E. Herfert: *J. Am. Chem. Soc.* **79** (1957), 3351.
94. R. G. Dickinson and L. Pauling: *J. Am. Chem. Soc.* **45** (1923), 1466.
95. R. J. Traill: *Can. Mineralogist* **7** (1963), 524.
96. J. J. Berzelius: *Pogg. Ann.* **7** (1826), 261.
97. L. Svanberg and H. Struve: *J. prakt. Chem.* **44** (1848), 257.
98. A. de Schulten: *Bull. Soc. Minéral. France* **12** (1889), 545.
99. A. E. van Arkel: *Rec. Trav. Chem.* **45** (1926), 442.
100. A. V. Nikolaev, A. A. Opalovsky, and V. E. Federov in: R. F. Schwenker and P. D. Garn (eds.), *Thermal Analysis*, vol. 2, Academic Press, 1969.
101. M. S. Silverman: *J. Inorg. Chem.* **6** (1967), 1063.
102. A. N. Zelikman, G. V. Indenbaum, and M. V. Teslitsaya: *Sov. Phys. Crystal.* **14** (5), (1970), 687.
103. A. A. Al-Hilli and B. L. Evans: *J. Crystal Growth* **15** (1972), 93.
104. H. Schäfer, T. Grofe, and M. Trenkel: *J. Sol. Stat. Chem.* **8** (1973), 14.

105. L. C. Towle, V. Oberbeck, B. E. Brown et al.: *Science* **154** (1966), 895.
106. B. L. Evans and R. A. Hazelwood: *Phys. Stat. Sol. (a)* **4** (1971), 181.
107. D. Puotinen and R. E. Newnham: *Acta Cryst.* **14** (1961), 691.
108. B. Brown: *Acta Cryst.* **20** (1966), 268.
109. E. Revolinsky and D. J. Beerntsen: *J. Phys. Chem. Sol.* **27** (1966), 523.
110. M. B. Vellinga, R. de Jong, and C. Haas: *J. Sol. Stat. Chem.* **2** (1970), 299.
111. E. Revolinsky and D. J. Beerntsen: *J. Appl. Phys.* **35** (1964), 2086.
112. M. Spiesser and J. Rouxel: *C. R. Acad. Sci. (Paris)* **265** (1967), 92.
113. O. Glemser, H. Sauer, and P. König: *Z. anorg. allg. Chem.* **257** (1948), 241.
114. E. Tiede and H. Lemke: *Ber. Deutsch. Chem. Ges.* **71** (1938), 584.
115. W. T. Hicks: *J. Electrochem. Soc.* **111** (1964), 1058.
116. M. Hansen in: *Constitution of Binary Alloys*, McGraw-Hill Book Co. Inc. New York, 1958, p. 1192.
117. L. H. Brixner: *J. Electrochem. Soc.* **110** (1963), 289.
118. O. Knop and H. Haraldsen: *Canad. J. Chem.* **34** (1956), 1142.
119. L. H. Brixner and G. Teufer: *Inorg. Chem.* **2** (1963), 992.
120. F. Hulliger: in *Structure and bonding*, vol. 4, Springer-Verlag, Berlin, Heidelberg, New York, 1968.
121. J. C. Wildervanck and F. Jellinek: *J. Less Comm. Metals* **24** (1971), 73.
122. N. W. Alcock and A. Kjekshus: *Acta Chem. Scand.* **19** (1965), 79.
123. S. Furuseth and A. Kjekshus: *Acta Chem. Scand.* **20** (1966), 245.
124. K. Meisel: *Z. anorg. allg. Chem.* **44** (1931), 243.
125. J. Lagrenaudie: *J. Phys Radium* **15** (1954), 299.
126. T. A. Catpaeva, P. A. Isakova, and T. P. Polyakova: *Trans. Onst. Geol. Nauk, Akad. Nauk Kaz. S.S.R.* (1963), 318.
127. G. Odent: *Rev. Chim. Minér.* **6** (1969), 933.
128. H. Haraldsen, F. Grönvold, and T. Hurlen: *Z. anorg. allg. Chem.* **283** (1956), 143.
129. S. Geller: *J. Am. Chem. Soc.* **77** (1955), 2641.
130. L. Wöhler, K. Ewald, and H. G. Kroll: *Chem. Ber.* **66** (1933), 1638.
131. W. Biltz: *Z. anorg. all. Chem.* **233** (1937), 282.
132. E. F. Hockings and J. G. White: *J. Phys. Chem.* **64** (1960), 1042.
133. S. Furuseth, K. Selte, and A. Kjekshus: *Acta Chem. Scand.* **19** (1965), 257.
134. F. Grönvold, H. Haraldsen, and A. Kjekshus: *Acta Chem. Scand.* **14** (1960), 1879.
135. L. Thomassen: *Z. Phys. Chem.* **B2** (1929), 349.
136. P. B. James and T. Lavik: *Acta Cryst.* **16** (1963), 1183.
137. K. C. Mills: *Thermodynamic Data for Inorganic Sulphides, Selenides and Tellurides*, Butterworths, London, 1974.

III–VI COMPOUNDS

R. M. A. LIETH

Chemical Physics Division, Solid State Group, Dept. of Physics, Technische Hogeschool, Eindhoven, The Netherlands

Table of Contents
1. Introduction
2. Description of the Crystal Structures
3. Properties of the Elements
    3.1. Thermodynamic Properties
    3.2. Purity and Purification of Starting Materials
4. Preparation and Purification of the Compounds
    4.1. Compound Preparation
    4.2. Purification of the Compounds
5. Phase Equilibrium
    5.1. Solid-Vapour Equilibrium
    5.2. Simultaneous Solid-Liquid-Vapour Equilibria
    5.3. Limitations in Crystal Preparation due to Phase Equilibria
6. Crystal Growth
    6.1. General Growth Methods Applicable to III-VI Compounds
        6.1.1. Melt Growth
        6.1.2. Vapour Growth
    6.2. Gallium Sulfide
        6.2.1. Melt Growth
        6.2.2. Vapour Growth
    6.3. Gallium Selenide
        6.3.1. Melt Growth
        6.3.2. Vapour Growth
    6.4. Gallium Telluride
        6.4.1. Melt Growth
        6.4.2. Vapour Growth
    6.5. Indium Selenide
        6.5.1. Melt Growth
7. Doping
8. Solid Solutions
References

1. Introduction

Among the more than twenty binary compounds which can be formed between the group III and the group VI elements, only four are known to possess a layered structure. The slab like structure of these compounds – GaS, GaSe, GaTe and InSe – consists of double-layers of metal atoms sandwiched between double layers of non-metal atoms.

These materials have received interest on account of their properties which lie in the field of semiconductivity, photoconductivity and luminescence. An important factor in their characterization is the ability to grow structurally and chemically pure crystals. This demand for purity has resulted from the recognition that the intrinsic properties of a material can be elucidated only if there are no foreign atoms present to interfere with the atoms of the base material and from the fact that measured quantities of known impurities can be added to a high purity material to obtain reproducible desirable and useful physical properties.

The need is therefore for both high purity materials as well as for materials with a controlled amount of impurity. In view of the lack of information on dopant addition to this family of compounds, more attention to this aspect is required in future.

In this article we shall approach preparation and crystal growth of the above mentioned compounds mainly from the practical side; for theories and studies of nucleation and controlled growth, the reader is referred to the existing literature on this topic.

It is a pity that only a meagre quantity of knowledge concerning InSe has accumulated in the past years, this article will therefore be concerned mainly with the three gallium compounds.

Since in volumes II and V the attention will be focused on crystal chemistry and crystallography here only an introductory description of the crystal structures is given in Section 2. In Section 3 we shall consider the thermodynamic properties of the component elements and the purification of starting materials. This will lead to the preparation and purification of compounds in Section 4.

The phase equilibria and the limitations on crystal preparation due to these phase relations are discussed in Section 5 and will be followed by a survey of general growth methods applicable to the III–VI compounds in Section 6.

The techniques applied with most success to each particular compound are described in Section 7. Much of the available literature will be cited, but this should not be construed as an attempt to provide an exhaustive bibliography.

Finally in Sections 8 and 9 a very brief account of doping procedures and the formation of solid solutions is given. Little data exists in either area.

2. Description of the Crystal Structures

The crystal structures of the four compounds will be presented in an introductory way to give the reader the necessary background information. For more detailed discussions the reader is referred to the current literature cited. Volumes II and V of this series will give a detailed discussion of this topic.

The three compounds GaS, GaSe and InSe have very similar structures [1–5], consisting of four-layered slabs each of which contains two close-packed metal layers and two close-packed anion layers. In the direction of the c-axis of the unit cell, the stacking sequence in such a four-fold slab is anion-metal-metal-anion (Figure 1a). The bonding between two adjacent slabs is of the Van der Waals type, while the bonding between the layers of a slab is predominantly covalent.

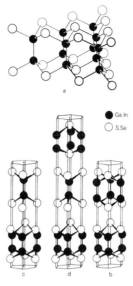

Fig. 1. Stacking of layers in GaS, GaSe, and InSe. (a) the arrangement of Ga (or In) atoms in one fourfold slab; (b) the stacking sequence in hexagonal (β)-GaS; (c) the stacking sequence in hexagonal (ε)-GaSe; (d) the stacking sequence in rhombohedral (γ)-GaSe. The stacking in InSe is still point of discussion, according to [7], it is that of GaS, according to [8] that of (γ) GaSe.

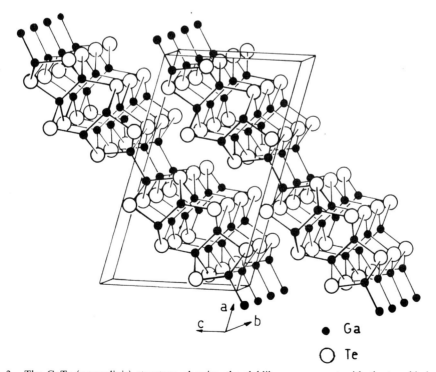

Fig. 2. The GaTe (monoclinic) structure, showing the slablike arrangement with the two kinds of Ga—Ga pairs. (Reference 1)

TABLE I

Polytypism, cell constants and colors of GaS, GaSe, GaTe and InSe

Compound	GaS	GaSe	GaTe	InSe
Color	greenish-yellow	red brown	metallic	black
Polytypism	no polytypism found yet	two polytypes known, five new recently discovered	no polytypism found yet	no polytypism reported yet
Cell constants	hexagonal stacking of four-fold sheets $a=3.585$ Å $c=15.50$ Å $D_{6h}^4(P6_3/mmc)$, two-layer repetition (so-called β-type) [5]	(1) hexagonal stacking of four-fold sheets $a=3.755$ Å $c=15.946$ Å $D_{3h}^1(P\bar{6}m2)$ two-layer repetition (2H) (so-called ε-type) [4, 5] and (2) rhombohedral stacking of four-fold sheets $a=3.755$ Å $c=23.910$ Å $C_{3v}^5(R3m)$ three-layer repetition (3R) (so-called γ-type) [4] Recently found, 9R, 12R and 15R, [67] and 4H and 6H [68,69]	monoclinic unit cell according to [11, 54] $a=23.81$ Å $b=5.076$ Å $c=10.48$ Å $\beta=45.4°$ but according to [11] $a=17.37$ Å $b=4.074$ Å $c=10.44$ Å $\beta=104°12'$	hexagonal stacking of four-fold sheets $a=4.05$ Å $\}$ [3] $c=16.93$ Å $D_{6h}^4(C6/mmc)$ [7] but according to [8] $a=4.01$ Å $c=16.63$ Å and according to [9] rhombohedral stacking $a=4.015$ Å $c=25.00$ Å and according to [70] $a=4.00$ Å $c=24.85$ Å

The difference between the various suggested structures arises from the different stacking of such slabs.

Thus in order to transform a four-fold slab into its neighbour in GaS, a rotation through 60° about the c-axis as well as a vertical translation is necessary. This results in the so-called (hexagonal) β-type of stacking (Figure 1b) with a two slab repetition of the type $A\beta\beta A$, $B\alpha\alpha B$.

Whereas no polytypism is found for GaS, GaSe has been found to be polytypic. In GaSe a horizontal translation in the [11.0] direction with respect to the adjacent slab is sufficient to give two structure types. One of them has a two slab repetition of the type $A\beta\beta A$, $B\gamma\gamma B$ and is the so-called (hexagonal) ε-type GaSe, while the other possibility which arises as a result of translation in the [11.0] direction is the so-called (rhombohedral) γ-type GaSe. This is a structure with a three slab repetition of the type $A\beta\beta A$, $B\gamma\gamma B$, $C\alpha\alpha C$ (Figures 1c and 1d). The symbols A, B and C indicate the position of the anions, α, β and γ those of the cations.

Unfortunately there has been some confusion in the past about the possibility of a β-type stacking for GaSe. Up to the present there has been no convincing evidence for the existence of such a stacking, which would be similar to the one found in GaS. For detailed discussions the reader is referred to recent literature on this subject [6], for polytypes see Reference 67. In the case of InSe there is some discrepancy between the published data. According to Semiletov [7] the hexagonal packing of layers is of the GaS-type. The disordered type of stacking found by this author is confirmed by detailed studies of Terhell [8], but his work leads him to reject the GaS type of stacking for InSe.

Schubert and co-workers [9] on the other hand proposed a rhombohedral structure with a large value for the c-axis. This is confirmed by Terhell [8] and Likforman [70].

As a consequence of the difference in stacking, the stacking faults which occur in these structures if part of a crystal slips along one of its layers, are different and it has been shown that the energy associated for example with a stacking fault in GaS is relatively large, while the stacking fault energy in GaSe is smaller [10].

The compound GaTe has a monoclinic structure [11] and the atoms are also arranged in layers, but of more complex type.

Portions of the structural units consisting of four planes of atoms, extending indefinitely in the direction of the b-axis, are joined together by other Te—Ga—Ga—Te strips in which the Ga—Ga bonds lie approximately at right angles to the other Ga—Ga bonds [11].

This is shown in Figure 2. Cell dimension data for GaTe published by different investigators show some difference as can be seen in Table I, where the cell constants, the colours and the possible polytypes are presented for the four compounds.

3. Properties of the Elements

In this section we shall be concerned with the elements of interest for the III–VI compounds. We shall present those thermodynamic properties which are of interest in phase equilibria and compound formation and also consider the availability of high purity elements and further purification.

3.1. THERMODYNAMIC PROPERTIES

The thermodynamic properties of the afore mentioned five elements Ga, In, S, Se and Te, are presented in Table II. They can be classed as relatively low melting. The metals listed have rather high boiling points and low vapour pressures at high temperatures (4×10^{-2} torr for In and about 6×10^{-3} torr for Ga at $1000\,°C$).

In general either the liquid or gaseous state will be of interest. Both metals Ga and In exists as monoatomic vapour species, the group VI elements are more complex, since they can exist in several forms depending on temperature and pressure [12].

TABLE II

Thermodynamic properties of the elements (Stull and Sinke [12])

Element	M.P.[a] (°C)	B.P.[b] (°C)	ΔH_m^c (kcal mole^{-1})	ΔH_v^d (kcal mole^{-1})
Ga	29.8	2,237	1.335	61.2
In	156.3	2,047	0.78	54.1
S	119	444.8	0.337	2.3
Se	217	685	1.3	6.29
Te	450	987	4.18	12.1

[a] Melting point at 1 atm pressure.
[b] Boiling point at 1 atm pressure.
[c] Heat of melting at the M.P.
[d] Heat of vaporization at 1 atm total pressure.

Sulphur vapours are very complex in composition; molecules from S_1 to S_8 are known to exist in equilibrium. In a recent report on the thermodynamics of sulphur vapour, Rau and co-workers [13] present a computer program for calculation of vapour densities and partial pressures of S_2, S_3, S_4, S_5, S_6, S_7 and S_8. Their results show the predominance of S_6 and S_8 – in saturated vapour at low temperatures (500–1000 °C), the decrease in the amount of these species with increasing temperature and the predominance of S_2 and S_3 at higher temperatures.

Selenium exists as Se_1, Se_2 and Se_6, while tellurium has been found to be stable as Te and Te_2 [12]. In a given temperature range, generally just one of the species mentioned will predominate. From the free energy changes listed in tables of thermodynamic properties of the elements one can readily determine the partial vapour pressure and the relative abundance of a particular form [12].

3.2. PURITY AND PURIFICATION OF STARTING MATERIALS

The five elemental components under consideration are commercially available to at least 99.999% purity and this eliminates in part the necessity for further purification. Sometimes however as in semiconductor work, further purification seems desirable; in such a case a careful check should be made to determine whether such treatment will effectively improve the situation. In handling or processing an already high purity semiconductor grade material, one runs the risk of introducing impurities. A potential source of contamination is the crucible used in the process. Furthermore, even though multiple purification steps may sufficiently decrease certain impurity concentrations, other less desirable impurity elements may at the same time be incorporated in the material. This may happen to such an extent that after the procedure the overall purity may be better but the material may be of lower quality with regard to specific applications.

Further purification can be attempted by distillation or vaporization or advanced techniques like zone melting, slow crystal growth and repeated or fractional crystallization (see for example Reference 14). Oxidation of metallic elements during the long periods of vacuum baking at about 10^{-5} torr is frequently observed. Removal of this surface oxide is readily achieved by firing the metal component in a stream of pure hydrogen at an appropriate temperature.

More detailed information about the various purification techniques and methods employed in detecting trace impurities can be found in the reports cited by Brooks and Kennedy [16] and the summary given by Lawson and Nielsen [17].

4. Preparation and Purification of the Compounds

In this section we shall be concerned with the preparation of the III–VI compounds in the form of powders and with further purification of the polycrystalline materials.

4.1. Compound preparation

The preparation of the four compounds was first reported by Klemm and Von Vogel [18]. The elements were melted together in sealed, evacuated vitreous silica ampoules. In recent work modifications of this method have been used. For the sulfide, ampoules of the two section type are employed; one section contains the metal component, the other section the sulphur. In heating this tube system, care is taken to keep the sulphur at a temperature not exceeding 600 °C, so that the sulphur vapour pressure does not exceed 5 atm. The metallic component can be kept at a higher temperature to ensure a rapid reaction, it should however not exceed 950 °C to avoid reaction between metal and tube wall with the resulting contamination of the compound with silicon and/or oxygen.

The need for a two-section ampoule which eliminates the danger of explosions, due to high sulphur vapour pressures at high temperatures, is absent in the case of the preparation of the selenides and tellurides [19, 20] since the vapour pressures of the more volatile components selenium and tellurium are lower at the reaction temperatures.

A discussion of the procedures employed in the synthesis of GaS is given in References 15 and 21. Some properties and preparation procedures are listed in Table III.

4.2. Purification of the compounds

Although element purification certainly is a recommended procedure it does not

TABLE III

Some properties and preparation procedures of the compounds GaS, GaSe, GaTe and InSe

Compound	M.P. [a] (°C)	$-\Delta H°_{298}$ [b] (kcal mole^{-1})	$S°_{298}$ [c] (cal mole^{-1} deg^{-1})	Specific density [d] (g cm^{-3})	Reaction of the components in a common ampoule	Reaction of the components in a two-section ampoule
					Ampoule is rotated to reduce reaction time	
GaS	962 ± 2 [d,e]	46.4 ± 3.0 [k] 34.5 ± 3.0 [j]	11 [h] 5 [j]	3.86	—	metal at ~950°C sulfur at ~600°C
GaSe	938 ± 5 [f]	35.0 ± 3.0 [k] 17.5 [n]	no data available	5.03	at ~800°C	—
GaTe	835 ± 5 [d,g]	28.6 ± 3.0 [k]	no data available	5.44	at ~850°C	—
InSe	660 ± 10 [d] <630 [m]	28.2 ± 3.0 [k]	no data available	5.55	at ~800°C	—

[a] Melting point at 1 atm pressure.
[b] Heat of formation or reaction at 25°C.
[c] Standard entropy.
[d] Klemm and Von Vogel [18].
[e] Lieth, Heijligers, and Van der Heijden [23].
[f] Dieleman and Engelfriet [24].
[g] Newman and co-workers [11].
[h] Lieth, Heijligers, and Van der Heijden [25].
[j] Nishinaga and Lieth [46].
[k] O. Kubaschewski, E. L. L. Evans, and G. B. Alcock [22].
[m] G. E. van Egmond [43], see also [70].
[n] Nishinaga et al. [66].

necessarily guarantee a high purity compound. Starting materials are always in contact with the container for a considerable length of time at the elevated temperatures needed for synthesis. Reduction of the long reaction time is a first step to obtain purer compounds. In the case of GaS this has been achieved by rotating the whole two-compartment tube system around its long axis.

This ensures a rapid rate of reaction and prevents the formation of a Ga_2S_3 crust on the liquid gallium, otherwise encountered, so that long reaction times are eliminated.

Aside from this source of contamination, the handling of elemental components prior to synthesis may introduce further impurities and therefore when feasible, compound purification should be given full consideration. Unfortunately, purification of the III-VI compounds is still an unexplored area. As a consequence only very little information is available.

In vapour growth, either by sublimation or by chemical transport processes with the aid of a halogen, there is often some purification effect. However, the degree of improvement depends, among other factors, on the concentration of impurities in the starting material, i.e. whether they are present above or below their high temperature solubility limit, the volatility of the impurities and the rate of crystal growth. Those impurities which are not volatile or do not form volatile intermediate compounds, will be effectively segregated; those which are volatile will probably be in equilibrium with source and seed. All impurities will be reduced at least to their solubility limit at the growth temperature, and one may therefore expect a difference in behaviour for different impurities. Purification effects observed in the course of GaS vapour growth experiments have been reported [21]. The same purification effect is observed for GaSe.

5. Phase Equilibrium

In this section we shall be concerned as far as data are available with the phase equilibria for the III–VI systems. Aside from their importance in crystal preparation, phase equilibrium considerations are of extreme value in the understanding of post heat treatments.

5.1. SOLID-VAPOUR EQUILIBRIUM

There are basically two types of equilibria of interest in the III–VI systems, viz., the equilibrium between the solid $AB(s)$ and the vapour phase (v) and the equilibrium between solid-liquid-vapour. Both types are of importance for crystal preparative work; the s-v equilibrium serves as a guide in the growth from the vapour phase, the s-l-v is of utmost importance in melt growth experiments.

Furthermore equilibria considerations help in understanding post heat treatments where they serve to specify the compositional state of the crystal. Only the gallium compounds have been studied more or less in detail (cf. References 11, 25 and 26) but it is not known whether the III–VI compounds dissociate at higher temperatures according to the reaction

$$AB(s) \rightleftharpoons A(v) + \tfrac{1}{2}B_2(v), \tag{5}$$

or whether molecular species $(AB)_x(v)$ are predominant.

As far as GaS is concerned, sublimation of the compound AB cannot be stopped by exerting a pressure of component B on the solid. The only effect such a sulphur vapour pressure has on the $AB(v)$ species is to reduce the rate of vapour diffusion and for GaS evidence therefore strongly favours the presence of molecular species in the vapour phase.

Experiments have indicated a very slight dissociation of GaS and small concentrations of elemental species [27]. A similar behaviour has been observed for GaSe in the temperature range 900–930 °C [26].

5.2. SIMULTANEOUS SOLID-LIQUID-VAPOUR EQUILIBRIA

The importance of these equilibria has already been emphasized and unfortunately enough, the amount of data is much more limited than that pertaining to solid-liquid equilibria.

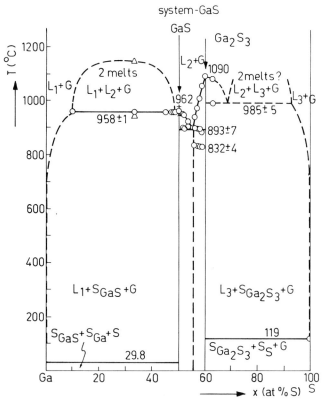

Fig. 3. The temperature-composition projection of the system Ga—S [23]., ($\triangle$); results of Spandau and Klanberg. [64]. ($+$); results of Brukl and Ortner, [65]; ($\bigcirc$); authors results 23.

For details about the determination of the various relationships and the construction of the $T-x$, $P-T$ and $P-x$ plots, the reader is referred to the book of Hume-Rothery and co-workers [28] and that of Prince [29].

$T-x$ diagrams of the four compounds under consideration have been reported; the earliest report is from Klemm and Von Vogel [18] on the system Ga—Te which was re-investigated by Newman and co-workers [11]. The system Ga—Se has been studied independently by Rustamov and co-workers [30] and Palatnik and Belova [31] and recently partly re-investigated by Dieleman and Engelfriet [24]. The system Ga—S was reported by Lieth and co-workers [23] and that of In-Se by Guliev and Medvedeva [32] and recently by Likforman et al. [70].

The $T-x$ plots are presented in Figures 3 to 6. There is a close similarity in the metal-rich parts of these diagrams, which show a region of liquid immiscibility. The sulphur-rich range of the system Ga—S is nearly inaccessible due to the large sulphur pressures and consequently the $T-x$ data are rather limited in this region [23].

$P-T$ and $P-x$ relationships have been reported for the system Ga—S and Ga—Te only; they are presented in Figures 7 to 11. For the system Ga—Se, the plot of P_{GaSe} vs $1/T$ is presented in Figure 12.

For GaS it can be seen that the pressure of the moleculair species P_{GaS} is considerably larger than the partical pressure of the most volatile component i.e. P_{S_2} [25, 27]; for GaSe and GaTe such a comparison cannot be made due to the absence of P_{GaTe} vs $(1/T)$ and P_{Se_2} vs $(1/T)$ data. It seems reasonable however, to assume a close resemblance between the behaviour of GaSe and GaS. See also [75].

Unfortunately the compound InSe has not been investigated to such an extent, and corresponding data are not available.

5.3. Limitations in Crystal Preparation due to Phase Equilibria

The two basic methods employed in crystal preparation are growth from a liquid phase and from a vapour phase. The limitations or advantages of each method will be examined, based on our knowledge of phase equilibria in the III–VI systems.

Successful growth from stoichiometric melts has been reported for all four compounds. The fact that this method is applicable is due to the fact that the melting points and the vapour pressure are not too high. This enables the use of easily sealable quartz, which softens above 1200°C, as the container material. Growth from off-stoichiometric melts has been reported for some III–VI compounds by Hársay [33]; the crystals grow at temperatures much lower than their melting points. However the need for growth at such lower temperatures does not seem so demanding as in the case of the II–VI or III–V compounds. Furthermore the basic shape of the four $T-x$ diagrams shows liquidus curves rising steeply with temperature in the vicinity of the pure metal. Common to all four systems is the region of liquid-liquid equilibrium on the metal-rich side. Hence crystal growth from off-stoichiometric melts at lower temperatures is confined to very diluted systems which in general does not lead to large and well-formed crystals.

For GaS and GaSe vapour phase growth is a convenient method. For GaTe the

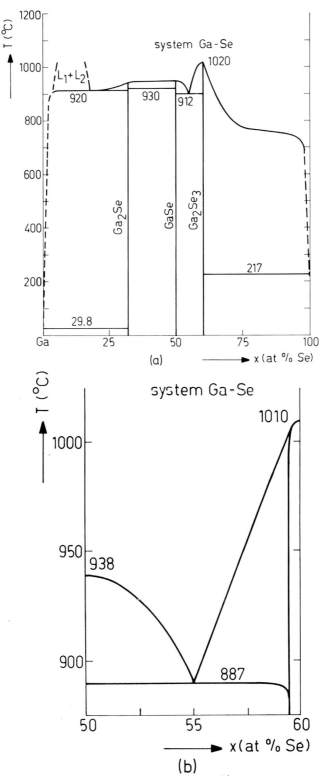

Fig. 4. The temperature-composition projection of the system Ga—Se, (a); as reported by [30], (b); as reported by [24].

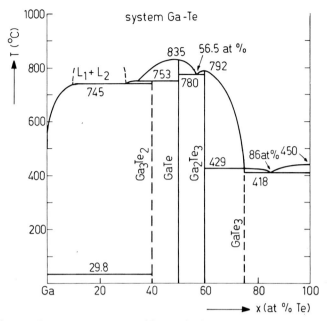

Fig. 5. The temperature-composition projection of the system Ga—Te [11].

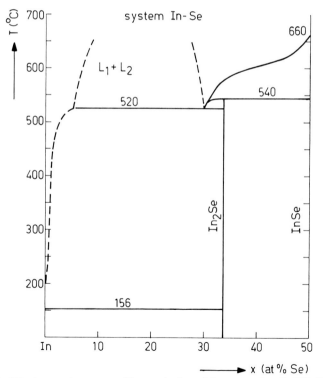

Fig. 6. Part of the temperature-composition projection of the system In—Se [32]. See also [70].

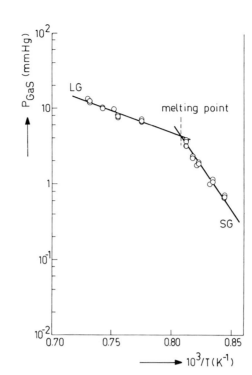

Fig. 7. The pressure-temperature relationship for the compound GaS in the range 910–1100 °C. (SG); represents the solid-vapour equilibrium. (LG); the liquid-vapour equilibrium, [27].

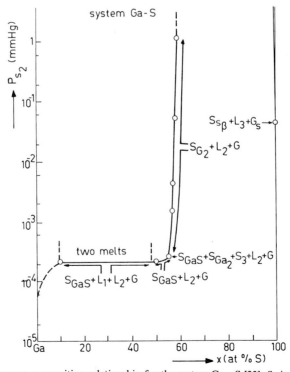

Fig. 8. The pressure-composition relationship for the system Ga—S [23]. S_β is monoclinic sulfur.

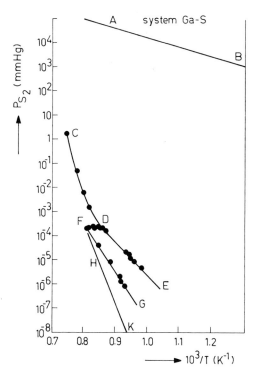

Fig. 9. The pressure-temperature relationship for the system Ga—S [27], (AB); the vapour pressure of pure sulfur, (HK); the 'p_{S_2}' of pure gallium, namely the vapour pressure of pure Ga translated into a sulfur pressure. (CDF); the pressure at the melting point, (FG); the pressure above solid GaS, (DE); the pressure above sulfur-rich GaS.

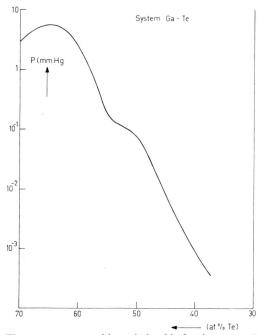

Fig. 10. The pressure-composition relationship for the system Ga—Te [11].

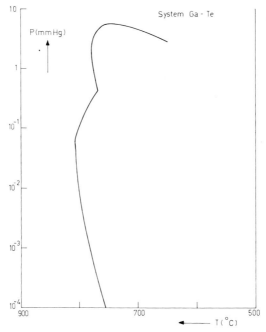

Fig. 11. The pressure-temperature relationship for the system Ga–Te [11].

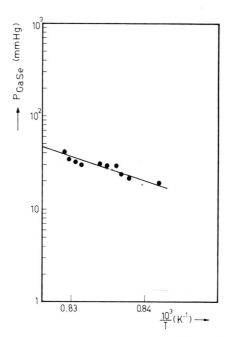

Fig. 12. The pressure-temperature relationship for the compound GaSe in the range 910–940 °C. [26].

partial Te-pressure is too small to give significant results. In the case of InSe attempts to prepare crystals from the vapour phase have been unsuccessful.

In some cases, the transport of vapour species can be increased either by using a forced flow of an inert carrier gas or by inclusion of another compound which, through reaction, changes the initial equilibrium. Such vapour growth experiments with the aid of iodine, which will be discussed later, have not been successful in the case of InSe.

6. Crystal Growth

In this section the general growth methods which have been utilized will first be mentioned briefly. The application of special techniques or slight modifications of commonly employed methods and the degree of success obtained will be discussed in a following section.

6.1. General Growth Methods Applicable to III–VI Compounds

6.1.1. *Melt Growth*

For the preparation of large volume single crystals, growth from the melt has been used with success for all four III–VI compounds [34, 35, 36]. Techniques like the Bridgman and Stockbarger methods predominate in all reports. Another arrangement employed successfully by Beck and Mooser [37] in the cases of GaS, GaSe and GaTe is a modified Czochralski technique; pulling and rotating of the seedholder which is enclosed in the evacuated quartz ampoule, which holds the melt, is carried out magnetically. The stirring effect which the rotating seed has in this set up, is a further advantage (Figure 13).

Crystal growth by way of zone melting has been successfully employed in growth and doping experiments of GaS [15].

6.1.2. *Vapour Growth*

The so-called chemical transport technique is the method most frequently used in vapour growth work. For the underlying principles the reader is referred to the work of Schäfer [38] and the article of Jeffes [39]. This technique aside from its simplicity has the advantage of operating at relatively low temperatures, leading to rather large crystals. It has however the distinct disadvantage of permitting possible contamination by the transport gas.

Another growth method employs a carrier gas in an open tube system. At the present it has only been reported for the growth of GaSe crystals (Paorici and Zuccalli [40]). A forced stream of argon gas carries the GaSe vapour down stream where the material is deposited in the form of many small crystals. A distinct disadvantage of this arrangement is the contamination of the GaSe crystals with gallium. Furthermore the dimensions of the crystals make them unsuitable for semiconductor work. Sublimation in a closed ampoule without the aid of a deliberately added agent or an inert carrier gas, is the third method utilized in vapour growth of crystals of the III–VI

6.2. GALLIUM SULFIDE

6.2.1. *Melt Growth*

Growth of sizeable crystals from the melt was first reported by Beck and Mooser [37]. Their technique has been mentioned briefly in Section 6.1.1. This modified Czochralski method, Figure 13, developed for the work with compounds containing one or more

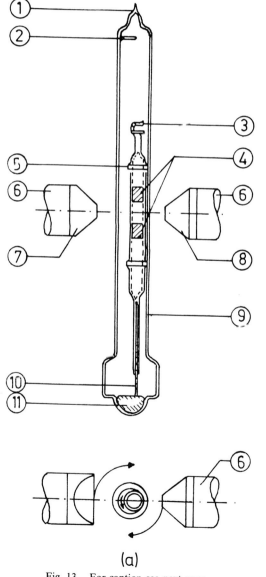

Fig. 13. For caption see next page.

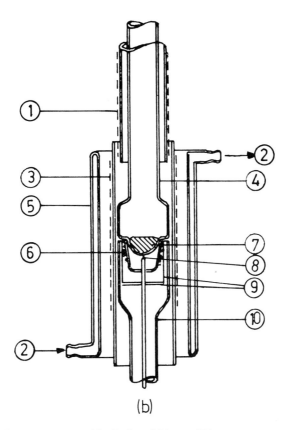

Fig. 13a-b. The pulling apparatus used by Beck and Mooser [37] to grow crystals from the melt. (a); seedholder and ampoule (b); ampoule and furnace.
(a): (1) sealing point, (2) pin, (3) hook, (4) hollow cylinders made of Hiperco (Curie point 970 °C), (5) sleeves to enhances smooth rotating along inner wall of ampoule, (6) magnet, (7) wedge shaped part of magnet, (8) conical shaped part of magnet, (9) sealed and calibrated ampoule, (10) seed crystal, (11) melt.
(b): (1) furnace 1, (2) water cooling, (3) furnace 2, (4) sealed ampoule, (5) cooling mantle, (6) furnace 3, (7) melt, (8) thermo couple, (9) radiation shield, (10) ampoule carrier.

highly volatile components, combines the advantage of magnetic seed pulling with that of magnetic rotation which ensures a smooth and jerk-free movement of the seed holder along the innerwall of the closed and evacuated vitreous silica capsule.

A further advantage of such a rotating seed is its stirring effect on the melt.

This arrangement has also been applied successfully in the case of GaSe and GaTe crystal growth.

The GaS crystals obtained were of the order of $6 \times 1 \times 0.3$ cm^3, see Figure 14.

Aulich, Brebner and Mooser have reported the growth of large monocrystalline samples by using the Bridgman method [41], while Patil and Tredgold [42] employed the Stockbarger method for the growth of GaS monocrystalline ingots.

In both procedures GaS powder is melted in evacuated capsules and pulled through an appropriate gradient, with the hot zone at about 1000 °C. The necessary pulling

Fig. 14. For caption see page 246.

Fig. 14. For caption see page 246.

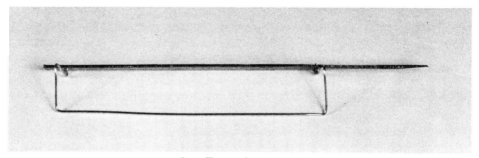

Ga Te edge-on

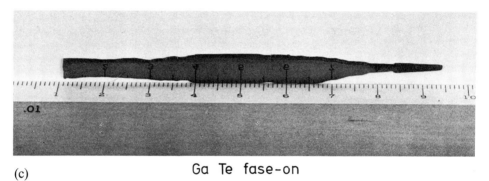

(c)　　　　　　　　　　　Ga Te fase-on

Fig. 14a-c. Melt grown single crystals of GaS, GaSe and GaTe obtained by Beck and Mooser with their set-up (Figure 13), (1.5×). (Courtesy by E. Mooser.)

rates vary between 10 and 15 mm h^{-1}. Still lower pulling rates down to 0.5 mm h^{-1} [43] seem to have the distinct advantage that no signs of intergrowth or twinning can be found in the large ingots. The absence of these defects in such slowly grown crystals is of particular interest for semiconductor work.

A horizontal zone-melting set-up has been used by Van der Leeden to prepare large ingots of GaS [15].

6.2.2. *Vapour Growth*

Chemical transport with iodine as transport agent (see Section 6.2), was first reported by Nitsche and co-workers [44]. The platelets grown at about 850 °C had dimensions of about $5 \times 5 \times 0.01$ mm^3, having their *c*-axis normal to the plane of the plate.

Ever since, this method has been employed successfully with or without modifications in tube length, tube diameter, halogen concentration and so on, to obtain large platelets. Excellent quality crystals have been obtained in tubes of 180 to 200 mm length having 20 mm I.D. or larger. Further conditions are a hot zone of 930 °C for the powder source, a growing zone of 850 to 870 °C and an iodine concentration in the range 2–5 mg cm^{-3} of tube volume.

The resulting crystals can reach dimensions of $10 \times 10 \times 0.01$ mm^3. Higher concentrations of the halogen have the disadvantage that the crystals are more intergrown, lower concentrations that an extended growth time is needed with corresponding increase in the diffusion of impurities into the growing crystals.

Other factors affecting the quality of the crystals are the tube diameter and the undercooling ΔT – the temperature difference between source zone and growth region. Too small a diameter hampers the outgrowth of the crystals, too large diameters cause strong convection which results in very fast growth of intergrown crystals. An undercooling ΔT which is too large also results in much intergrown material. A detailed discussion of the influence of these parameters on GaS growth is given in Reference 45, thermodynamic calculations for this system have been reported by Nishinaga et al. [46].

The use of a preformed compound instead of elemental components in these transport reactions, seems to have a certain advantage. This is best illustrated by the trivial error encountered in some GaS growth experiments with iodine, in which the crystals obtained turned out to be Ga_2S_3. Investigations showed that during the filling procedure, the sulphur powder added to the metal component, completely covered the gallium. After addition of the halogen, the tube was carefully heated to the appropriate temperature and the sulphur could react with the top surface of the metal only. This resulted in the subsequent formation of a layer of Ga_2S_3. In this way the metal bulk was closed-off for any further reaction. As a result the iodine could only transport Ga_2S_3 from the source zone to the growth zone.

The sublimation technique has been used with variable success to grow crystals from the vapour phase. The crystals obtained in this way have the shape of small platelets (with the c-axis normal to the plane of the plate), needles and twinned needles (having their c-axis parallel to the growth axis of the needle). From the point of view of crystal growth, this habit variation is interesting [45, 47], for semiconductor work these needle crystals probably enable specific physical measurements to be carried out in a direction normal to the c-axis.

On the other hand, this sublimation technique is rather unfavourable with respect to the dimensions of the growing plates and needles; it is of minor interest compared with the chemical transport technique or the melt growth methods. Variables like tube geometry and temperature gradient are for this sublimation work the same as discussed earlier for the halogen transport method.

Of particular interest was the investigation of the concentration of impurities incorporated in vapour phase grown GaS and GaSe and the degree of contamination with the halogen used as transporting agent. This was determined with the aid of neutron activation analysis. The amount of iodine incorporated in GaS crystals lies between 0.2 and 10 ppm. For data about their impurities, see Reference 48.

From what is known today it appears that the preparation of single crystals of GaS is equally successful both from the vapour phase and from the melt and that the optimum conditions in the various techniques are sufficiently well established. The melt growth technique has the advantage that iodine free crystals can be produced, which are p-type, while the material transported with the aid of iodine is n-type.

6.3. GALLIUM SELENIDE

6.3.1. *Melt Growth*

The properties of GaSe are sufficiently similar to GaS that as far as crystal growth is concerned, the same considerations apply to both compounds. Consequently some of the particulars will be mentioned more briefly here since they have been discussed under GaS (Section 6.2). With their special apparatus Beck and Mooser [37] also obtained GaSe crystals of centimeter dimensions and it is not surprising that the Bridgman method also proved to be of great value for GaSe (Aulich *et al.* [41]). The procedure is similar to that followed in the case of GaS, and as GaSe has a melting point (938 °C) which is only 25 °C lower than that of GaS, all other variables are identical to those mentioned in Section 6.2.

6.3.2. *Vapour Growth*

An analogous situation exists for growth experiments from the vapour phase. The conditions for best growth with the aid of iodine as transporting agent are in general identical to those presented for GaS, see Reference 49 and also Reference 26. Several investigations concerned with different areas of crystal preparation have been reported; Terhell [6] for instance studied the effect of sample preparation and growth techniques on the structure and habit, Van Egmond [26] reported on the growth rate as a function of variables like halogen concentration, tube geometry and growth temperature and Mooser and Schlüter [50] investigated high angle twist in vapour grown platelets. Employment of the sublimation technique has also led to the production of GaSe needle crystals. This technique has the same disadvantage in comparison to the other growth techniques as discussed for GaS. However, among the needle crystals Terhell found pure γ-GaSe [51] and discovered several new polytypes [67], which have not been obtained with the iodine technique.

Another investigation reported by Paorici and Zuccalli [40], is concerned with the application of the open tube technique for GaSe growth. As mentioned earlier (Section 6.1.2) argon is used as a carrier gas, in their set up. It flows over the mass of polycrystalline GaSe, which is kept at about 920 °C, and carries the vapour down stream where it is deposited at a lower temperature in the form of very thin and small platelets, thin ribbons and very thin small needles.

Microscopic- and X-ray examinations show the majority of the needles to be twinned [52]. Most of the material is contaminated with free gallium in the form of microscopically small droplets. In analogy with a similar effect observed in GaS transport experiments with an open tube system [27] this gallium condensate can be explained by the fact that in the vapour phase GaSe is very slightly dissociated into its components, so that at too low a temperature in the crystallization region the back reaction is hampered and the metal condenses as small droplets. (See reaction 1.)

From the foregoing one may conclude that in the case of GaSe, also single crystals can be prepared equally well from the melt as from the vapour phase.

6.4. GALLIUM TELLURIDE

6.4.1. *Melt Growth*

Gallium telluride is the lowest melting of the three gallium compounds. Its melting point of 835 °C (Table III) makes it quite suitable for melt growth, which in fact is the only technique yielding large crystals of this particular compound. The previously mentioned Beck and Mooser pulling method and the Bridgman/Stockbarger technique are most frequently employed for GaTe growth [35, 36, 53].

6.4.2. *Vapour Growth*

Experiments with iodine transport are reported by Klosse [54]. His work shows the need for a two-phase source to grow GaTe crystals by way of iodine transport in the vapour phase. This is due to the low vapour pressure of tellurium in the system gallium-tellurium. He mentions growth experiments with gradients of 750–730 °C, and 600–560 °C respectively, both employing a source consisting of a mixture of GaTe and Ga_2Te_3.

In the case of growth at 730 °C, the crystals have the shape of ribbons about 10 mm in length and 1 mm in width. When the lower growth temperature of 560 °C was used, the growth rate was considerable reduced and fragile whiskers up to 8 mm in length were obtained and even screw like whiskers appeared in the low temperature batches. Mooser [55] succeeded in growing large GaTe platelets only when rare-earth elements – like Gd – were added as a dope. Without dopant he obtained long flat needles which formed a mass like steelwool and failed to get large platelets. Very recently Mancini and co-workers [56], reported the growth of single crystals both by way of sublimation in a closed ampoule and by the iodine transport process. In their experiments the largest platelike crystals ($8 \times 6 \times 0.3$ mm^3) were obtained in the sublimation procedure, the iodine process yielding largely needles and ribbons and only occasionally a few platelets.

6.5. INDIUM SELENIDE

6.5.1. *Melt Growth*

For this compound the amount of information on preparation, purification, crystal growth techniques and other procedures is rather meager.

No experiments are known where vapour phase growth seems to have been successful. In all reports on InSe, the melt growth method is mentioned in connection with growth of single crystals. The material is either slowly cooled with a constant temperature gradient along the ingot, or slowly pulled in a Bridgman set up at a rate of 0.5 mm h^{-1} [8]. This technique gives monocrystalline regions which are very easily cleaved. [8, 34, 57] The phase diagram reported by Likforman *et al.* [70], shows InSe to have an incongruent meltingpoint. This explains why large pure InSe single crystals cannot be grown by melt-growth techniques when stoichiometric mixtures of In and

Se are employed. The resulting product in such experiments is mainly a mixture of InSe and other phases existing in the system In-Se. The report of Čelustka et al. [71] who observed p- and n-type conductivity in different parts of the same ingot and the work of Nagpal and Ali [72] who could not improve their X-ray powder patterns by prolonged heating at 600 °C are in agreement with the above mentioned property of InSe.

The use of a nonstoichiometric composition will in this case result in a better product as has been shown recently by the experiments of the Paris group of Kuhn [73] and by the investigations of Terhell and van Egmond of our laboratory [74].

Very pure monocrystalline InSe ingots were obtained by using nonstoichiometric source material in Bridgman experiments. In Terhell's work, In and Se were used in the ratio 55 at. % In and 45 at. % Se; this yielded centimeters long pure InSe ingots. Segregation could only be observed in the last 0.5 cm of the top end, that part which is the last to cool in this procedure. Extensive structural work [74] confirmed the results of the single crystal work of Likforman et al. and of Nagpal et al. [70, 72], i.e., that the InSe unit cell is rhombohedral. Powdering the crystals however, totally destroys the rhombohedral lattice, resulting in a disordered polytype. The introduction of such stacking faults in layered materials – by grinding the crystals – affects the intensities in different ways. Especially, in layered structures with very weak bonding between adjacent slabs – as in GaSe and InSe – this will lead to an error in the structure determination (GaS has a stronger bonding between the slabs).

This fact has already been discussed in connection with the structure-work on GaSe [6]. Such a behaviour can explain the controversial structure data reported in the literature on InSe.

The methods used for preparation of single crystals of the four compounds are listed in Table IV.

7. Doping

Doping of the III–VI compounds has not been investigated to such an extent as for example in the case of the elemental semiconductors Si and Ge or of the III–V compounds GaP and GaAs.

In case when crystals are to be grown from the melt or from the vapour phase in closed ampoules the dopant is added to the starting material, if a dynamic system is used in vapour growth work, the dopant is added in a gaseous form to the carrier gas.

Attempts to incorporate various metals in GaS are discussed in detail by Van der Leeden [15]. In his experiments the dopant was added to GaS and the material was subsequently zonemelted. The dopant concentrations were afterwards detected spectroscopically.

Fivaz and Mooser reported attempts to incorporate Zn, Ge and Sn in GaSe grown by the iodine process [58].

TABLE IV
Reported growth techniques as employed for III–VI compounds

Compound	Melt growth	Growth from the vapour phase		
		Open tube system with carrier gas	Sublimation in closed ampoules	Chemical transport in closed ampoules with iodine
GaS	T_h is $>1000\,°C$ Beck-Mooser method [37] Bridgman technique [35, 36, 41, 43] Stockbarger method [42] Zone-melting technique [15]	no report known so far	$T_h = 920\,°C$ $T_l = 870$ to $900\,°C$ [27, 45, 46, 47]	$T_h = 930\,°C$ $T_l = 850$ to $870\,°C$ [44, 45, 46]
GaSe	T_h about $1000\,°C$ Same methods as used for GaS, see References 35, 36, 37, 41, 43	Argon is used as carrier gas [40] $T_h = 920\,°C$ Flow rate $= 250\,cm^3\,min^{-1}$	$T_h = 920\,°C$ $T_l = 840\,°C$ to $890\,°C$ [26]	$T_h = 920\,°C$, $T_l = 850$ to $870\,°C$ [26] $T_h = 870\,°C$, $T_l = 840\,°C$ or $T_h = 850\,°C$, $T_l = 760\,°C$ [58]
GaTe	T_h about $900\,°C$ see References 35, 36, 37, 53	no report known so far	$T_h = 820\,°C$ $T_l = 770\,°C$ [55]	$T_h = 750\,°C$, $T_l = 730\,°C$ [54] ($T_h = 600\,°C$, $T_l = 560\,°C$) $T_h = 800\,°C$, $T_l = 785\,°C$ [55]
InSe	$T_h > 660\,°C$, see References 70, 73, 74	no report known so far	no report known so far	no report known so far

8. Solid Solutions

We shall consider in this section what combination of compounds form complete solid solutions and which have miscibility gaps. The only combination so far known which shows complete miscibility in all proportions is that of GaS with GaSe, although the work of Rustamov seems to be in disagreement with this statement [59]. The number of chemical, crystallographical and physical investigations of this combination is large and still growing. A rather complete account of all work reported on gallium and its compounds is presented in the annual information bulletin on Gallium; volume 10 of 1971 gives information on III–VI solid solutions and presents a complete list of references on this topic [60]. The combinations GaS—GaTe, GaSe—GaTe and GaSe—InSe show miscibility gaps. The combination GaS—GaTe [61] has two narrow regions of solid solubility on either side of the diagram, one up to 10 mole % GaTe in GaS at 770 °C and one of about 10 mole % GaS in GaTe at 770 °C, while the eutectic point lies at 45 mole % GaTe at 770 °C. For the combination GaSe—GaTe, Rustamov [62] reports the ternary compound Ga_2SeTe with a melting point at 820 °C and the existence of two narrow regions of solid solubility.

The combination GaSe—InSe has been studied by Mushinskii [63]. It contains two ranges of solid solution, one between 100 and 80 mole % GaSe and one between 10 and 0 mole % GaSe.

Single crystals of mixed composition $GaS_{1-x}Se_x$ can easily be obtained either by the chemical transport process (Section 6.1.2) or with the Bridgman technique [41].

References

1. W. B. Pearson: *The Crystal Chemistry and Physics of Metals and Alloys*, Wiley-Interscience, New York, 1972.
2. J. C. Philips: *Phys. Rev.* **188** (1969), 1225.
3. R. W. G. Wyckoff: *Crystal Structures*, Vol. 1, Wiley-Interscience Inc. New York, 1965.
4. K. Schubert, E. Dörre, und M. Kluge: *Z. Metallkunde* **4**, Heft 3, (1955), 216.
5. H. Hahn und G. Frank: *Z. Anorg. allgem. Chemie* **278** (1955), 340.
6. J. C. J. M. Terhell and R. M. A. Lieth: *Phys. Stat. Sol. (a)* **10** (1972), 529.
7. S. A. Semiletov: *Kristallografiya* **3** (1958), 288.
8. J. C. J. M. Terhell: private communications.
9. K. Schubert, E. Dörre, und E. Günzel: *Naturwiss.* **41** (1954), 448.
10. Z. S. Basinski, D. B. Dove, and E. Mooser: *Helv. Phys. Acta* **34** (1961), 373.
11. P. C. Newman, J. C. Brice, and H. C. Wright: *Philips Res. Rep.* **16** (1961), 41.
12. D. R. Stull and G. C. Sinke: 'Thermodynamic Properties of the Elements', *Advances in Chemistry*, Series No. 18, Amer. Chem. Soc., Washington D. C., 1956.
13. H. Rau, T. R. N. Kutty, and J. R. F. Guedes de Carvalho: *J. Chem. Thermodynamics* **5** (1973), 833.
14. W. D. Lawson and S. Nielsen: J. J. Gilman (ed.), *The Art and Science of Growing Crystals*, John Wiley and Sons, New York, London, 1963.
15. G. A. van der Leeden: Thesis, Eindhoven, 1973.
16. M. S. Brooks and J. K. Kennedy (eds.); *Ultrapurification of Semiconductor Materials*, The MacMillan Co., New York, Ontario, 1962.
17. W. D. Lawson and S. Nielsen: *Preparation of Single Crystals*, Butterworths Scientific Publications, London, 1958.
18. W. Klemm und H. U. von Vogel: *Z. Anorg. allg. Chemie* **219** (1934), 45.
19. P. Fielding, G. Fischer, and E. Mooser: *J. Phys. Chem. Solids* **8** (1959), 434.

20. K. K. Mamedov, I. G. Kerimov, V. N. Konstryukov, and G. D. Guseinov, *Russ. J. Phys. Chem.* **41** (1967), 691.
21. R. M. A. Lieth, C. W. M. van der Heijden, and J. W. M. van Kessel: *J. Crystal Growth* **5** (1969), 251.
22. O. Kubaschewski, E. L. L. Evans, and G. B. Alcock; *Metallurgical Thermochemistry*, 4th edition, Pergamon Press, Oxford, London, New York, 1967.
23. R. M. A. Lieth, H. J. M. Heijligers, and C. W. M. van der Heijden: *J. Electrochem. Soc.* **113** (1966), 798.
24. J. Dieleman and H. R. C. Engelfriet: *J. Less Common Metals* **25** (1971), 231.
25. R. M. A. Lieth, H. J. M. Heijligers, and C. W. M. van der Heijden: *Mat. Sci. Eng.* **2** (1967), 193.
26. G. E. van Egmond and R. M. A. Lieth: *Mat. Res. Bull.* **9** (1974), 763.
27. R. M. A. Lieth: Thesis Eindhoven, 1969.
28. W. Hume-Rothery, J. W. Christian, and W. B. Pearson: *Metallurgical Equilibrium Diagrams*, Chapman and Hall Ltd., Londen, 1953.
29. A. Prince: *Alloy Phase Equilibria*, Elsevier Publ. Co., Amsterdam, London, New York, 1966.
30. P. G. Rustamov, B. K. Babaeva, and N. P. Luzhnaya: *Inorg. Materials* **1** (1965), 775.
31. L. S. Palatnik and E. K. Belova: *Inorg. Materials* **2** (1966), 657.
32. T. N. Guliev and Z. S. Medvedeva: *Inorg. Materials* **1** (1965), 777.
33. M. Hársay: *Mat. Res. Bull.* **3** (1968), 483.
34. R.W. Damon and R. W. Redington: *Phys. Rev.* **96** (1954), 1498.
35. R. H. Williams and A. J. McEvoy: *Phys. Stat. Sol. (a)* **12** (1972), 277.
36. A. J. Niilisk and J. J. Kirs: *Phys. Stat. Sol.* **31** (1969), K91.
37. A. Beck and E. Mooser: *Helv. Phys. Acta* **34** (1961), 370.
38. H. Schäfer: *Chemische Transportreaktionen*, Verlag Chemie, 1962.
39. J. W. E. Jeffes: *J. Cryst. Growth* **3/4** (1968), 13.
40. C. Paorici and G. Zuccalli: *J. Cryst. Growth* **15** (1972), 240.
41. E. Aulich, J. L. Brebner, and E. Mooser: *Phys. Stat. Sol.* **31** (1969), 129.
42. S. G. Patil and R. H. Tredgold: *J. Phys. C., Solid State Physics* **4** (1971), 3199.
43. G. E. van Egmond: unpublished results, 1972.
44. R. Nitsche, H. U. Boelsterli, and M. Lichtensteiger: *J. Phys. Chem. Solids* **21** (1961), 199.
45. R. M. A. Lieth: *Phys. Stat. Sol. (a)* **12** (1972), 399.
46. T. Nishinaga and R. M. A. Lieth: *J. Cryst. Growth* **20** (1973), 109.
47. E. Mooser, M. Schlüter, and R. M. A. Lieth: *J. Cryst. Growth* **16** (1972), 62.
48. J. C. J. M. Terhell and R. M. A. Lieth: *Phys. Stat. Sol. (a)* **5** (1971), 719.
49. S. Nikitine, R. Nitsche, M. Sieskind, and J. Vogt: *J. de Chemie Physique* (1963), 667.
50. E. Mooser and M. Schlüter: *Philos. Mag.* **23** (1971), 811.
51. J. C. J. M. Terhell and R. M. A. Lieth: *J. Cryst. Growth* **16** (1972), 54.
52. J. C. J. M. Terhell and C. Paorici: to be published.
53. J. L. Brebner, G. Fisher, and E. Mooser: *J. Phys. Chem. Solids* **23** (1962), 1417.
54. K. Klosse: Thesis, Utrecht, 1973.
55. E. Mooser: private communications.
56. A. M. Mancini, C. Manfredotti, A. Rizzo, and G. Micocci: *J. Cryst. Growth* **21** (1974), 187.
57. G. A. Akhundov and T. G. Kerimova: *Phys. Stat. Sol.* **16** (1966), K15.
58. R. Fivaz and E. Mooser: *Phys. Rev.* **163** (1967), 743.
59. P. G. Rustamov: *Chem. Abstr.* **63**, 5029[e]; also in *Gallium* **5**, 1966.
60. *Gallium*: Bulletin d'information vol. 10, Alusuisse France S.A., Marseille, 1971, editor Pierre de la Bretèque.
61. P. G. Rustamov and T. A. Dzhalilzade: *Chem. Abstr.* **63**, 15963[e]; *Gallium* **9**, 1970.
62. P. G. Rustamov, V. B. Cherstvova, and M. G. Safarov: *Chem. Abstr.* **64**, 4324[b]; *Gallium* **5**, 1966.
63. V. P. Mushinskii and N. M. Pavlenko: *Kristall und Technik* **4** (1969), K5; *Gallium* **9**, 1970.
64. N. Spandau und F. Klanberg: *Z. anorg. allgem. Chem.* **295** (1958), 300.
65. A. Brukl und G. Ortner: *Naturwiss.* **18** (1930), 393; *Monatshefte* **56** (1930), 358.
66. T. Nishinaga, R. M. A. Lieth, and G. E. van Egmond: *Japan. J. Appl. Phys.* **14** (1975), 1659.
67. J. C. J. M. Terhell, R. M. A. Lieth, and W. C. van der Vleuten: *Mat. Res. Bull.* **10** (1975), 577.
68. A. Kuhn, R. Chevalier, and A. Rimsky: *Acta Cryst.* **B31** (1975), 2841.
69. J. C. J. M. Terhell and W. C. van der Vleuten: *Mat. Res. Bull.* **11** (1975), 101.
70. A. Likforman and M. Guittard: *C.R. Acad. Sci. Paris* **279** (1974), 33.
71. B. Čelustka and S. Popović: *J. Phys. Chem. Sol.* **35** (1974), 287.

72. K. C. Nagpal and S. Z. Ali: *Indian J. Pure Appl. Phys.* **13** (1975), 258.
73. A. Kuhn and A. Chevy: to be published in the Proceedings of the E.C.C.G.-1 Zurich 1976.
74. J. C. J. M. Terhell and G. E. van Egmond: to be published.
75. V. Piacente, G. Bardi, V. Di Paolo and D. Ferro: *J. Chem. Thermodynamics* **8** (1976), 391.

IV–VI COMPOUNDS

P. BUCK

Kristallographisches Institut der Universität Freiburg i. Br., West-Germany

Table of Contents
1. Introduction
2. Crystallographic Data
   2.1. Lead Monoxide PbO
      2.1.1. Polymorphism of Lead Monoxide
      2.1.2. Crystal Structures of Red and Yellow Lead Monoxide
   2.2. Siliconsesquitelluride Si_2Te_3
   2.3. Tin Monoxide SnO
   2.4. Tin Disulfide SnS_2 and Tin Diselenide $SnSe_2$
3. Thermodynamic Data
   3.1. Lead Monoxide
   3.2. Siliconsesquitelluride
   3.3. Tin Monoxide
   3.4. Tin Disulfide and Tin Diselenide
4. Preparation Methods
   4.1. Lead Monoxide
      4.1.1. Preparation of Lead Monoxide by Thermal Decomposition of Lead Compounds
      4.1.2. Preparation of Lead Monoxide by Reaction of Lead Salts with Alkalies
   4.2. Siliconsesquitelluride
   4.3. Tin Monoxide
   4.4. Tin Disulfide and Tin Diselenide
5. Growth of Single Crystals
   5.1. Lead Monoxide
   5.2. Siliconsesquitelluride
   5.3. Tin Monoxide
   5.4. Tin Disulfide and Tin Diselenide
6. Solid Solutions
   6.1. PbO–SnO Solid Solutions
   6.2. SnS_2–$SnSe_2$ Solid Solutions
References

1. Introduction

The group IV elements Si, Ge, Sn and Pb form about thirty binary compounds with the group VI elements O, S, Se and Te. Among these, five compounds are known to have pronounced layer structures: lead monoxide PbO, siliconsesquitelluride Si_2Te_3, tin monoxide SnO, tin disulfide SnS_2 and tin diselenide $SnSe_2$.

Lead monoxide has found interest because of its photoelectric properties [1–6]. Lead monoxide layers are used as photosensitive elements in television pick-up tubes ('plumbicons') [7, 8]. Siliconsesquitelluride Si_2Te_3, the crystal structure of which has not yet been determined in full detail, shows semiconducting and photoconductive properties [9–11]. Tin monoxide is iso-structural with the tetragonal modification of lead monoxide. The conductivity in SnO is used in solid solutions with lead monoxide [12]. Tin disulfide, tin diselenide and their solid solutions also find interest because of their semiconducting and photoconductive properties [13–18].

This article on the IV–VI compounds comprises six sections. The following Section 2 gives the necessary information on the crystallography of the IV–VI compounds with layered structures. This item will be treated in more detail in a following volume of this series. In Section 3 thermodynamic data like heats of formation, standard entropy values, free energies of formation and heat capacities are presented in condensed form. Section 4 describes in detail methods for the preparation of the polycrystalline compounds. In Section 5, the growth of single crystals is treated. Section 6 deals with the formation of solid solutions.

2. Crystallographic Data

2.1. LEAD MONOXIDE PbO

2.1.1. *Polymorphism of Lead Monoxide*

Lead monoxide exists in two polymorphic forms. Red, tetragonal lead monoxide (litharge) is the thermodynamically stable modification at room temperature. At temperatures about 500 °C it transforms into the yellow orthorhombic form (massicot) which is stable up to the melting point at about 885 °C. The reversible transition $PbO_{red} \rightleftharpoons PbO_{yellow}$ has been subject of many investigations. The reported values for the transition temperature vary from 488.5 to 587 °C. Jaeger and Germs [19] report 587 °C, Le Blanc and Eberius [20] 586 °C, Cohen and Addink [21] 488.5 °C, Peterson [22] and Pamfilov and Ponomareva [23] 489 °C. Coussaert et al. [24] state that the transition starts at 489 °C and is not completed before 600 °C. White et al. [25] studying the pressure – temperature curve of the transition found a transition temperature of 550 °C at atmospheric pressure. Rooymans [26] and Rooymans and Langenhoff [27] found a broad transition range from 552° for stoichiometric crystals to 529° for oxygen-rich crystals. They suggest that these variations are due to differences in the Pb/O ratio. Drogomiretskii and Ivancheva [28] report that impurities affect the transition temperature.

The yellow form of lead monoxide is metastable at room temperature. Depending on the method of preparation (see Section 4) pure red or pure yellow PbO can be obtained. Commercial lead monoxide is usually a mixture of both modifications. The transformation of yellow to red lead monoxide is rather slow. Lewis et al. [29] studied the phase transformation induced by strain.

They found that either of the two polymorphs can be partially converted to the other by means of ball milling at temperatures in the range – 196 °C to 200 °C. The high-pressure, high-temperature polymorphism of lead monoxide is considered in more detail in the section on crystal growth.

2.1.2. *Crystal Structures of Red and Yellow Lead Monoxide*

Many X-ray investigations on the two modifications of PbO have been reported

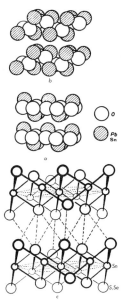

Fig. 1. The crystal structures of tetragonal and orthorhombic lead monoxide, tin monoxide, tin disulfide and tin diselenide. a: PbO$_{red}$, SnO. b: PbO$_{yellow}$. c: SnS$_2$, SnSe$_2$.

[21, 30–40], Table I gives the crystallographic data together with those for the other IV–VI compounds with layered structures.

The red tetragonal form of lead monoxide belongs to the space group $P4/nmm$ with $a=3.796$, $c=5.023$ Å and $Z=2$ [37], the yellow orthorhombic one to the space group $Pbma$ with $a=5.489$, $b=4.755$, $c=5.891$ Å and $Z=4$ [37]. The crystal structure of the former was determined by Dickinson and Friauf [30] and confirmed by Moore and Pauling [34] and Leciejewicz [40] by neutron diffraction.

The structure of the latter was determined by Byström [35] and by Kay [38] and Leciejewicz [39] by neutron diffraction. Both crystal structures are shown schematically in Figure 1, PbO$_{red}$ consists of planes of lead and oxygen atoms parallel to (001). Each oxygen atom is surrounded tetrahedrally by four lead atoms and each lead atom is bonded to four oxygen atoms which form a square to one side of it. The oxygen atoms are sandwiched between two lead layers. Adjacent Pb—O—Pb sandwiches are hold together by van der Waals forces. The structure of PbO$_{yellow}$ also consists of Pb—O—Pb layers stacked in the c direction. However, the oxygen layers between two lead planes are puckered. A detailed study of the bonding in red and yellow lead monoxide has been carried out by Dickens [41].

2.2. SILICONSESQUITELLURIDE Si$_2$Te$_3$

Weiss and Weiss [42] took X-ray powder diagrams of 'silicon telluride', the formula of which they stated to be SiTe$_2$. Later studies of the Si—Te phase diagram [9, 43] however showed that Si$_2$Te$_3$ is the only compound to form in the solid state in this system. Klein Haneveld *et al.* [44] partially determined the crystal structure of

TABLE I
Crystallographic data for PbO$_{red}$, PbO$_{yellow}$, Si$_2$Te$_3$, SnO, SnS$_2$, SnSe$_2$

Compound	PbO$_{red}$	PbO$_{yellow}$	Si$_2$Te$_3$	SnO	SnS$_2$	SnSe$_2$
Space group	$P4/nmm$	$Pbma$	—	$P4/nmm$	$P\bar{3}m1$	$P\bar{3}m1$
a [Å]	3.976 [37]	5.489 [37]	7.429 [44]	3.802 [50]	3.639 [15]	3.811 [13]
b	3.976	4.755	7.429	3.802	3.639	3.811
c	5.023	5.891	13.471	4.836	5.884	6.137
Z	2	4	4	2	1	1
ρ_{exp} [g cm^{-3}]	9.34 (30°C) [21]	9.63 (30°C) [21]	4.50 [9]	6.25 [45]	4.47 [101]	6.01 [101]
ρ_{X-ray}	9.35 (30°C) [21]	9.64 (30°C) [21]	4.53 [44]	6.40 [50]	4.50 [15]	6.01 [101]

siliconsesquitelluride. Their X-ray powder diagrams agreed with those given by [42]. They found a hexagonal unit cell with $a=7.429$ and $c=13.471$ Å, $Z=4$ and $\rho=4.53$ g cm^{-3}. However, no space group is given. The X-ray intensities showed that the Te atoms are hexagonally close-packed. Probably, six of the silicon atoms are in the tetrahedral holes of this packing, the other two are octahedrally surrounded by the Te atoms.

2.3. Tin monoxide SnO

Crystallographic data on various forms of tin monoxide SnO have been reported. The blue-black modification of SnO has a layered structure and is iso-structural with the tetragonal form of lead monoxide [34, 45]. At higher temperatures it becomes unstable. Spandau and Kohlmeyer [46] state that SnO disproportionates at about 400°C according to: $4\,SnO \rightarrow Sn + Sn_3O_4$. Other authors however report the decomposition of SnO into Sn and SnO_2. [45, 47] give 400°C and Platteeuw and Meyer [48] showed that SnO is already unstable at 300°C. [46] and [48] even suppose that SnO is at room temperature in a metastable state. [49] reports that the disproportionation starts at 175°C and is intensive at 460° to 500°C. According to Swanson et al. [50] the lattice constants for SnO are $a=3.802$ and $c=4.836$ Å.

Serebryanaya et al. [51] showed that the tetragonal form of SnO undergoes a reversible phase transition at high pressures (15 kbars) and room temperature. The resulting modification has no longer a layered structure but can be described as an orthorhombic deformed sphalerite structure. Another red, orthorhombic modification of SnO, the crystal structure of which is not yet known has been reported by Partington and Moser [52] and Donaldson et al. [53]. Kwestroo and Vromans [54] recently described the preparation of another, yet unknown red modification of SnO.

2.4. Tin disulfide SnS_2 and tin diselenide $SnSe_2$

Tin disulfide and tin diselenide have the CdI_2 structure (Figure 1). The crystal structure of SnS_2 was determined by Oftedal [55], that of $SnSe_2$ by Busch et al. [13]. The lattice constants are $a=3.639$, $c=5.884$ Å [15] and $a=3.811$, $c=6.137$ Å [13] respectively, the space group is $P\bar{3}m1$. [56–58] report the occurrence of SnS_2 polytypes. Moh [59] studied the tin-sulfur system and found that SnS_2 transforms at 692°C into a high-temperature form which is stable in the presence of excess sulfur and melts congruently at 865°C.

3. Thermodynamic Data

In this section the following thermodynamic data are listed: standard heat of formation of the compound from the elements $\Delta H°_{298}$, standard free energy of formation $\Delta G°_{298}$, the temperature dependence of the free energy of formation $\Delta G°_T$, standard entropy values $S°_{298}$ and heat capacity data C_p. $\Delta H°_{298}$, $\Delta G°_{298}$ and $S°_{298}$ values for the IV–VI compounds with layered structures are presented in Table II.

3.1. Lead monoxide

Table II gives the $\Delta H°_{298}$, $\Delta G°_{298}$ and $S°_{298}$ values of different authors for the two forms of PbO.

Alcock and Belford [67] calculated the standard molar free energy of formation of PbO at elevated temperatures from heat capacity, entropy and enthalpy data. According to these authors the standard free energy change for the reaction $Pb(s, l) + \frac{1}{2}O_2(g) = PbO$(yellow, s, l) may be represented by the equations:

$$\Delta G°(298\ 600\ K) = (-53010 + 39.02\ T + 0.2935 \times 10^{-3}\ T^2 -$$
$$- 5.090\ T \log T + 44145\ T^{-1}) \cdot 10^{-3}\ [\text{kcal mole}^{-1}]$$
$$\Delta G°(600-1170\ K) = (-53260 + 24.64\ T - 1.4665 \times 10^{-3}\ T^2 +$$
$$+ 0.622\ T \log T + 32645\ T^{-1}) \cdot 10^{-3}$$
$$\Delta G°(1170-1200\ K) = (-50420 + 52.09\ T - 0.110 \times 10^{-3}\ T^2 -$$
$$- 9.63\ T \log T - 10000\ T^{-1}) \cdot 10^{-3}.$$

Charette and Flengas [68] more recently directly determined the free energy of formation at elevated temperatures by electrochemical measurements. They obtained for the reaction $Pb(l) + \frac{1}{2}O_2(g) = PbO(s)$ over the temperature range 772 to 1160 K:

$$\Delta G° = (-51400 + 23.037\ T) \cdot 10^{-3}\ [\text{kcal mole}^{-1}].$$

Detailed ΔH and ΔG values are given by [69] and [70].

King [66] measured the heat capacities of the red and yellow form of lead monoxide over the temperature range 51–298 K. The C_p values rise from 4.095 cal deg^{-1} mole^{-1} (PbO$_{red}$) at 53.4 K to 10.95 at 298 K. The values for the yellow modification do not differ significantly from those for PbO$_{red}$. The values of King agree fairly well with those given by [65]. Knacke and Prescher [71] give the following expression for the heat capacity over the temperature range 298 to 1158 K:

$$C_p = 10.712 + 0.003946\ T\ [\text{cal deg}^{-1}\ \text{mole}^{-1}].$$

According to Spencer and Spicer [72] the heat capacity of PbO$_{red}$ can be described by

$$C_p = 17.331 - 0.02098\ T + 2.175 \times 10^{-5}\ T^2\ (298-823\ K),$$

and that of PbO$_{yellow}$ by

$$C_p = 2.981 + 0.02760\ T - 1.708 \times 10^{-5}\ T^2\ (298-923\ K).$$

3.2. SILICONSESQUITELLURIDE

According to Exsteen et al. [73] the standard heat of formation $\Delta H°_{298}(Si_2Te_3) = -18.6 \pm 4$ kcal mole^{-1}. The standard entropy was estimated to be $S°_{298} = 40 \pm 4$ cal deg^{-1} mole^{-1}. Estimated heat capacity values at 298 and 1165 K (melting point) are 6.35 and 7.25 cal deg^{-1} mole^{-1}, respectively.

Brebrick [43] determined the free energy change for the reaction $\frac{2}{5}Si(s) + \frac{3}{5}Te(l) = Si_{2/5}Te_{3/5}(s)$. He gives the following expression: $\Delta G°[Si_{2/5}Te_{3/5}(s)] = -5.754 + 3.904 \times 10^{-3}\ T\ [\text{kcal g-atom}^{-1}]\ (604-892°C)$.

3.3. TIN MONOXIDE

Data for the standard heat of formation, standard free energy of formation and standard entropy from several authors are listed in Table II.

TABLE II

$\Delta H°_{298}$, $\Delta G°_{298}$ and $S°_{298}$ data for PbO$_{red}$, PbO$_{yellow}$, Si$_2$Te$_3$, SnO, SnS$_2$ and SnSe$_2$

Compound	PbO$_{red}$	PbO$_{yellow}$	Si$_2$Te$_3$	SnO	SnS$_2$	SnSe$_2$
$-\Delta H°_{298}$	52.360 [60]	52.07 [61]	18.6 [73]	68.4 [61]	40.0 [80]	21.4 [83]
(kcal mole^{-1})	52.40 [61]	50.860 [64]		69.1 [74]		
	51.720 [64]			68.35 [75]		
				68.3 [76]		
$-\Delta G°_{298}$	45.050 [60]	45.05 [61]		61.5 [61]	46.231 [80]	
(kcal mole^{-1})	45.25 [61]	44.956 [62]		62.1 [74]		
	45.101 [62]	45.360 [64]		61.4 [76]		
	45.100 [63]			60.95 [77]		
	45.530 [64]					
$S°_{298}$	15.62 [65]	16.40 [65]	40 [73]	23.3 [74]	20.9 [80]	
(cal deg^{-1} mole^{-1})	15.6 [66]	16.1 [66]		13.56 [65]		

Detailed data for the temperature dependence of ΔH and ΔG are given by [69] and [70]. Belford and Alcock [78] calculated the standard free energy of formation from heat capacity (estimated), entropy and enthalpy data. $\Delta G°$ may be represented by the equation:

$$\Delta G°(\text{SnO}) = (-69670 + 18.37\,T - 1.50 \times 10^{-3}\,T^2 - 10000\,T^{-1} + 3.06\,T \log T) \cdot 10^{-3} \,[\text{kcal mole}^{-1}] \,(505-1273\,\text{K}).$$

Heat capacities of tin monoxide from 70 K to room temperature have been measured by Millar [65]. He reports C_p values from 4.067 cal deg^{-1} mole^{-1} (69.6 K) to 10.52 (292.5 K). Over the temperature range 298 to 1273 K Kelley [79] gives the equation $C_p = 9.55 + 3.5 \times 10^{-3}\,T$.

3.4. TIN DISULFIDE AND TIN DISELENIDE

Only few thermodynamic data of SnS$_2$ and SnSe$_2$ are available. For SnS$_2$ the standard heat of formation is reported to be $\Delta H°_{298} = 40.0$ kcal mole^{-1}, $\Delta G°_{298} = -46.231$ kcal mole^{-1} and the standard entropy $S°_{298} = 20.9$ cal deg^{-1} mole^{-1} [80]. King and Todd [81] measured heat capacities from 53 K (4.139 cal deg^{-1} mole^{-1}) to 296 K (16.76 cal deg^{-1} mole^{-1}). Orr and Christensen [82] studied the high temperature heat content of SnS$_2$. According to them the heat capacity over the temperature range 298–1000 K is represented by the equation: $C_p = 15.51 + 4.20 \times 10^{-3} T$. The only value for SnSe$_2$ is $\Delta H°_{298} = 21.4$ kcal mole^{-1} [83].

4. Preparation Methods

4.1. Lead monoxide

A review of the preparation methods for the two modifications of lead monoxide shows that there are two fundamental methods. Firstly, the thermal decomposition of lead compounds and secondly the reaction of lead salt solutions with alkali hydroxides. Whether red or yellow lead monoxide is formed depends on the experimental conditions.

4.1.1. *Preparation of Lead Monoxide by Thermal Decomposition of Lead Compounds*

Kohlschütter and Roesti [84] decomposed lead carbonate at 400 °C and obtained red lead monoxide. According to [20], the decomposition of PbCO$_3$ into yellow lead monoxide is possible at 250 °C in vacuum. [22] found under the same experimental conditions a mixture of red and yellow PbO. He states a dependence of the yield of yellow oxide on the particle size of the lead carbonate. Small particles give predominantly yellow PbO, the coarsest ones the red modification. Clark and Tyler [85] obtained red PbO by heating neutral lead carbonate in vacuum at temperatures between 350 and 500 °C, but yellow PbO by heating basic lead carbonate to about 700 °C. At temperatures below 500 °C basic lead carbonate yields a mixture of both forms. King [66] reports that 80 h heating of lead carbonate at 560–580 °C in vacuum was insufficient to decompose the carbonate completely or to prevent the formation of some red oxide. After grinding the product, heating it for 10 h at 725 °C and subsequently quenching to room temperature, yellow PbO was formed.

[20] prepared red lead monoxide by thermal decomposition of hydrated lead oxide PbOxH$_2$O which was obtained by reacting a lead acetate solution with sodium-hydroxide solution. When hydrated PbO is dehydrated in vacuum at about 120°, varied fractions of the resulting oxide are obtained in the yellow form, depending on the original state of the hydrate [22]. The most finely precipitated hydrates yield almost pure yellow, the coarsest nearly red oxide. [85] observed that the amount of yellow PbO increases with the decomposition temperature.

Jones and Rotschild [86] obtained red PbO by heating lead nitrate in air at about 500 °C for 30 min, Katz [87] reports the formation of yellow PbO platelets when heating Pb(NO$_3$)$_2$ in air or in vacuum up to 600 °C. Heating of the higher lead oxides PbO$_2$ and Pb$_3$O$_4$ in vacuum to 560 °C results also in the formation of red PbO [87].

Thermal decomposition of lead oxalate in vacuum at temperatures of about 300 °C yields red lead monoxide [88].

4.1.2. Preparation of Lead Monoxide by Reaction of Lead Salts with Alkalies

Applebey and Reid [89] prepared lead monoxide by the action of strong potassium hydroxide solution on lead hydroxide near the boiling point. On cooling, the oxide is deposited in crystallized form. The modification which forms depends on the KOH concentration: 15 N – KOH gives the red form and 10 N – KOH the yellow one. The rate of cooling also effects the product. Rapid cooling always results in yellow PbO.

Garrett et al. [90] obtained yellow PbO by first precipitating hydrated lead oxide from lead nitrate with carbonate-free alkali hydroxide. Dissolving this material in hot 5 M – NaOH and subsequent cooling yielded large flakes of yellow PbO. In accordance with the observations of [89] higher concentrations of NaOH (10 M) resulted in the formation of red PbO. Cohen and Addink [21] give a detailed description for the preparation of red lead monoxide from lead nitrate and potassium hydroxide. Pure KOH is dissolved in H_2O (1:1) and heated to 100°C. Under nitrogen atmosphere (in order to prevent the formation of PbO_2) a lead nitrate solution, saturated at 100°C, is added. The mixture is continuously stirred. A yellow-greenish precipitate is formed which disappears soon. After some time a precipitate of yellow PbO forms which changes to red PbO after some minutes. The red powder dissolves completely in HNO_3. This means that no traces of PbO_2 are present in the reaction product. The yellow form of PbO was prepared by the same authors by heating red PbO under N_2 – atmosphere up to 650–700 °C for about 30 min and quenching the product to liquid nitrogen temperature. Katz [87] obtained red lead monoxide by the action of a boiling sodium hydroxide solution on lead acetate.

More recently, Kwestroo and Huizing [91] described the preparation of ultra-pure lead monoxide from lead acetate with ammonia at room temperature. In the following their method will be treated in more detail.

The first step of the preparation is the purification of lead acetate. Commercially available 'analytical reagent' quality is very pure. The amounts of impurities are close to the limit of detection of spectrochemical analysis. Foreign ions more noble than lead, as for example Cu, Ag and Bi, are removed from the lead acetate solution by treatment with metallic Pb. After Kwestroo and Huizing, 165 g of lead acetate $Pb(CH_3COO)_2 \cdot 3H_2O$ are dissolved in 350 ml of water in a polyethylene bottle. The solution passes through a Pb-column at a rate of 75 ml min^{-1}. A silica tube (10 × ⌀2.5 cm) is filled with 325 g of pure lead in the form of small balls (diameter 10–100 μ). The solution is collected in another polyethylene bottle. A pure ammonia solution is prepared from NH_3 gas and deionized water. 1650 ml of 10 N ammonia solution are added to the purified lead acetate solution while vigorously stirring. A white gelatinous precipitate of basic lead acetate, $2Pb(OH)_2 \cdot Pb(CH_3COO)_2$ is formed. After some minutes, it transforms into a greenish crystalline precipitate which is found to be orthorhombic PbO. If the transformation does not occur, the precipitate has to be washed several times by decanting with concentrated ammonia. The greenish precipitate is washed five times with 2 N ammonia by decantation. Then 200 ml water are added and the mixture is kept at 80°C for 10 h. The resulting red crystalline mass is washed five times with 1 N ammonium hydroxide and dried at 95 °C in a platinum dish. Finally the dry red PbO is heated to 150 °C. The yield is about 80–85 g ultra-pure PbO which contains 1 ppm silica and about 10 ppm carbon.

During their investigations Kwestroo and Huizing made an interesting observation which they used for the preparation of the yellow form of lead monoxide. If more than 20 ppm Si are present in the white precipitate of basic lead acetate, the spontaneous recrystallization in water from orthorhombic to tetragonal PbO is prevented. This is the case if the preparation is carried out not in polyethylene vessels but in silica vessels. The procedure is then as follows:

A purified solution of 120 g of lead acetate in 250 g of water is introduced into a silica vessel. 1500 ml ammonium hydroxide (15 N) are added while stirring. The white precipitate of basic lead acetate forms and transforms after stirring for one or two hours into greenish looking orthorhombic (yellow) PbO. After drying at 70–95 °C the precipitate is heated to 150 °C and about 60 g of yellow PbO are obtained. Spectrochemical analysis showed silicon (100 ppm) and carbon (0.1%), from acetate residues, to be the main impurities.

In another paper Kwestroo et al. [92] studied the influence of impurities on the course of reaction between lead acetate and ammonium hydroxide. They found that the elements Si, Ge, P, As, Sb, Se, Te, Mo and W when present in the precipitation solution as anions at concentrations as low as 10–100 ppm prevent the transformation of the orthorhombic into the tetragonal modification.

4.2. SILICONSESQUITELLURIDE

Weiss and Weiss [42] heated a stoichiometric mixture of thoroughly purified, finely powdered Si and Te (1:2) for 24 h in an evacuated quartz ampoule at 1050–1070 °C. They thought to have synthesized the compound $SiTe_2$. However, as a comparison of their X-ray powder diagrams with the more recent results of other authors shows, they obtained Si_2Te_3. Indeed, they report that small amounts of Te had deposited in the tip of the quartz ampoule. The small red platelets decompose in air into H_2Te and SiO_2. Due to the fast decomposition of H_2Te, a metallic grey layer of Te forms on the surface, retarding the further decomposition of Si_2Te_3. [44] prepared Si_2Te_3 by heating the mixed elements in an evacuated quartz tube at 900 °C. Vennik and Callearts [93] describe the formation of hexagonal lamellae (about 5×5 mm^2, 2–50 μ thick) of siliconsesquitelluride by the action of Te vapour on a silicon single crystal. At 77 K the platelets are transparent, at room temperature of red colour and at 370 K opaque. Bailey [9] prepared polycrystalline Si_2Te_3 by high-temperature reaction of mixtures of 40 at.% Si and 60 at.% Te. The starting materials (semiconductor grade silicon of $10^2 \Omega$ cm electrical resistivity and tellurium of 99.999% purity) were sealed under vacuum in quartz capsules and prereacted at 1100 °C for about 20 h in a rocking furnace and then held at 1250 °C for at least 1 h. Subsequently the capsule was quenched in water and annealed for 90 h at 790 °C. The result was polycrystalline Si_2Te_3. Brebrick [43] reports a similar way of preparation. Exsteen et al. [73] placed very pure Si and excess Te at the ends of an evacuated quartz ampoule (25 cm long) which was inserted into a horizontal two zone furnace with a temperature gradient from 900 (Si) to 700 °C (Te). Red platelets of Si_2Te_3 formed in the middle of the ampoule.

4.3. Tin monoxide

The tetragonal modification of tin monoxide may be prepared by a method which was first described by Ditte [94] and successively used by other authors [95, 96, 48]. A sodium carbonate solution is added to a solution of tin chloride $SnCl_2$. If the mixture is alkaline (pH 5–6.5) a white precipitate of tin hydroxide is formed. After some hours of heating at about 110 °C the precipitate turns into blue-black SnO.

This method may be varied by the use of ammonium hydroxide instead of a sodium carbonate solution [54, 65, 75]. In the following the method of Kwestroo and Vromans [54] for the preparation of spectrochemical pure tin monoxide by the action of ammonium hydroxide on $SnCl_2$ will be described in detail. Pure ammonium hydroxide is prepared by saturation of deionized water with NH_3-gas in a polyethylene bottle. The $SnCl_2$ solution is obtained by dissolving high purity tin (99.9999%) in pure hydrochloric acid (made by dissolving HCl-gas in deionized water in a polyethylene bottle). The tin (II) chloride solutions have to be freshly prepared in order to avoid tin (IV) formation. 30 g of tin are dissolved in 120 ml HCl (6 N) and 200 ml concentrated ammonium hydroxide are added while stirring. A white precipitate forms at pH 9. The precipitate is washed three times with 2 N ammonium hydroxide and heated overnight with 50 ml 2 N ammonium hydroxide at about 60–70 °C. The precipitate transforms into black crystalline SnO, which is washed ten times with deionized water and dried at 90 °C. The yield is about 30 g very pure, black SnO in which about 50–100 ppm chloride are still present. Spectrochemical analysis shows that Si, Mg, Fe, Al, Cu and Pb impurities are ≤ 2 ppm.

SnO may also be prepared by thermal decomposition of tin oxalate. SnC_2O_4 decomposes at temperatures around 320 °C into $SnO + CO + CO_2$ [45].

4.4. Tin disulfide and tin diselenide

King and Todd [81] prepared tin disulfide from reagent grade $SnCl_2 \cdot 2H_2O$, S and NH_4Cl. These ingredients were mixed in the molar proportion of 1:3:2 and heated in an open-end tube at 150 °C until the water had volatilized and then for three hours at 300–350 °C. Then the tube was closed and cooled. The reaction product was shaken and centrifuged with water, then with alcohol and dried at 60 °C. Finally the product was mixed with 10% by weight of S and heated in vacuum for five hours at 300 °C where the excess S distilled off.

Busch et al. [13] describe the preparation of $SnSe_2$. The spectroscopically pure elements were melted together in sealed and evacuated quartz tubes. The same procedure is given by Asanabe [14]. Greenaway and Nitsche [15], in the course of growing single crystals of SnS_2, synthesized the compound by reacting the high-purity elements in sealed quartz ampoules in the presence of 5 mg iodine cm^{-3}. Domingo et al. [97] also prepared SnS_2 and $SnSe_2$ by melting stoichiometric amounts of the elements (S, Se:5 – nine, Sn:6 – nine purity) in evacuated quartz tubes. For SnS_2 a 3 mm wall tubing was used because of the high vapour pressure of S (about 40 at.) at the melting point of SnS_2 (869 °C). Temperature was increased slowly over 24 h to 980 °C, held for 72 h and slowly cooled down. Traces of unreacted S and SnS where still present in the reaction product. $SnSe_2$ has a lower melting point (657 °C) so that the Se vapour pressure is not dangerous. In the case of $SnSe_2$ the temperature was raised to 730 °C, held constant for 20 h and cooled down. Nakata et al. [17] describe a similar method for the preparation of SnS_2, SnSSe and $SnSe_2$.

5. Growth of Single Crystals

5.1. LEAD MONOXIDE

For a better understanding of the methods for growing single crystals of red and yellow lead monoxide the $p-T$ diagram of PbO has to be considered. The high-pressure, high temperature polymorphism of PbO was studied by White et al. [25] and Rooymans [26]. The former authors report that no unequivocal data for the $p-T$ dependence of the transition $Pb_{red} \rightleftharpoons PbO_{yellow}$ were obtained. The results depended on starting materials and experimental conditions. The equilibrium pressure-temperature relation shown in Figure 2 was deduced from many different experiments. The latter author also emphasizes that the broad transition region may be due to deviations from stoichiometry in the starting material. Figure 2 shows the negative slope of the $p-T$ diagram of PbO. The high-temperature, high-pressure phase is the orthorhombic modification which has a higher density than the tetragonal form.

The polymorphism of lead monoxide prevents crystal growth from the melt or by

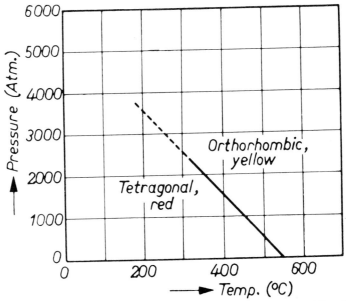

Fig. 2. The $p-T$ diagram of lead monoxide. After White, Dachille and Roy [25].

flux methods. However, hydrothermal techniques can be successfully used. Rooymans and Langenhoff [27] report on the hydrothermal growth of single crystals of tetragonal lead monoxide in diluted lithium hydroxide solutions in autoclaves. The PbO starting material, prepared after the method given in [91] was very pure, LiOH of Merck p.a. grade was used. In order to prevent contamination with Si, the lithium hydroxide solution was prepared in PVC bottles. The $p-T$ conditions for the growth of tetragonal lead monoxide are limited as Figure 2 shows. In most experiments the autoclaves were filled only to about 60%. At 450 °C this corresponds to a pressure of about 1000 atmospheres. The best results were obtained under the following conditions: the autoclave is filled up to 60% with 1 N-LiOH, heated up to a bottom temperature of 450 °C (top temperature about 430 °C) and subsequently cooled down to 250 °C with a cooling rate of 1–2 °C h^{-1}. The maximum size of the crystals was $8 \times 8 \times 0.5$ mm^3. Impurities of 0.08 wt.% of Li were spectrochemically detected.

Keezer et al. [3, 6] grew single crystals of both modifications of lead monoxide also by hydrothermal techniques. Lead monoxide powder (99.999%) and analytical grade 1 N-NaOH were sealed in thin-walled silver, gold or platinum tubes which were put into the pressure vessel. The temperature maintained during growth was 400 °C. Orange-red tetragonal crystals resulted at pressures below 1500 atm, yellow-green orthorhombic crystals above 2040 atm. The maximum size of the platelets was $3 \times 3 \times 0.1$ mm^3. The highest growth rates were about 1 mm day^{-1} for red and 3 mm day^{-1} for yellow PbO. After several days of growth the entire pressure vessel was quenched to 0 °C before lowering the pressure. This was found necessary to prevent transformation during cooling. Elongated spots on X-ray Laue patterns indicated that deformation of the crystals had occurred during growth or upon subsequent quenching.

Bordovskii and Izvozchikov [5] used a 16 N solution of sodium hydroxide for the hydrothermal growth of single crystal platelets ($6 \times 1 \times 0.1$ mm^3) of orthorhombic lead monoxide.

5.2. SILICONSESQUITELLURIDE

Single crystals of Si_2Te_3 were grown from the vapour phase. Bailey [9] placed a mixture of Si and Te (58 at.%) at the end of an evacuated (10^{-3} torr) quartz ampoule which was brought into a furnace with a temperature gradient from 830 to 700 °C. Crystals (maximum size $20 \times 20 \times 0.2$ mm^3) formed at the walls of the quartz tube at temperatures between 800 and 750 °C. After 70–80 h about 30% of the charge had converted into single-crystalline platelets of Si_2Te_3.

Rau and Kannewurf [10] obtained single-crystalline platelets ($10 \times 10 \times 0.5$ mm^3) of Si_2Te_3 (they thought them to be $SiTe_2$) by vapour transport both in vacuum as well as in the presence of iodine. The experiments were performed in a two-zone furnace. The presynthesized compound was placed in a quartz tube (15 cm length, 15 mm diam). Optimum temperatures for vacuum transport were 850–750 °C. If iodine was used as transport agent, a temperature gradient of 650 to 590 °C yielded the best crystals. Within 5 days about 5 g of polycrystalline Si_2Te_3 were transported into the cooler region of the ampoule.

5.3. TIN MONOXIDE

Until now, no attempt to grow single crystals of the tetragonal modification of tin monoxide has been reported. This may be due to the thermal instability of this compound (see Section 2.3.).

5.4. TIN DISULFIDE AND TIN DISELENIDE

Busch et al. [13] succeeded in growing single crystals of $SnSe_2$ from the melt using the Bridgman method. They obtained specimens from 20–30 mm length and 10 mm in diameter. However, more recently crystals of SnS_2 and $SnSe_2$ have been grown from the vapour phase by chemical transport reactions using iodine as transport agent. The principle of the vapour transport technique consists in the volatilization of the polycrystalline compound forming gaseous iodides and sulfur (or selenium) vapour at a temperature T_1 and the back-reaction of the gas mixture at a temperature T_2. In the case of tin disulfide and tin diselenide T_1 is higher than T_2. If T_1 and T_2 are properly chosen, single crystals may be formed at T_2. Greenaway and Nitsche [15] were the first to apply this method to the growth of SnS_2 single crystals. They synthesized SnS_2 by reacting the high-purity elements in sealed quartz ampoules (length 100 mm, diameter 16 mm) in the presence of 5 mg iodine cm^{-3}. The ampoules were placed in a two-zone furnace with $T_1 = 800\,°C$ and $T_2 = 700\,°C$. After about 60 h transparent yellow hexagonal plates with a size of about 4×4 mm^2 had formed. Domingo et al. [97] report on the vapour growth of SnS_2 ($T_1 = 700$, $T_2 = 600\,°C$) and $SnSe_2$ single crystals under similar conditions. Crystals of $SnSe_2$ were obtained in a temperature gradient 500 to 400 °C. The same temperatures were chosen by [17]. Said and Lee [18] describe the vapour transport growth of SnS_2 crystals with an area up to 5 cm^2 and 0.1 mm thickness in a temperature gradient 690° to 650 °C. Rimmington et al. [98] grew single crystals of SnS_2 ($T_1 = 680$, $T_2 = 620\,°C$) and $SnSe_2$ ($T_1 = 650$, $T_2 = 610\,°C$) also by the iodine vapour transport technique. It should be mentioned that all temperatures given above refer to the temperatures present in the furnace and not in the ampoule.

6. Solid Solutions

6.1. PbO–SnO SOLID SOLUTIONS

The tetragonal modifications of lead and tin monoxide are iso-structural. As their lattice constants do not differ significantly, solid solutions between them may be expected. Kwestroo et al. [99] studied the possibility of preparing solid solutions between PbO and SnO. The decomposition of tin monoxide at temperatures about 300 °C or lower (see Section 2.3.) prevents the preparation of solid solutions from molten mixtures. Even if SnO would not decompose, the PbO transformation at about 490 °C would make this way of preparation impossible. The authors of [99] therefore used a low temperature method. As mentioned in Sections 4.1.2 and 4.3 pure lead monoxide is prepared by the action of ammonium hydroxide on lead acetate and pure tin monoxide by the action of ammonium hydroxide on tin (II) chloride. A reasonable method seemed to be to prepare the solid solutions from mixtures of lead acetate and tin (II) chloride. However, this is impossible because of the formation of insoluble $PbCl_2$. Kwestroo et al. consequently used mixtures of $Sn(ClO_4)_2$ and lead acetate solutions in varying ratios and treated them with concentrated ammonia. White precipitates were formed immediately. They consisted of solid solutions of Pb^{2+} and Sn^{2+} hydroxides ($Pb(OH)_2$ and $Sn(OH_2)$ are also iso-structural) with the composition $3Pb_{1-x}3Sn_xO \cdot H_2O$ ($0 < x < 1$). After heating at 80 °C in water black precipitates formed. These precipitates were dried in air or in an inert atmosphere. X-ray powder diagrams gave the following results. If the ratio Sn^{2+} to Pb^{2+} in the starting solution is 1:4 or less, stable solid solutions $Pb_{1-x}Sn_xO$ ($0 < x < 0.25$) are formed. With increasing Sn^{2+} content the lattice constants decrease and the colour changes from light-brown to black-brown. From mixtures of solutions with 25–60

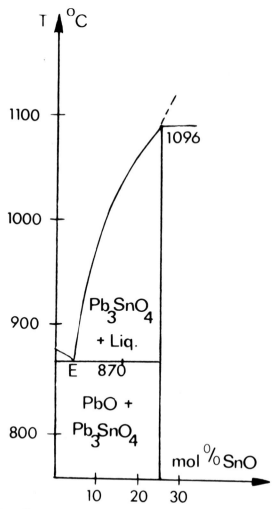

Fig. 3. Partial $T-x$ diagram of the system PbO—SnO. After Fournier and Kohlmuller [100].

mol % SnO, yellow products resulted when the black precipitates are dried in air. No PbO—SnO solid solutions are detected but $PbSnO_3$ and PbO are formed. Furthermore, if more than 40 mol % SnO are present in the starting mixture an unidentified compound forms. However, if the precipitates are dried in a nitrogen atmosphere, the $Pb_{1-x}Sn_xO$ solid solutions are stable up to $x=0.4$. Starting solutions with more than 60 mol % SnO yield grey-green to black coloured products, which are mixtures of $PbSnO_3$, SnO and the unknown compound. No solid solutions form.

The authors of [99] give values of a- and c-axis of the solid solutions with $Pb_{1-x}Sn_xO$ composition ($0 \leq x < 0.34$). Despite of the low accuracy of the X-ray measurements due to line broadening, the decrease of the lattice constants with increasing x was obvious.

Fournier and Kohlmuller [100] studied the system PbO—SnO by differential thermal analysis. They give a $T-x$ diagram up to $x=0.25$ which is shown in Figure 3. Mixtures with less than 25 mol % SnO were prepared from the oxides. Samples with more than 25 mol % SnO were prepared by mixing a solution of SnO in 6 N-HCl with a solution of PbO in 8 N-NaOH. At pH$=7.5$ a white precipitate consisting of the desired mixture of the hydroxides resulted. After heating 5 to 10 min to 1050 °C the reaction between PbO and SnO took place. These authors state that the following compounds exist in the system PbO—SnO: $3PbO \cdot SnO$, $2PbO \cdot SnO$, $PbO \cdot SnO$ and $PbO \cdot 6SnO$. X-ray powder data for these compounds are given.

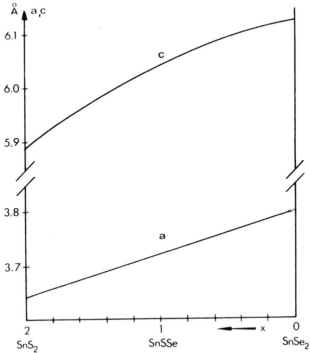

Fig. 4. Variation of lattice constants a and c with composition in SnS_xSe_{2-x} solid solutions. After Rimmington and Balchin [101].

6.2. SnS_2–$SnSe_2$ SOLID SOLUTIONS

Solid solutions SnS_xSe_{2-x} are possible in the whole range from $0 < x < 2$ [101, 102]. Rimmington and Balchin [101] grew mixed crystals by the above (Section 5.4) described iodine vapour transport technique. Pure SnS_2 crystals were obtained in a temperature gradient from 690 to 650 °C, for $SnSe_2$ the corresponding temperatures are 550 and 500 °C. Intermediate growth conditions were chosen for the mixed compositions. The colour changes from yellow, transparent (SnS_2) through red, transparent to metallic opaque ($SnSe_2$). The length of the a-axis decreases linearly with increasing x. The variation of the c-axis, however, is non-linear. This is shown in Figure 4.

References

1. L. Heijne: *Philips Res. Rept. Suppl.* (1961), No. 4.
2. V. A. Izvozchikov and G. A. Bordovskii: *Soviet Phys. Doklady* **7** (1962), 740.
3. R. C. Keezer, D. L. Bowman, and J. H. Becker: *Bull. Am. Phys. Soc.* **10** (1965), 534.
4. B. J. Mulder and J. de Jonge: *Solid State Comm.* **4** (1966), 293.
5. G. A. Bordovskii and V. A. Izvozchikov: *Krist. Tech.* **3** (1968), 219.
6. R. C. Keezer, D. L. Bowman, and J. H. Becker: *J. Appl. Phys.* **39** (1968), 2062.
7. L. Heijne, P. Schagen, and H. Bruining: *Nature* **173** (1954), 220.
8. E. E. de Haan, A. van der Drift, and P. P. M. Schampers: *Philips Tech. Rev.* **25** (1963/64), 133.
9. L. G. Bailey: *J. Phys. Chem. Solids* **27** (1966), 1593.
10. J. W. Rau and C. R. Kannewurf: *J. Phys. Chem. Solids* **27** (1966), 1097.
11. G. G. Roberts and E. L. Lind: *Phys. Letters* **A33** (1970), 365.
12. J. van den Broek and A. Netten: *Philips Res. Rept.* **25** (1970), 145.
13. G. Busch, C. Fröhlich, F. Hulliger, and E. Steigmeier: *Helv. Phys. Acta* **34** (1961), 359.
14. S. Asanabe: *J. Phys. Soc. Japan* **16** (1961), 1789.
15. D. L. Greenaway and R. Nitsche: *J. Phys. Chem. Solids* **26** (1965), 1445.
16. B. L. Evans and R. A. Hazelwood: *Brit. J. Appl. Phys.* **2** (1969), 1507.
17. R. Nakata, M. Yamaguchi, S. Zembutsu, and M. Sumita: *J. Phys. Soc. Japan* **32** (1972), 1153.
18. G. Said and P. A. Lee: *Phys. Stat. Sol. (a)* **15** (1973), 99.
19. F. M. Jaeger and H. C. Germs: *Z. Anorg. allgem. Chem.* **119** (1921), 145.
20. M. Le Blanc and E. Eberius: *Z. Physik. Chem.* **160** (1932), 69.
21. E. Cohen and N. W. H. Addink: *Z. Physik. Chem.* **168** (1934), 188.
22. M. Petersen: *J. Am. Chem. Soc.* **63** (1941), 2617.
23. A. V. Pamfilov and M. N. Ponomareva: *Zh. Obshch. Khim.* **16** (1946), 963.
24. R. Coussaert, J. M. Leroy, D. Thomas, and G. Tridot: *Compt. Rend.* **266** (1968), 1496.
25. W. B. White, F. Dachille, and R. Roy: *J. Am. Cer. Soc.* **44** (1961), 170.
26. C. J. M. Rooymans: *Philips Res. Rept. Suppl.* (1968), No. 5.
27. C. J. M. Rooymans and W. F. Th. Langenhoff: *J. Cryst. Growth* **3, 4** (1968), 411.
28. P. V. Drogomiretskii and E. G. Ivancheva: *Ukr. Khim. Zh.* **36** (1970), 1001; *Chem. Abstr.* **74** (1966), 80617x.
29. D. Lewis, D. O. Northwood, and R. C. Reeve: *J. Appl. Cryst.* **2** (1969), 156.
30. R. G. Dickinson and J. B. Friauf: *J. Am. Chem. Soc.* **46** (1924), 2457.
31. G. R. Levi and G. Natta: *Nuovo Cimento* **3** (1926), 114.
32. F. Halla and F. Pawlek: *Z. Physik. Chem.* **128** (1927), 49.
33. J. A. Darbyshire: *J. Chem. Soc.* (1932), 211.
34. J. W. Moore and L. Pauling: *J. Am. Chem. Soc.* **63** (1941), 1392.
35. A. Byström: *Arkiv Kemi, Mineral. Geol.* **B17**, No. 8 (1943), 1.
36. A. Byström: *Arkiv Kemi, Mineral. Geol.* **A20**, No. 11 (1945), 1.
37. H. E. Swanson and R. K. Fuyat: *Nat. Bur. Std. (U.S.)*, Circ. 539 (1953), Vol. 2, 1.
38. M. I. Kay: *Acta Cryst.* **14** (1961), 80.

39. J. Leciejewicz: *Acta Cryst.* **14** (1961), 66.
40. J. Leciejewicz: *Acta Cryst.* **14** (1961), 1304.
41. B. Dickens: *J. Inorg. Nucl. Chem.* **27** (1965), 1495, 1503.
42. A. Weiss and A. Weiss: *Z. Anorg. allgem. Chem.* **273** (1953), 124.
43. R. F. Brebrick: *J. Chem. Phys.* **49** (1968), 2584.
44. A. J. Klein Haneveld, W. van der Veer, and F. Jellinek: *Rec. Trav. Chim.* **87** (1968), 255.
45. M. Straumanis and C. Strenk: *Z. Anorg. allgem. Chem.* **213** (1933), 301.
46. H. Spandau and E. J. Kohlmeyer: *Z. Anorg. Chem.* **254** (1947), 65.
47. J. Trillat, L. Tertian, and M. Plattard: *Compt. Rend.* **240** (1955), 526.
48. J. C. Platteeuw and G. Meyer: *Trans. Faraday Soc.* **52** (1956), 1066.
49. L. G. Berezkina, N. I. Ermakova, and D. M. Chizhikov: *Zh. Neorgan. Khim.* **9** (1964), 1760, *Chem. Abstr.* **61** (1964), 6622.
50. H. E. Swanson, R. K. Fuyat, and G. M. Ugrinic: *Nat. Bur. Std. (U.S.)*, Circ. 539 (1955), Vol. 4, 1.
51. N. R. Serebryanaya, S. S. Kabalkina, and L. F. Vereschagin: *Soviet Phys. Doklady* **14** (1970), 672.
52. J. R. Partington and W. Moser: *Nature* **154** (1944), 643.
53. J. D. Donaldson, W. Moser, and W. B. Simpson: *J. Chem. Soc.* (1961), 839.
54. W. Kwestro and P. H. G. M. Vromans: *J. Inorg. Nucl. Chem.* **29** (1967), 2187.
55. I. Oftedal: *Z. Phys. Chem.* **134** (1928), 301.
56. J. R. Guenter and H. R. Oswald: *Naturwiss.* **55** (1968), 177.
57. A. H. Clark: *Naturwiss.* **59** (1972), 361.
58. A. H. Clark: *Neues Jahrb. Mineral. Monatsh.* (1969), 426.
59. G. H. Moh: *Neues Jahrb. Mineral. Abhandl.* **111** (1969), 227.
60. D. F. Smith and H. K. Woods: *J. Am. Chem. Soc.* **45** (1923), 2632.
61. F. D. Rossini, D. D. Wagman, W. H. Evans, S. Levine, and I. Jaffe: *Nat. Bur. Std. (U.S.)*, Circ. 500, (1952), 154, 158.
62. H. M. Spencer and J. H. Mote: *J. Am. Chem. Soc.* **54** (1932), 4618.
63. I. V. Andrews and D. J. Brown: *J. Am. Chem. Soc.* **56** (1934), 388.
64. J. A. Makolkin: *Zh. Fiz. Khim.* **16** (1942), 13, *Chem. Abstr.* (1943), 2641.
65. R. W. Millar: *J. Am. Chem. Soc.* **51** (1929), 207.
66. E. G. King: *J. Am. Chem. Soc.* **80** (1958), 2400.
67. C. B. Alcock and T. N. Belford: *Trans. Faraday Soc.* **60** (1964), 822.
68. G. G. Charette and S. N. Feengas: *J. Electrochem. Soc.* **115** (1968), 796.
69. J. P. Coughlin: *U.S. Bur. Mines, Bull.* **542** (1954), 1.
70. C. E. Wicks and F. E. Block: *U.S. Bur. Mines, Bull.* **605** (1963), 1.
71. O. Knacke and K. E. Prescher: *Z. Erzbergbau Metallhüttenw.* **17** (1964), 28.
72. H. M. Spencer and W. M. Spicer: *J. Am. Chem. Soc.* **64** (1942), 617.
73. G. Exsteen, J. Drowart, and A. Vander Auwera-Mahieu: *J. Phys. Chem.* **71** (1967), 4130.
74. L. Brewer: *Chem. Rev.* **52** (1953), 1.
75. G. L. Humphrey and C. J. O'Brien: *J. Am. Chem. Soc.* **75** (1953), 2805.
76. D. D. Wagman, W. H. Evans, V. B. Parker, I. Halow, S. M. Bailey, and R. H. Schumm: *Nat. Bur. Std. (U.S.)*, Tech. Note 270-3 (1968), 181.
77. A. B. Garrett and R. E. Heiks: *J. Am. Chem. Soc.* **63** (1941), 562.
78. T. N. Belford and C. B. Alcock: *Trans. Faraday Soc.* **61** (1965), 443.
79. K. K. Kelley: *U.S. Bur. Mines, Bull.* **476** (1949), 1.
80. I. Barin and O. Knacke: *Thermochemical Properties of Inorganic Substances*, Springer-Verlag, Berlin 1973.
81. E. G. King and S. S. Todd: *J. Am. Chem. Soc.* **75** (1953), 3023.
82. R. L. Orr and A. U. Christensen: *J. Phys. Chem.* **62** (1958), 124.
83. M. I. Karakhanova, A. S. Pashinkin, and A. V. Novoselova: *Izv. Nauk SSSR, Neorg. Mater.* **3** (1967), 1550; *Chem. Abstr.* **68** (1968), 33780 m.
84. V. Kohlschütter and H. Roesti: *Ber. Deut. Chem. Ges.* **56** (1923), 275.
85. G. L. Clark and W. P. Tyler: *J. Am. Chem. Soc.* **61** (1939), 58.
86. R. O. Jones and S. Rotschild: *J. Electrochem. Soc.* **105** (1958), 4460.
87. T. Katz: *Ann. Chim. (Paris)* **5** (1950), 5.
88. L. L. Bircumshaw and J. Harris: *J. Chem. Soc.* (1939), 1637.
89. M. P. Applebey and R. D. Reid: *J. Chem. Soc.* **121** (1922), 2129.
90. A. B. Garrett, S. Vellenga, and C. M. Fontana: *J. Am. Chem. Soc.* **61** (1939), 367.

91. W. Kwestro and A. Huizing: *J. Inorg. Nucl. Chem.* **27** (1965), 1951.
92. W. Kwestro, J. de Jonge, and P. H. G. M. Vromans: *J. Inorg. Nucl. Chem.* **29** (1967), 39.
93. J. Vennik and R. Callaerts: *Compt. Rend.* **260** (1965), 496.
94. A. Ditte: *Ann. Chim. Phys.* [5] **27** (1882), 145.
95. C. G. Fink and C. L. Mantell: *J. Phys. Chem.* **32** (1928), 103.
96. W. Fraenkel and K. Snipischski: *Z. Anorg. allgem. Chem.* **125** (1922), 235.
97. G. Domingo, R. S. Itoga, and C. R. Kannewurf: *Phys. Rev.* **143** (1966), 536.
98. H. P. B. Rimmington, A. A. Balchin, and B. K. Tanner: *J. Cryst. Growth* **15** (1972), 51.
99. W. Kwestroo, J. H. van den Biggelaar, and C. Langereis: *Mat. Res. Bull.* **5** (1970), 307.
100. J. Fournier and R. Kohlmuller: *Bull. Soc. Chim. France* (1970), 4283.
101. H. P. B. Rimmington and A. A. Balchin: *Phys. Stat. Sol. (a)* **6** (1971), 47.
102. P. A. Lee, G. Said, R. Davis, and T. H. Lim: *J. Phys. Chem. Solids* **30** (1969), 2719.

INDEX OF SUBJECTS

Acids 26-28
Adsorbed adducts
 – halogen 19
 – LEED and 18
 – $MoCl_5$ 18
 – order-disorder 18
 – structure of 18
α-forms of metal hydroxides 75
Aluminium chloride 19
Anhydrous metal halides 37, 42
Antimony 2, 6, 8
Antimony triiodide 46
Arsenic 2, 6, 8
Arsenic triiodide 46

Barium fluorobromide 64
Barium fluorochloride 47, 64
Barium fluoroiodide 64
Basic salts of bivalent metals 79, 80
Bismuth 2, 6, 8
Bismuth triiodide 46, 64
Bivalent metal halides 35
Bivalent metal hydroxides 71
Boiling points of metal halides 47, 48, 51
Bond distances in metal halides 40
Bonding
 – covalent 161, 226
 – ionic 41, 161
 – metallic 161
 – molecular 41
 – Van der Waals 226
Boron nitride 3, 31
Boron trichloride 15
Bromine 12, 14, 23
Brucite lattice type 73, 82

Cadmium hydroxide 129
 – properties of 132
Cadmium iodide type of lattice 73
Calcium hydroxide 86
 – properties of 87
 – thermal decomposition 89
Carbon, forms of 3
Carbons, polycrystalline 9
 – black 9
 – columnar 10
 – fibers 10
 – glassy 10
 – hard 9
 – porylytic 10
 – soft 9

Cesium 22, 23
Chemical transport 54, 173, 174, 177, 178,
  195–198, 211–213, 220, 241, 246, 248, 251
Chlorine 14, 19–21
Chromium tribromide 46
Chromium trichloride 46
Chromyl chloride 28–30
Cobalt (II) hydroxides 100
 – properties of 103
 – thermal decomposition 106
Cobalt (III) oxide hydroxide 107
Colors 146–156, 228
Controlled freezing 55
Coordination-number 37
      -octahedral 158, 161, 162
      -trigonal priomatic 144, 158, 161,
        162
Copper (II) hydroxides 118
 – properties of, structures of 120
Covalency 41
Crystal growth techniques 53–56, 241
Crystal growth
 – of Ag_2F 57
 – of AsI_3 61
 – of BaFBr 64
 – of BaFCl 62, 64
 – of BaFI 64
 – of BaI_3 61, 64
 – of CaFBr 64
 – of CaFCl 64
 – of CaFI 64
 – of CaI_2 58
 – of $CdBr_2$ 60
 – of $CdCl_2$ 60
 – of CdI_2 58
 – of $CoBr_2$ 59
 – of $CoCl_2$ 59
 – of CoI_2 59
 – of EuFBr 64
 – of EuFCl 64
 – of EuFI 64
 – of $FeBr_2$ 59
 – of $FeCl_2$ 61
 – of $FeCl_3$ 61
 – of GaS 242, 246, 251
 – of GaSe 248, 251
 – of GaTe 249, 251
 – of HfS_2 170, 174
 – of $HfSe_2$ 171, 174
 – of $HfTe_2$ 172, 174
 – of InSe 249, 250, 251

Crystal growth (continued)
- of $MnBr_2$ 59
- of MnI_2 59
- of MoS_2 202, 211
- of $MoSe_2$ 205, 211
- of $MoTe_2$ 206, 212
- of $NiBr_2$ 61
- of $NiCl_2$ 61
- of NbS_2 182, 195
- of $NbSe_2$ 184, 195
- of $NbTe_2$ 186, 196
- of PbFBr 63
- of PbFCl 63
- of PbFI 63
- of PbI_2 59
- of PbO 266
- of ReS_2 218, 220
- of $ReSe_2$ 218, 220
- of SbI_3 61
- of Si_2Te_3 267
- of SnO 268
- of SnS_2 268
- of $SnSe_2$ 268
- of SrFBr 64
- of SrFCl 62, 64
- of SrFI 64
- of TaS_2 188, 196, 197
- of $TaSe_2$ 193, 197
- of $TaTe_2$ 194, 198
- of TcS_2 216, 220
- of $TcSe_2$ 216, 220
- of $TcTe_2$ 216, 220
- of TiS_2 163, 173
- of $TiSe_2$ 165, 173
- of $TiTe_2$ 166, 173
- of VCh_2 180, 195
- of WS_2 208, 212
- of WSe_2 209, 213
- of WTe_2 210, 213
- of ZrS_2 167, 173
- of $ZrSe_2$ 169, 174
- of $ZrTe_2$ 170, 174

Crystal structures 2, 4, 6, 7, 8, 36, 37, 40, 144–156, 158, 159, 160, 161, 226, 227, 229, 256–259
Crystalline layer 53
Crystallization of thin films 54, 57–59, 61
Crystallization temperature 53
C_6 type of lattice 73

Decomposition potential 51
Dehydration 43
Densities 146–156, 232
Dichlorine heptoxide 30
Disproportionation 54
Distorted layers 145
Doping 250
Double layer lattice 81, 103, 104, 124, 131

Electrical conductivity, anisotropy 37
Electrolysis 55
Electronegativity 161
Elements 1
Enthalpy, see thermodynamic data
Epitaxy 18, 19
Europium fluorobromide 64
Europium fluorochloride 64
Europium fluoroiodide 64
Evaporation growth 53, 55, 56, 59

Feitknechtite 90
- synthetic 95
Fluorine 14
Free energy of formation, see thermodynamic data
Free enthalpy of formation, see thermodynamic data

Gallium selenide 231, 232
Gallium sulfide 231, 232
Gallium telluride 231, 232
Gel growth 56, 59, 60
Germanium diiodide 44
Graphite
- defects in 9
- hexagonal 2, 4
- natural 8
- nickel system 11
- oxidation 28
- oxide 29
- Pauling theory of 3
- rhombohedral 2
- turbostratic 8
Green ferroferrites 98
Green rusts 99

Hafnium diselenide 171, 174
Hafnium disulfide 170, 174
Hafnium ditelluride 171, 172, 174
Halides of bivalent metals 35
Halides of trivalent metals 35
Halogens 14, 19
Heat capacity, see thermodynamic data
Heat of formation, see thermodynamic data
Hydrogen bonding 75
Hydrolysis 43
Hydroxides of bivalent metals 71
Hydroxy-salts of bivalent metals 79, 101, 119
Hygroscopic metal halides 36, 44–47

Indium selenide 225
Indium trichloride 46
Insulator 143
Intercalation in graphite
- adducts list of 12
- C—C distance in a layer 23
- capillary condensation during 12

Intercalation in graphite (continued)
- conductivity, electrical 13
- expansion during 16
- experimental 12
- flexibility of layers 17, 18, 28
- general behaviour 11
- hysteresis 12
- isotherm 12
- preparation of compounds 12
- splitting during 17
- stages 12, 16, 18
- superconductivity 14
- theory of 16
- threshold pressure 12, 16

Interhalogens 14
Iodine 14
Ionic
- bond 41, 161
- character 40
- radius ratio 37, 40

Ionicity 40, 161
Iridium ditelluride 219, 220
Iridium trichloride 46
Iron dibromide 44
Iron dichloride 45, 64
Iron diiodide 44
Iron (II) hydroxide 96
- colloidal solutions of 96
- properties of 98
Iron tribromide 46
Iron trichloride 46

Lattice energy of metal hydroxides 74, 77
Layer structures
- of antimony 2, 6, 8
- of arsenic 2, 6, 8
- of bismuth 2, 6, 8
- definition of 2
- of Group IV 3
- of Group V 3
- of Group VI 8
- of Group VII 8
- of phosphorus 2–7
Lead fluorobromide 47
Lead fluorochloride 47
Lead fluoroiodide 47
Lead iodide 36
Lead monoxide 255, 263, 264, 266
Ligandfield stabilization energy 75
Litharge 256
Lithium 25
Lutetium trichloride 46

Magnesium dibromide 44
Magnesium dichloride 45, 64
Magnesium diiodide 44
Magnesium hydroxide 83
- properties of
- thermal decomposition of 85

Manganese dibromide 44
Manganese dichloride 45, 64
Manganese diiodide 44
Manganese (II) hydroxide 90
- oxidation of 95
- properties of 93
- thermal decomposition of 95
Massicot 256
Melt grown crystals 55, 57–63, 241, 242, 248, 249, 251
Melting points
- of gallium chalcogenides 232
- of indium selenide 232
- of metal halides 40, 47, 48, 51
Metal atoms
- double layers of 225
Metal chlorides 15, 16
Metal oxides 30
Metal sulfides 30
Metals 22
Metals in solvents 25
Mixed crystals, see solid solutions
Molecular bond 41
Molybdenum diselenide 159, 200, 204, 211
Molybdenum disulfide 159, 200, 201
Molybdenum ditelluride 159, 200, 205, 206, 212
Mössbauer spectroscopy 21

Nemalite (Nematolite) 83
Nickeldibromide 45
Nikkel-Cadmium storage batteries 108, 129
Nickel dichloride 45, 64
Nickel diiodide 45
Nickel hydroxide 109
- high purity preparation 112
- thermal decomposition 116
- in higher valence states 117
Nickel ditelluride 219, 220
Niobium diselenide 183, 195
Niobium disulfide 181, 195
Niobium ditelluride 186, 195
Nitroethane, solvent 15, 21
Nitrogen pentoxide 30
Nitromethane, solvent 15–17, 21

Optical absorption 36, 37, 53
Oxidation state 42, 43

Paladium ditelluride 219, 220
Phase diagrams 234–240
Phase equilibria 233, 234
Phosphorus 2–7, 9
Platinum diselenide 219, 220
Platinum disulfide 219, 220
Platinum ditelluride 219, 220
Polarizability 40, 42
Polarization 37, 42
Polymorphism 157, 256
Polytypism 37, 42, 60

Polytypism (continued)
- of CdI_2 58
- of GaSe 228, 229
- of MoS_2 159
- of $MoSe_2$ 159
- of $MoTe_2$ 159
- of NbS_2 159, 160
- of $NbSe_2$ 159, 160
- of $NbTe_2$ 159, 160
- of PbI_2 60
- of TaS_2 159, 160
- of $TaSe_2$ 159, 160
- of WS_2 159
- of WSe_2 159

Portlandite 74, 86
Potassium 22, 23
Preparation
- of Ag_2F 44
- of $AlCl_3$ 46
- of AsI_3 46
- of BaFBr 64
- of BaFCl 47, 64
- of BaFI 64
- of BiI_3 46, 64
- of CaFBr 64
- of CaFCl 64
- of CaFI 64
- of $Ca(OH)_2$ 87
- of CaI_2 44
- of $CdBr_2$ 45
- of $CdCl_2$ 45
- of $Cd(OH)_2$ 129
- of CdI_2 44
- of $CoBr_2$ 44
- of $CoCl_2$ 45, 64
- of $Co(OH)_2$ 101
- of CoOOH 107
- of CoI_2 44
- of $CoTe_2$ 218, 220
- of $CrBr_3$ 46
- of $CrCl_3$ 46
- of $Cu(OH)_2$ 119
- of EuFBr 64
- of EuFCl 64
- of EuFI 64
- of $FeBr_2$ 44
- of $FeBr_3$ 46
- of $FeCl_2$ 45, 64
- of $FeCl_3$ 46
- of $Fe(OH)_2$ 97
- of FeI_2 44
- of GaS 231, 232
- of GaSe 231, 232
- of GaTe 231, 232
- of GeI_2 44
- of HfS_2 170, 174
- of $HfSe_2$ 171, 174
- of $HfTe_2$ 172, 174
- of $InCl_3$ 46

Preparation (continued)
- of InSe 231, 232
- of $IrCl_3$ 46
- of $IrTe_2$ 219, 220
- of $LuCl_3$ 46
- of $MgBr_2$ 44
- of $MgCl_2$ 45, 64
- of $Mg(OH)_2$ 83
- of MgI_2 44
- of $MnBr_2$ 44
- of $MnCl_2$ 45, 64
- of $Mn(OH)_2$ 90
- of MnI_2 44
- of MoS_2 201, 211
- of $MoSe_2$ 204, 211
- of $MoTe_2$ 206, 211
- of $NiBr_2$ 45
- of $NiCl_2$ 45, 64
- of $Ni(OH)_2$ 109
- of NiI_2 45
- of $NiTe_2$ 219, 220
- of NbS_2 181, 195
- of $NbSe_2$ 183, 195
- of $NbTe_2$ 186, 195
- of PbFBr 47
- of PbFCl 47
- of PbFI 47
- of PbI_2 44
- of PbO 262-264
- of $PdTe_2$ 219, 220
- of PtS_2 219, 220
- of $PtSe_2$ 219, 220
- of $PtTe_2$ 219, 220
- of ReS_2 217, 220
- of $ReSe_2$ 217, 220
- of $RhCl_3$ 46
- of $RhTe_2$ 218, 220
- of SbI_3 46
- of $ScCl_3$ 46
- of Si_2Te_3 264
- of SnO 265
- of SnS_2 265
- of $SnSe_2$ 265
- of SrBFr 64
- of SrFCl 47, 64
- of SrFl 64
- of TaS_2 187, 196, 197
- of $TaSe_2$ 190, 197
- of $TaTe_2$ 194, 197, 198
- of TeS_2 216, 220
- of $TeSe_2$ 216, 220
- of $TeTe_2$ 216, 220
- of ThI_2 44
- of $TiBr_2$ 44
- of $TiBr_3$ 46
- of $TiCl_2$ 44
- of $TiCl_3$ 46
- of TiI_2 44
- of TiS_2 162, 173

Preparation (continued)
- of TiSe$_2$ 164, 173
- of TiTe$_2$ 165, 173
- of TlCl$_3$ 46
- of TmI$_2$ 44
- of VBr$_2$ 44
- of VCl$_2$ 44
- of VCl$_3$ 46
- of VI$_2$ 44
- of VI$_3$ 46
- of VS$_2$ 180, 195
- of VSe$_2$ 180, 195
- of VTe$_2$ 180, 195
- of WS$_2$ 207, 212
- of WSe$_2$ 208, 213
- WTe$_2$ 210, 213
- of YbI$_2$ 44
- of YCl$_3$ 46
- of ZnBr$_2$ 45
- of Zn(OH)$_2$ modifications 124
- of ZnI$_2$ 45
- of ZrS$_2$ 166, 173
- of ZrSe$_2$ 166, 173
- of ZrTe$_2$ 166, 174

Purification 230, 231, 233, 263, 264
Pyrochroite 74, 90
- synthetic 93

Reversible electromotive force 51
Rhenium diselenide 217, 220
Rhenium disulfide 217, 220
Rhodium ditelluride 218, 220
Rhodium trichloride 46

Silicon sesquitelluride 255, 257, 264, 267
Scandium trichloride 46
Slabs
- bonding in, bonding between 145
- four fold 226
- number of, in a unit cell 157

Sodium 24
Solid Solutions 64, 177, 178, 200, 215, 252, 268, 271
Solubility products of Me(OH)$_2$ 74, 77, 78
Special forms of hydroxides
- of Cd(OH)$_2$ 131
- of Co(II)(OH)$_2$ 102
- of Mg(OH)$_2$ 84
- of Mn(II)(OH)$_2$, morphology 92
- of Ni(II)(OH)$_2$, gels, Sols 112, 113
- of Zn(OH)$_2$ 122

Stacking facults 42
Stacking sequence 4-7, 27, 39, 42, 227-229
Strontium fluorobromide 64
Strontium fluorochloride 47, 64
Strontium fluoroiodide 64
Structural types of bivalent
   metal hydroxides 74

Sublimation 53, 59-61, 247-249, 251
Sublimation point 47, 48, 51
Sulfur trioxide 30
Superlattices 21, 22
Supersaturation 55, 56

Tantalum diselenide 190, 197
Tantalum disulfide 187, 196, 197
Tantalum ditelluride 194, 197, 198
Techneticum diselenide 216, 220
Techneticum disulfide 216, 220
Techneticum ditelluride 216, 220
Thallium trichloride 46
Thermal decomposition 43, 85, 89, 95, 106, 116
Thermochemical properties of MeX$_2$ and MeX$_3$ 45, 47, 48
Thermodynamic data 45, 47, 48, 50, 74, 86, 89, 96, 100, 107, 117, 128, 131, 146-156, 232, 259-262
Thionyl chloride, solvent 16
Thorium diiodide 44
Thullium diiodide 44
Tin diselenide 255, 259, 265, 268
Tin disulfide 255, 259, 265, 268
Tin monoxide 255, 259, 265, 268
Titanium dibromide 44
Titanium dichloride 44
Titanium diiodide 44
Titanium diselenide 164, 173
Titanium disulfide 162, 173
Titanium ditelluride 165, 173
Titanium tribromide 46
Titanium trichloride 46
Topotaxy, topotactic reactions 99, 119, 122, 129, 134
Transition metal dichalcogenides 142
Transition metal halides 36
Tungsten diselenide 208, 213
Tungsten disulfide 207, 212
Tungsten ditelluride 210, 213
Turbostratic Ni(OH)$_2$ 113

Vanadium dibromide 44
Vanadium dichalcogenides 180, 195
Vanadium dichloride 44
Vanadium diiodide 44
Vanadium trichloride 46
Vanadium triiodide 46
Vapour phase growth 53, 54, 58-60

Ytterbium diiodide 44
Yttrium trichloride 46

Zink dibromide 45
Zink diiodide 45
Zink hydroxide 122
- α-, β-, γ-, ε-, modifications 123
- properties of 127
- solubilities 128

Zirconium diselenide 168, 174
Zirconium disulfide 166, 173

Zirconium ditelluride 169, 174